S0-AUG-115

Applied Linear Programming

For the Socioeconomic and Environmental Sciences

OPERATIONS RESEARCH AND INDUSTRIAL ENGINEERING

Consulting Editor: J. William Schmidt

CBM, Inc., Cleveland, Ohio

Applied Statistical Methods, *I. W. Burr*

Mathematical Foundations of Management Science and Systems Analysis, *J. William Schmidt*

Urban Systems Models, *Walter Helly*

Introduction to Discrete Linear Controls: Theory and Application, *Albert B. Bishop*

Integer Programming: Theory, Applications, and Computations, *Hamdy A. Taha*

Transform Techniques for Probability Modeling, *Walter C. Giffin*

Analysis of Queueing Systems, *J. A. White, J. W. Schmidt, and G. K. Bennett*

Models for Public Systems Analysis, *Edward J. Beltrami*

Computer Methods in Operations Research, *Arne Thesen*

Cost-Benefit Analysis: A Handbook, *Peter G. Sassone and William A. Schaffer*

Modeling of Complex Systems, *V. Vemuri*

Applied Linear Programming: For The Socioeconomic and Environmental Sciences, *Michael R. Greenberg*

Applied Linear Programming

For the Socioeconomic and Environmental Sciences

MICHAEL R. GREENBERG

Departments of Urban Studies and Geography
Rutgers University
New Brunswick, New Jersey

ACADEMIC PRESS New York San Francisco London 1978

A Subsidiary of Harcourt Brace Jovanovich, Publishers

ACADEMIC PRESS, INC.
111 Fifth Avenue, New York, New York 10003

United Kingdom Edition published by
ACADEMIC PRESS, INC. (LONDON) LTD.
24/28 Oval Road, London NW1 7DX

Library of Congress Cataloging in Publication Data

Greenberg, Michael R
Applied linear programming for the socioeconomic and environmental sciences.

(Operations research and industrial engineering series)
Includes bibliographies.
1. Linear programming. 2. Social sciences--Linear programming. 3. Environmental engineering--Linear programming. I. Title.
T57.74.G73 001.4'24 78-6295
ISBN 0-12-299650-X

PRINTED IN THE UNITED STATES OF AMERICA

TO GWEN

Contents

Chapter 3 Beyond the Global Optimum

Chapter 4 Computer Algorithms

PART TWO APPLICATIONS

Chapter 5 Mathematical Models for Policy Analysis

Chapter 6 Solid Waste

Chapter 7 Water Resources Management

Chapter 8 Health, Education, and Law Enforcement

Chapter 9 Intraregional Land Use and Transportation

Chapter 10 Economic Development and Transportation

Chapter 11 A Linear Programming Case Study

Preface

One of the shelves in my office is lined with books that have linear or nonlinear programming and optimization in the title. Why then write another linear programming book? Because I think it is useful to include the basic methods and a broad survey of applications under a single cover. Most of the dozen or so books I have used in my course since 1969 stress the mathematics of linear programming (e.g., Gass, Hadley, Simmons, Dantzig, Smythe and Johnson, and Spivey and Thrall). None emphasizes the range of linear programming applications that are relevant to students of socioeconomic and environmental management. Most books acknowledge the existence of only transportation applications. Those that do contain applications tend to present the proven applications, a few applications in depth, or linear programming and other models (e.g., Stark, Helly, Salkin and Saha, Driebeck, and DeNeufville and Marks). The major objective of this book is to stimulate new applications of linear and related programming by specialists and nonspecialists by providing many of the already reported uses of linear and related optimization techniques and the basic methods under a single cover.

This volume was molded on two other perspectives which, while not novel, are geared toward the nonspecialist student. Optimization models have become a tool that must be understood not only by the industrial engineer, operations research analyst, economist, and mathematician, but also by the planner, geographer, regional scientist, civil engineer, public servant, and other people without extensive mathematics backgrounds. Therefore, the second major goal of this book is to prevent the mathematics of linear programming from acting as a barrier to the potential user. While all users or readers of linear programming should understand the mathematics, they can understand it without corollaries and lemmas and mathematical presentations that assume backgrounds which many students do not possess.

A third goal of this book is to quickly move the student from the book to the computer. This goal is derived from years of teaching and talking with people who had not taken linear programming but were determined to look over the analyst's shoulder, people who had a course in linear programming but were never exposed to computer codes, and people who were unfamiliar with the plethora of information one can acquire beyond the optimum solution.

The three goals of this book are accomplished in eleven chapters. The first three chapters present the theory and mathematics of linear programming. Chapter 1 introduces the concept of optimization, a geometric interpretation, the mathematical formulation and solution, and a brief history of standard and new applications. Chapter 2 presents the essential matrix mathematics for solving linear programming problems. The simplex procedure is presented as the most useful technique for the potential user. Diagrams are used to demonstrate the feedback between the single- and two-phase solutions. The third chapter highlights special properties of linear programming (the dual and postoptimal analysis) and other optimizations methods (e.g., mixed integer, branch and bound, dynamic, and heuristic).

The fourth chapter presents MPS/360 and MPSX/370 computer codes for solving linear programming problems. Methods of preparing job control, system, and data cards are presented. System options are explained, including postoptimal analysis. Sample printouts are displayed and interpreted.

Chapters 5–11 provide applications of interest to planners, managers, economists, environmental and regional scientists, geographers, civil engineers, and public servants. Each chapter has three components: a broad overview of linear programming applications, one or more examples to be solved by the reader on the computer, and a bibliography. The author has chosen to emphasize breadth rather than depth in the application chapters. Five areas of application are presented: solid waste in Chapters 6 and 11; water resources in Chapter 7; health, education, and law enforcement in Chapter 8; intraregional land use and transportation in Chapter 9; and economic development and transportation in Chapter 10. Within each chapter, as much breadth in applications as possible is presented.

The breadth approach has three important advantages. First, the user should see the great extent to which applications vary. For example, in some applications optimizing behavior seems like a reasonable assumption;

in others this is dubious. Some applications require a readily available and a narrowly defined data set; others have data requirements that must be met with estimates. Numerous examples are necessary to highlight the subtle and sometimes not so subtle differences between problems. Second, with the exception of a few review articles, the vast majority of the applications in this book are available only in technical literature that may be difficult to locate and/or to readily comprehend for the nonspecialist. More than sixty percent of the applications in this book have been published during the 1970s. By bringing together many of the recent applications under a single cover, it is hoped that the inventory in and of itself will spawn new applications.

The breadth approach to applications has drawbacks. The most important drawback is overselling optimization. Optimization is only one of many mathematical and nonmathematical approaches used in the socioeconomic and environmental sciences. In many of the disciplines presented in the book, linear programming is not the most important mathematical tool. To mitigate against misrepresenting the role of linear programming, Chapter 5 compares optimization and other approaches and presents guidelines for mathematical programming applications. Chapter 11 expands on Chapter 5 through a case study.

Another important decision was to aggregate the applications on the basis of disciplines rather than by type of application. The latter includes such groupings as personnel allocation, transportation, siting, and capital replacement. A discipline classification appeals to the discipline specialist who may be primarily interested in solid-waste and water-resources applications, or land-use and transportation applications. This book gives the discipline specialist the opportunity to select one or two of the applications chapters to read in depth.

The discipline approach will not appeal to the reader who perceives applications as multidisciplinary. To this reader, sitings of a landfill, a hospital, a school, or other facilities are similar. The author decided on discipline aggregation because the vast majority of people with whom he spoke favored the discipline classification.

Recapitulating, the three major goals of this volume are to teach the mathematics of linear programming in as simple a manner as possible, to move the user to the computer as soon as possible, and especially to stimulate interest in new applications by presenting numerous examples of the use of linear and related programming. Overall, this book is intended to serve the large body of readers who are primarily interested in applications, not mathematics.

Acknowledgments

Many friends have contributed to this volume. Bill Spencer helped me learn goal programming. Tom Fitzgerald, Jim Gallagher, and John Caruana helped set up and run some of the problems at the ends of the chapters. Nancy Neuman made detailed comments on three chapters. Numerous colleagues contributed suggestions on specific chapters, in particular, George Carey and Don Krueckeberg of Rutgers University and Ethan Smith of the United States Geological Survey.

I thank the many typists who have struggled with my handwriting: Vera Lee, Barbara Swan, Gayle Youngman, Marilyn Watterson, and Gail Scher.

I thank the Center for Urban Policy Research of Rutgers University for allowing me to abstract and reproduce portions of my solid waste management book in Chapter 11.

Finally, I thank my many students at Rutgers, Columbia and Cornell Universities for motivating me to write this book and for reading it and offering suggestions that have greatly improved it.

Applied Linear Programming

For the Socioeconomic and Environmental Sciences

PART ONE

THEORY AND METHODS

Chapter 1

Theory, Methods, and Applications

Linear programming is a specialized mathematical decisionmaking aid. In and of itself, it has no socioeconomic or environmental content. Like other widely used mathematical methods (e.g., regression, input–output analysis, and the calculus), it can only help us interpret data and to examine theories about the way things work or should work. If the data are spurious and incomplete, or if the match of model and theory is inappropriate, mathematical decisionmaking aids such as linear programming are more likely to confuse rather than clarify a decision. Given these important caveats, linear programming was developed for specific classes of applications which assume optimal behavior, linear relationships among different pieces of information, and partially restrictive specifications.

A. OPTIMAL BEHAVIOR

You must be able to assume optimal behavior if you hope to learn by applying linear programming to a problem. Optimal behavior implies that decisionmakers are aware of all the important information and will choose the most favorable decision, or at least would like to know the best decision. Not everyone can agree on the criteria for the best decision. Some people regard economic gain as something to be maximized; others prefer to minimize the use of resources. Still others seek to minimize political turmoil and others have additional "most favorable" criteria. Different most favorable criteria are based on the different philosophies and roles of the decisionmakers and the different information supplied to them.

Optimal behavior has been one of the cornerstones of economic theory and operations research.[1] Economic theorists, operations analysts, engineers, and others have assumed optimal behavior in modeling many private and sometimes public decisionmaking processes. However, it seems apparent that decisions are influenced by political, sociological, and psychological factors. While proponents of optimization assume that economic decisions are made on the basis of carefully reasoned debate, other theorists argue that decisions are more like psychological dramas in which the players, loaded with personal perspectives and axes to grind, arrive at clearly suboptimal decisions.

Most institutional decisions probably are made with explicit notions of optimal behavior and implicit psychological motivations. When the weight of the decisionmaking process is more on the side of information analysis than personal motivations, then optimal behavior is a possible behavioral theory from which to operate. However, if data are manipulated to support a pet intuition or past practice, or both, then optimal behavior cannot be assumed or may be assumed to occur only over a narrow range of decisions, most likely routinized operating decisions. On the whole, optimal behavior is most likely to be found when decisionmakers want to consider a relatively limited range of policies and choose from a broad range of specific alternatives drawn from each major policy. The limited range of major policies enables decisionmakers to grope toward specific tradeoff and contingency choices which can be modeled to seek optimality.

When optimal behavior is a reasonable-looking assumption, then the following mathematical conditions should exist if linear programming is to be used: constraints that partially restrict the range of the solution, nonnegative solutions, and linear relationships among the components of the problem.

B. MATHEMATICAL CONDITIONS FOR A SOLUTION

Linear programming may not be the appropriate decisionmaking aid, even when optimal behavior is assumed. Calculus is used to solve many optimization problems. Calculus methods cannot solve optimization problems with imprecise specifications and with nonnegativity requirements.

Imprecise specifications typify the socioeconomic and environmental planning sciences. For example, a water supply manager seeks to find a least-

[1] For a spectrum of viewpoints on optimality and decisionmaking, see Clarkson (1968), Ansoff (1969), Pugh (1971), and Starbuck (1971).

cost solution to meeting a public potable water demand of 10 million gallons per day (mgd). He may draw his supply from the local reservior or from a pipeline to an adjacent town. A solution that supplies 9.9 or less mgd or 10.1 mgd or more is not acceptable. The resource manager therefore seeks a least-cost solution while constrained by a specific demand requirement. If all the restrictions are as specific as the 10.0 mgd demand restriction, calculus can usually be used to solve the problem. Typically, however, other constraints are present but are not precise. The water supply manager's local reservoir and pipeline are likely to have imprecise constraints. The local reservoir has a daily yield of 5 mgd, which may not be exceeded. The manager can pump 5 mgd, but he may also pump less than 5 mgd. The reservoir restriction is an inequality rather than an equality. The pipeline also has two inequality restrictions. On the one hand, the community can only take 10 mgd because of the size of the pipeline, on the other, by contractual agreement it must take out at least 6 mgd. These typical inequality constraints may be treated with a linear programming formulation.

Another constraint is that the manager is not allowed to transfer negative amounts of water from the sources of supply to the areas of demand. This constraint is obvious to people but not to the computer. To the computer, it is cheaper to ship 5 mgd than 10 mgd, then why not ship -5 mgd? Linear programming problems are written with ≥ 0 constraints to prevent negative allocations.

Linear programming is one of a number of mathematical programming tools which seek a least-cost solution to meeting imprecise specifications. What distinguishes linear programming from the other mathematical programming techniques, which are briefly reviewed in the third chapter, is that linear programming requires linearly proportional relationships. The resources to be consumed by an activity must be linearly proportional to the level of the activity. For example, if the planner wants to double the number of garden apartments of a particular design in the community, then the amount of land and public services required for the previous garden apartments would have to double.

Similarly, the contribution of the activities to the goal that is to be optimized must be linearly proportional to the activity level. If 100 garden apartment units will cost the region $100,000 to service, then 200 garden apartment units will cost $200,000 to service. Economies and diseconomies of scale are not recognized. Finally, when added together, all activities must obey a materials balance. The sum of the resource inputs must equal the sum of the product outputs. Summarizing, linear programming is a mathematical tool for obtaining optimum solutions that do not violate imprecise constraints, that cannot have negative activities, that require linearly proportional relationships, and that account for all inputs and outputs within the system.

C. AN ILLUSTRATION OF DIFFERENT SOLUTIONS

At this point, it is appropriate to illustrate the theory and mathematical requirements of linear programming. Let us review the water manager's problem. The water department can utilize reservoir water which costs $300 for a million gallons or pipeline water which costs $500 for a million gallons. The reservoir water will be called activity X_1; the pipeline water will be called activity X_2. The community wishes to minimize the cost. The total cost and objective of the manager can be written as the following objective function:

Minimize cost: $$Z = 300X_1 + 500X_2 \tag{1}$$

We remember that the total demand for water is 10 mgd. Accordingly, the following equation states that activities X_1 (reservoir) and X_2 (pipeline) must supply 10 mgd:

$$X_1 + X_2 = 10 \tag{2}$$

Equation (2) is an example of a rigid specification, or equality constraint. Next, the water manager is permitted to take only 5 mgd from the reservoir:

$$X_1 \leq 5 \tag{3}$$

Equation (3) exemplifies imprecise specifications, or inequality constraints. Next, we need two constraints that specify that the pipeline can yield only 10 mgd, but that it must be used for more than 6 mgd:

$$X_2 \leq 10 \tag{4}$$

$$X_2 \geq 6 \tag{5}$$

Finally, we cannot supply negative amounts of water, so we need two non-negativity constraints:

$$X_1 \geq 0 \tag{6}$$

$$X_2 \geq 0 \tag{7}$$

The entire linear programming problem is summarized in the following:

Minimize: $$Z = 300X_1 + 500X_2 \tag{1}$$

subject to the constraints

$$\begin{aligned}
X_1 + X_2 &= 10 && (2)\\
X_1 \qquad &\leq 5 && (3)\\
X_2 &\leq 10 && (4)\\
X_2 &\geq 6 && (5)\\
X_1 \qquad &\geq 0 && (6)\\
X_2 &\geq 0 && (7)
\end{aligned}$$

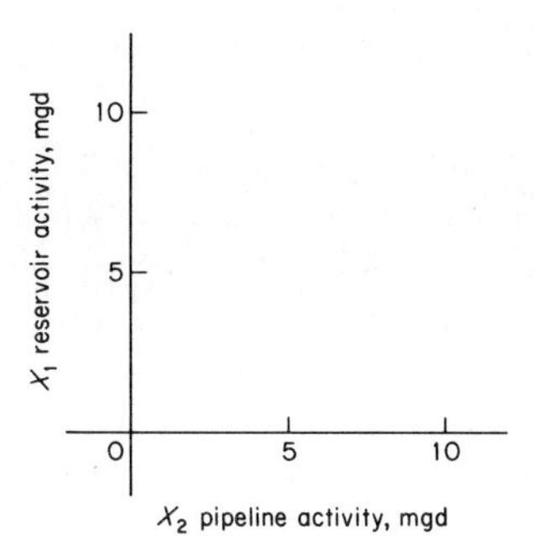

Fig. 1

This is the standard form for a linear programming problem. Let us briefly review the theory and mathematical conditions in this problem. The objective function assumes that the optimal least-cost choice is being sought. Equations (3)–(7) exemplify the imprecise inequality restrictions, and Eqs. (6) and (7) represent nonnegativity requirements. The proportional linear relationships are obvious in the objective function and the constraints (e.g., \$300 for 1 million gallon from X_1, \$600 for 2 million gallons, \$900 for 3 million gallons, etc.). Finally, the outputs of water from the two sources of water equal the inputs of water to the areas of demand.

1. A GRAPHICAL METHOD

In this section, we will review a graphical solution to a linear programming problem. Graphical solutions are impractical for real-world applications since most decisions involve many possible activities and many constraints. Nevertheless, the graphical solution is a device that may be used to explain how the more abstract algebraic methods are used to solve a linear programming problem and the possible solutions that can be obtained from a linear programming problem.

To picture the solution graphically (Fig. 1), we label one axis as activity X_1 (reservoir) and the other axis as activity X_2 (pipeline).

2. FEASIBLE SOLUTION

First, we seek a feasible solution. When a feasible solution has been found, then an optimal solution is sought. We find a feasible solution by drawing each constraint on the graph and locating the area of feasible solutions. The first constraint [Eq. (2)] is graphed in Fig. 2.

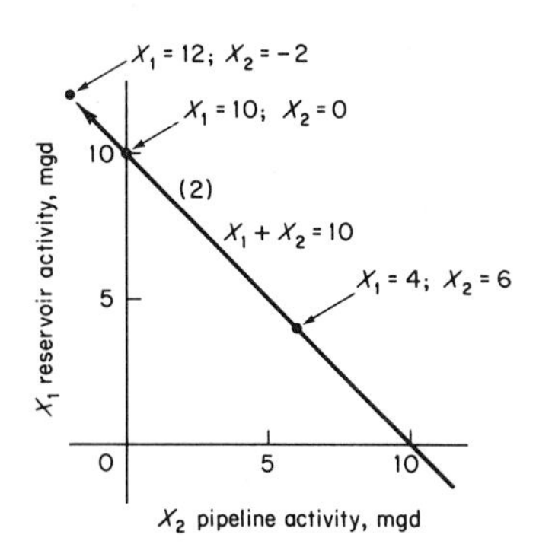

Fig. 2

The solution space is the line $X_1 + X_2 = 10$. Any combination of X_1 and X_2 is possible. For example, $X_1 = 10\ X_2 = 0$; $X_1 = 12$, $X_2 = -2$; $X_1 = 4$, $X_2 = 6$. At this point, as long as the sum of X_1 and X_2 is 10, the solution is feasible. If the value of X_1 is 10 and X_2 is 0, then the cost of the solution is \$3000 = (300(10) + 500(0)). If X_1 is 12 and X_2 is -2, then the value is \$2600 (300(12) + 500(-2)). Finally, the cost of supplying water is \$4200 if X_1 is 4 and X_2 is 6 (400(4) + 500(6)). The important conclusion to be drawn from using the first constraint is that, in the absence of the nonnegativity constraints, a computer solution to this linear programming problem would be an illogical solution in which the X_2 activity would be negative.

The second constraint [Eq. (3)] requires that no more than 5 mgd be supplied from the reservoir. To graphically portray the constraint, we use arrows pointing downward to indicate that the $X_1 \leq 5$ line forms an upper boundary (Fig. 3). The feasible solution space has now been reduced. Since $X_1 \leq 5$, the feasible space starts at point $A(5,5)$ and runs along the Eq. (2) line, where it intersects the X_2 axis and continues into coordinate space; there, the X_2 values are positive and the X_1 values are negative.

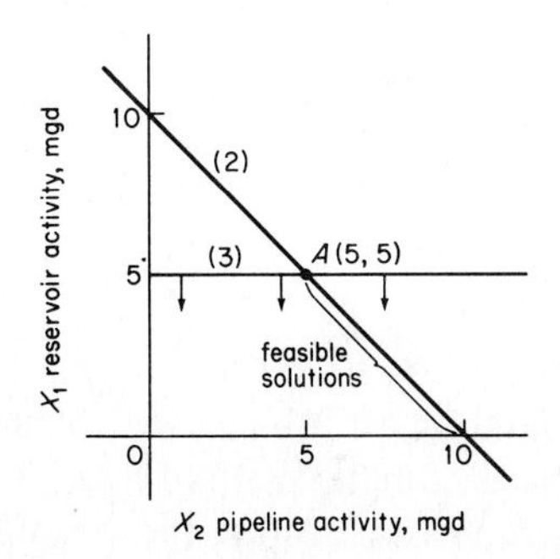

Fig. 3

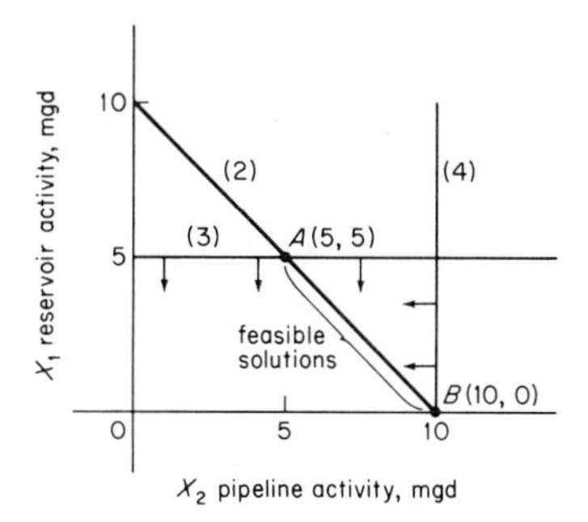

Fig. 4

The third constraint [Eq. (4)] demands that no more than 10 mgd be supplied from the pipeline. In Fig. 4, it intersects the X_2 axis at point $B(10{,}0)$. The feasible solution now lies between points $A(5{,}5)$ and $B(10{,}0)$. At $A(5{,}5)$, the cost of the solution is $\$4000 = (300(5) + 500(5))$. At $B(10{,}0)$, the solution costs $\$5000 = (300(0) + 500(10))$.

The fourth constraint [Eq. (5)] requires that at least 6 mgd be supplied through the pipeline. In Fig. 5, the possible solutions have now been reduced to the line between points $C(6{,}4)$ and $B(10{,}0)$. At $C(6{,}4)$, the value of the problem is $\$4200 = (300(4) + 500(6))$.

Finally, we add the nonnegativity constraints [Eqs. (6) and (7)]. The final solution lies on or between points $C(6{,}4)$ and $B(10{,}0)$.

The feasible solution space to the water supply linear programming problem is a line. The equality constraint [Eq. (2)] has restricted the feasible solution to a line. All inequality constraints will lead to a polygonal solution space. This difference between the equality and inequality constraints may be illustrated by changing the equality constraint [Eq. (2)] $X_1 + X_2 = 10$ to an inequality constraint $X_1 + X_2 \geq 10$. As Fig. 6 demonstrates, when the

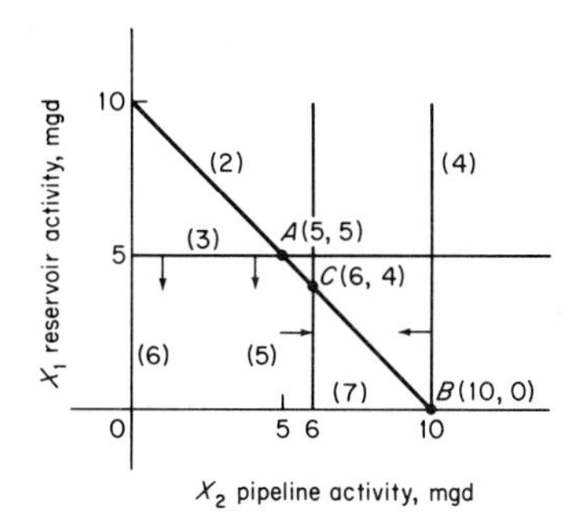

Fig. 5

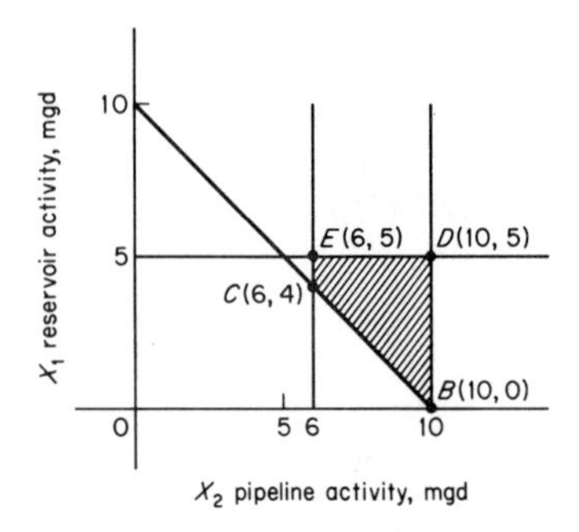

Fig. 6

inequality replaces the equality constraint, the line of feasible space (CB) is replaced by the polygon ($CBDE$).

3. OPTIMAL SOLUTION

Our next step is to determine at which point along the feasible solution line between points C and B the objective function ($X_1, X_2 = 300X_1 + 500X_2$) is a minimum. We begin the search for the optimum point by solving the objective function equation for different values of the cost Z. The different Z values are chosen by examining the Z values produced by different values of X_1 and X_2.

$$Z = 300X_1 + 500X_2 \tag{1}$$

We specify different values of Z and X_1 and solve for the values of X_2.

	Given		Solved
	Z	X_1	X_2
(a)	4200	2	7.2
(b)	4200	4	6.0
(c)	4200	6	4.8
(d)	4200	8	3.6

(a)
$$\begin{aligned} 4200 &= 300(2) + 500X_2 \\ 4200 &= 600 + 500X_2 \\ 3600 &= 500X_2 \\ 7.2 &= X_2 \end{aligned}$$

(b)
$$\begin{aligned} 4200 &= 300(4) + 500X_2 \\ 4200 &= 1200 + 500X_2 \\ 3000 &= 500X_2 \\ 6 &= X_2 \end{aligned}$$

(c)
$$\begin{aligned} 4200 &= 300(6) + 500X_2 \\ 4200 &= 1800 + 500X_2 \\ 2400 &= 500X_2 \\ 4.8 &= X_2 \end{aligned}$$

(d)
$$\begin{aligned} 4200 &= 300(8) + 500X_2 \\ 4200 &= 2400 + 500X_2 \\ 1800 &= 500X_2 \\ 3.6 &= X_2 \end{aligned}$$

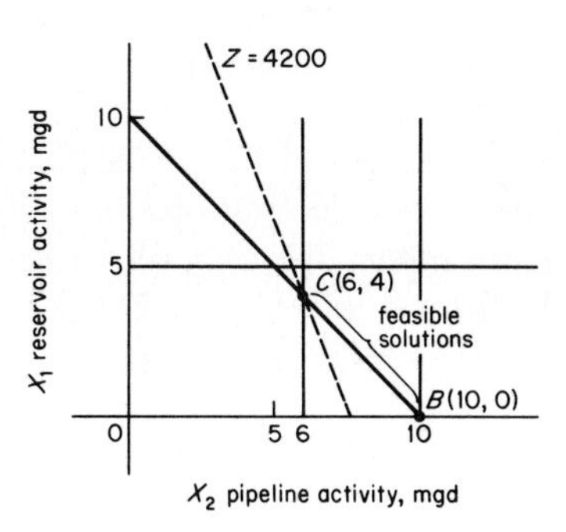

Fig. 7

Next, we plot the resulting equation for $Z = 4200$ (Fig. 7). The line passes through one feasible solution point $C(6,4)$. Next, we solve for X_2, given the cost values of \$3200 and \$5200, and graph the results (Fig. 8).

Z	X_1	X_2	Z	X_1	X_2
3200	2	5.2	5200	2	9.2
3200	4	4.0	5200	4	8.0
3200	6	2.8	5200	6	6.8
3200	8	1.6	5200	8	5.6

The $Z = 3200$ line is a less expensive but an infeasible solution that does not intersect the feasible solution line CB. The $Z = 5200$ line comes closer to intersecting the feasible line and is more expensive than the $Z = 4200$ line. The $Z = 4200$ objective function line is a less expensive solution than the $Z = 5200$ line and therefore would be selected even if $Z = 5200$ was feasible. Indeed, the $Z = 4200$ line is the optimum solution. Any objective function line that we derive from $Z = 300X_1 + 500X_2$ and that we pass through the

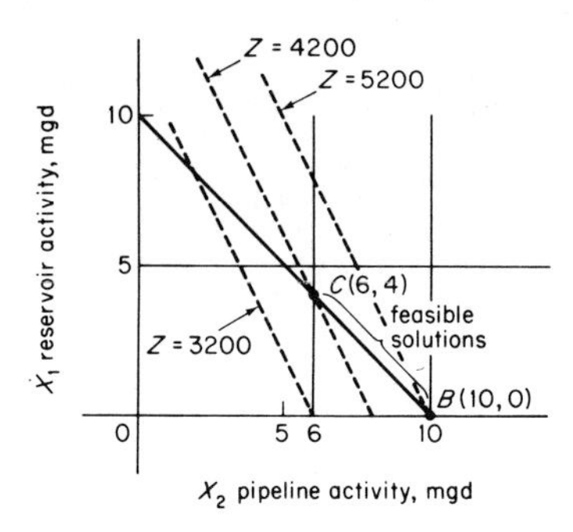

Fig. 8

line of feasible solution will be a more expensive one than $Z = 4200$, $X_1 = 4$, $X_2 = 6$.

Summarizing, a graphical procedure has produced an optimum and feasible solution. We found a line of feasible solutions by a step-by-step introduction of the constraints that narrowed the area of feasibility. Then we mapped alternative objective function solutions onto the feasible solutions to find a least-cost solution.

4. INFEASIBLE SOLUTION

We may use graphs to give geometrical interpretations of three other possible linear programming solutions: infeasible, unbounded, and multiple optima. An infeasible solution is one in which one or more of the constraints is not satisfied. If we change Eq. (4) from $X_2 \leq 10$ to $X_2 \leq 4$, we have an infeasible solution because 10 mgd are demanded [Eq. (2)] and only 9 mgd are available: reservoir [Eq. (3)] plus pipeline [Eq. (4)]. Graphically, the infeasible solution is apparent from the fact that there is no common point at which all six constraints are satisfied (see Fig. 9). Equations (3) and (4) do not have a common intersection with Eq. (2), demand. Equation (4), pipeline restriction, and Eq. (5), amount that must come through the pipeline, are contradictory. When you work with a real problem, the computer will clearly identify infeasible solutions and flag the specific constraints that are not being satisfied.

5. UNBOUNDED SOLUTION

An unbounded solution has an unconstrained infinite solution. For example, in the following problem, the objective function [Eq. (8)] crosses the four constraints [Eqs. (9)–(12)] in such a way as to make it impossible to

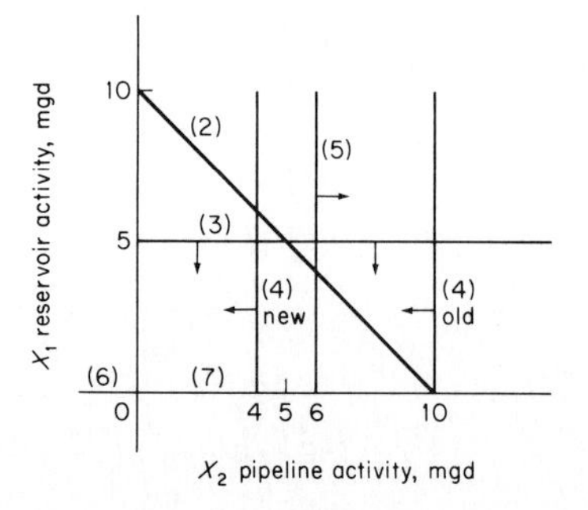

Fig. 9

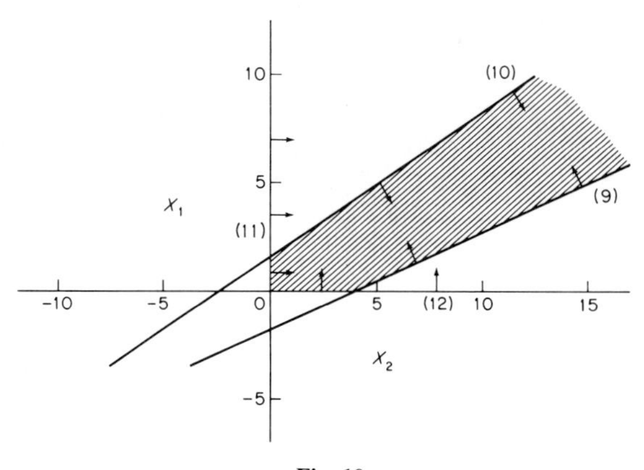

Fig. 10

find a bounded maximum solution (see Fig. 10). We can continue to construct an infinite number of higher and higher valued objective function values that will be feasible. If an additional constraint is added, then an optimum feasible solution can be found within the hatched polygon (Fig. 11). An unbounded solution usually results from poor preparation of constraint equations and is identified by the computer solutions.

Maximize: $$2X_1 + 4X_2 \tag{8}$$

subject to

$$-2X_1 + X_2 \leq 4 \tag{9}$$

$$-X_1 + X_2 \geq -2 \tag{10}$$

$$X_1 \geq 0 \tag{11}$$

$$X_2 \geq 0 \tag{12}$$

Additional constraint: $$X_1 + X_2 \leq 3 \tag{13}$$

One of the problems at the end of Chapter 2 has an unbounded solution.

6. MULTIPLE-OPTIMA SOLUTION

An unbounded solution has an unbounded infinite solution which is not useful. A multiple-optima solution has a bounded infinite set of solutions which are normally useful. A multiple-optima solution is produced by the objective function equation paralleling one or more of the boundary lines of the feasible solution space, or by redundant constraints, or both. The result is an infinite combination of solutions within a precise solution space. For example, if we change the objective function of the water supply problem

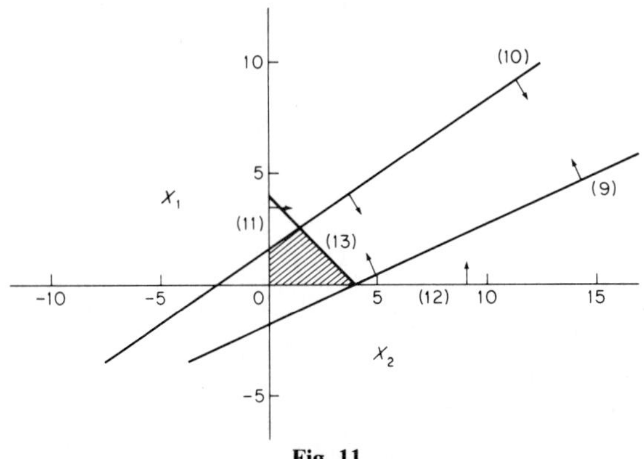

Fig. 11

from minimize: $300X_1 + 500X_2$ to minimize: $300X_1 + 300X_2$, we will have a multiple-optima solution. At a cost of \$3000, the objective function equation $300X_1 + 300X_2$ directly overlaps the feasible solution line *CB*. An infinite combination of optimal solutions between *C* and *B* are possible (see Fig. 12). In contrast to an unbounded solution, a multiple-optima solution may be a useful result since it presents decisionmakers with an infinite variety of options within a bounded area. One of the exercise problems for Chapter 2 has multiple optima.

Summarizing, the graphical method has helped us illustrate that the following solutions are possible:

(1) one optimum feasible solution,
(2) no feasible solution, and
(3) infinitely many feasible solutions.

Chapter 3 will explore the more abstract concepts of a dual solution and in the process will reexamine the multiple-optima solution and the associated degenerate solution.

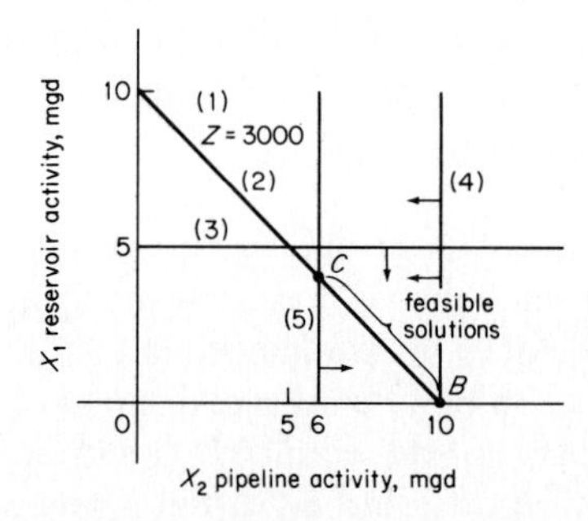

Fig. 12

D. THE CORNER POINT, THE OPTIMAL SOLUTION, AND THE COMPUTER

Perhaps you noticed that the optimal solution to the water supply problem was found at the corner point of the line *CB* and at a corner point of the polygon *CBDE*. With the exception of the multiple-optima case, the optimal solution will always be one of the corner points, and the corner points will be feasible solutions.

This fortunate mathematical characteristic is the basis for solving linear programming problems. The simplex method that is detailed in Chapter 2 begins by finding any corner point of the feasible solution. The remainder of the mathematical process consists of systematically evaluating corner points to determine if the solution is in the direction of increased optimality. If one of the adjacent corner points is more optimal, then a move is made to that adjacent corner point; if not,we have found the optimum.

In essence, the mathematical solution to linear programming problems is a systematic trial-and-error procedure which finds a corner point and then moves toward optimality. Mathematical rules instruct the computer how and when to shift corner points. At each step, the computer moves closer to the optimal solution, if it exists, and the computer codes have been constructed to yield the optimum result in a finite number of steps.

The computer codes are critical. The computer program reviewed in Chapter 4 has solved linear programming problems developed by the author that contain more than 200 constraint equations and 200 possible activities. These problems usually took less than 30 seconds of computer time and cost less than $5.00. Had we been so foolish as to have tried to solve these programs by hand calculations, we would have quickly realized such an effort was beyond the capability of a squad of the most fervent people. Specifically, given liberal assumptions about our abilities to add, subtract, multiply, and divide quickly, a single solution to one of the author's linear programming problems would have taken a full-time worker more than 1 million years.

E. APPLICATIONS: HISTORICAL AND RECENT PERSPECTIVES

The Second World War provided the interest and funds for the development of solutions to many operations research problems, including the problem solving of a system of equations with inequality constraints. Large-scale military planning such as fleets of cargo ships and convoys required efficient resource allocation. While linear programming problems had been formulated in the early nineteenth century by Fourier, a French mathematician, and shortly before the Second World War by Kantorovich (1960), a Russian mathematician, these works were basically ignored because of a lack of serious enough application to warrant research funds. The Second World War provided the urgency and the funds.

George Dantzig developed the simplex method in 1947. T. C. Koopmans (1951) developed a special linear programming solution for transportation problems and organized a major conference. During the early 1950s, Charnes *et al.* (1953) wrote the first linear programming text and industrial applications had begun.

Industry and business have been the major users of linear programming. The biggest users are petroleum refineries. They seek, for example, to maximize the value of the oil inputs subject to constraints on refinery equipment and gasoline blend requirements, and to find the best locations for pipelines and best routes and schedules for tankers. Other major industrial users are the steel, dairy, and paper industries. A major agricultural application has been the least-cost solution to choosing cattle feed. Overall, the four most common applications of linear programming are finding the least-cost means of meeting product specifications, determining optimal production processes and/or products, finding least-cost transportation routes, and determining the best schedules for production and sales.

This book emphasizes recent applications of linear programming in the fields of the socioeconomic and environmental sciences. Most of these applications are products of the 1970s. Specifically, about 60% of the applications reviewed in this book were published in 1970 or later. Another 30% appeared between 1965 and 1969. Few applications in these fields appeared during the 1950s, when the groundwork was laid for most industrial and business applications. In short, the vast majority of linear programming applications in the socioeconomic and environmental sciences are very recent.

Even while linear programming is being applied to the disciplines which constitute the core of this book, many for the first time, and some without success, it has become apparent that the proportionality and linearity assumptions of linear programming are not always tenable. For example, a linear programming solution can suggest that 2.4 new schools be constructed. Obviously, a fraction of a school, bridge, water pipe, or other fixed facility will be of no use. The need for integer solutions has led to the formulation of mathematical tools and computer codes for solving linear programming problems which require integer solutions. Other optimization problems have nonlinear objective functions or constraints, or both. These problems are much more difficult to solve and are reviewed in Chapter 3.

F. SUMMARY

The purpose of this chapter has been to provide an overview of the theory, methods, and applications of linear programming. We began by introducing the basic theories and mathematical assumptions of linear programming. Next, a graphical solution was used to illustrate the following possible solutions: feasible, optimal, infeasible, unbounded, and multiple optima. The relationship between the corner point, a solution, and the computer was reviewed. Finally, a historical and recent perspective on applications was provided.

REFERENCES

Ansoff, H. I., ed. (1969). "Business Strategy." Penguin, Baltimore.

Charnes, A., Cooper, W., and Henderson, A. (1953). "An Introduction to Linear Programming." Wiley, New York.

Clarkson, G. P., ed. (1968). "Managerial Economics." Penguin, Baltimore.

Kantorovich, L. V. (1960). "Mathematical Methods of Organizing and Planning Production." Translated from the Russian and reprinted in *Management Sci.* **6**, No. 4, 366–422.

Koopmans, T. C., ed. (1951). "Activity and Analysis of Production and Allocation." Wiley, New York.

Pugh, D. S., ed. (1971). "Organization Theory." Penguin, Baltimore.

Starbuck, W. H., ed. (1971). "Organizational Growth and Development." Penguin, Baltimore.

Chapter 2

Algebraic Methods for Solving Linear Programming Problems

A. ESSENTIAL MATRIX METHODS

Linear programming may be discussed as a theory of optimization and graphically illustrated with two-variable examples. The notion of simplicity conveyed by the relatively simple theory and graphical solutions gives way to the harsh reality of real-world applications that include hundreds of activities, hundreds of constraints, and thousands of decision choices. Meaningful problems can only be solved by computer algorithms. Fortunately, the solutions are based on elementary algebra. This chapter will demonstrate that, by and large, the rules for manipulating data sets duplicate the rules used for individual numbers.

1. DEFINITIONS AND NOTATION[1]

As in most specialized fields, the manipulation of large data sets has brought forth a technical jargon that dominates the literature. The pivotal term in large-data set analysis is *matrix* (plural: *matrices*). A matrix is a set of

[1] A simple introduction to matrix algebra is provided by the School Mathematics Study Group (1960). See Almon (1969) for an interesting treatment. Formal mathematical presentations are found in Hohn (1964) and in Marcus and Minc (1965).

symbols systematically arranged by rows and columns. The symbols may be numbers or any other notation that can be stored as part of a data bank.

The accompanying tableau is an example of a matrix containing information about the socioeconomic characteristics of a population in three counties.

	Variable		
Geographical unit	Median family income $ thousands	Median education completed	Median family size
County 1	12	12	2.80
County 2	10	11	3.00
County 3	8	9	2.90

The nine pieces of information are called *elements* of the matrix. Each element in the matrix is identified by the intersection of a row and column. The first subscript identifies the row, the second identifies the column. The median education completed in county 2 is 11 years. The element is at the intersection of row 2 (county 2) and column 2 (median education completed). It is represented as a_{22}. The element a_{23} is 3.00, the element a_{31} is 8, and so forth.

The matrix of which the elements are a part of may be represented by any of the following symbols: A, $\bar{A}$, $[A]$, and $[a_{ij}]$. The last form expresses that the matrix $\bar{A}$ has i rows and j columns. Matrix $\bar{A}$ has 3 rows and 3 columns and may be represented in the following way: $\bar{A} = [a_{ij}]$, $i = 1, 2, 3$, $j = 1, 2, 3$. Matrix $\bar{A}$ is said to have the *order* of 3 by 3. The matrix $\bar{B}$ with 4 rows and 2 columns has the order $[b_{42}]$ and is represented as $\bar{B} = [b_{ij}]$, $i = 1, 2, 3, 4$, $j = 1, 2$. In the most general case, the matrix $\bar{C}$ with m rows and n columns may be represented as $\bar{C} = [c_{ij}]$, $i = 1, 2, \ldots, m$, $j = 1, 2, \ldots, n$.

Eight special types of matrices should be distinguished for the purposes of this book: square, transpose, equal, diagonal, identity, null, symmetric, and vector. A matrix in which the number of rows equals the number of columns as in $\bar{A}$ above is called *square*. Square matrices are not especially important in linear programming; however, they are the standard operating matrix in closely related fields such as interindustry analysis, correlation, and factor analysis. A *transpose* matrix has the property that the column elements of one matrix are identical to the row elements of the second matrix. In essence, the rows in one matrix become the columns in the second. The upper right-hand element changes places with the lower left-hand element.

$$\bar{A} = \begin{bmatrix} 2 & 2 & 6 \\ 4 & 1 & 1 \\ 8 & 3 & 2 \end{bmatrix}; \qquad \bar{B} = \begin{bmatrix} 2 & 4 & 8 \\ 2 & 1 & 3 \\ 6 & 1 & 2 \end{bmatrix}; \qquad \bar{A}^{\mathrm{T}} = \bar{B}$$

Matrix $\bar{B}$ is the transpose of matrix $\bar{A}$. For example, element a_{23} equals element b_{32}; element a_{21} equals element b_{12}. The transpose matrix is used as an aid in solving some matrix problems.

Equal matrices have the property that all corresponding elements of both matrices are equal:

$$\bar{C} = \begin{bmatrix} 1 & 4 & 5 \\ 2 & 1 & 3 \end{bmatrix}; \quad \bar{D} = \begin{bmatrix} 1 & 4 & 5 \\ 2 & 1 & 3 \end{bmatrix}; \quad \bar{E} = \begin{bmatrix} 1 & 4 & 5 & 1 \\ 2 & 1 & 3 & 2 \end{bmatrix}$$

where

$$\bar{C} = \bar{D}; \qquad \bar{C} \neq \bar{E}; \qquad \bar{D} \neq \bar{E}$$

A square matrix in which the principal diagonal elements do not all equal zero and the nondiagonal elements equal zero is called a *diagonal* matrix. Elements f_{11}, f_{22}, and f_{33} in matrix $\bar{F}$ do not equal zero:

$$\bar{F} = \begin{bmatrix} 2 & 0 & 0 \\ 0 & 1 & 0 \\ 0 & 0 & 3 \end{bmatrix}$$

An *identity* matrix is a special case of a diagonal matrix in which the diagonal values are 1:

$$\bar{G} = \begin{bmatrix} 1 & 0 & 0 \\ 0 & 1 & 0 \\ 0 & 0 & 1 \end{bmatrix}$$

The identity matrix $\bar{I}$ is a key tool in matrix algebra. It serves the same purpose as the number 1 does in algebra. As we will see in the next section, the identity matrix is an important tool in linear programming.

A *null* matrix contains all 0 elements and is equivalent to the value 0 in algebra:

$$\bar{H} = \begin{bmatrix} 0 & 0 & 0 \\ 0 & 0 & 0 \end{bmatrix}$$

A *symmetric* matrix has the property that the row values equal the corresponding column values in the same matrix. A correlation coefficient matrix is symmetric. The transpose and the initial matrix are equal:

$$\bar{J} = \bar{J}^{\mathrm{T}} = \begin{bmatrix} 1 & 2 & 3 \\ 2 & 2 & 6 \\ 3 & 6 & 4 \end{bmatrix}$$

Finally, a *vector* is a special matrix which possesses only one row or one column:

$$\bar{K} = \begin{bmatrix} 2 \\ 1 \\ 3 \end{bmatrix}; \qquad \bar{L} = [2 \quad 4 \quad 7 \quad 9 \quad 3]$$

2. RULES FOR MATRIX ADDITION, SUBTRACTION, AND MULTIPLICATION

Two rules apply to adding and subtracting matrices. First, the matrices must be of the same order to be conformable for addition and subtraction. Second, the elements of one matrix are added to or subtracted from the corresponding elements of the second matrix.

$$\bar{A} = \begin{bmatrix} 1 & 3 \\ 2 & 2 \\ 3 & 1 \end{bmatrix}; \qquad \bar{B} = \begin{bmatrix} 4 \\ 1 \\ 6 \end{bmatrix}; \qquad \bar{C} = \begin{bmatrix} 2 & -3 \\ 1 & 4 \\ 2 & -1 \end{bmatrix}$$

Matrices $\bar{A}$ and $\bar{C}$ may be added because they are of the same 3 by 2 order. The order of matrix $\bar{B}$ is 3 by 1. The matrix sum of $\bar{A}$ and $\bar{C}$ is formed by adding the corresponding elements.

$$\bar{A} + \bar{C} = \bar{D}$$

$$\begin{bmatrix} 1 & 3 \\ 2 & 2 \\ 3 & 1 \end{bmatrix} + \begin{bmatrix} 2 & -3 \\ 1 & 4 \\ 2 & -1 \end{bmatrix} = \begin{bmatrix} 1+2 & 3-3 \\ 2+1 & 2+4 \\ 3+2 & 1-1 \end{bmatrix} = \begin{bmatrix} 3 & 0 \\ 3 & 6 \\ 5 & 0 \end{bmatrix}$$

$$\bar{A} - \bar{C} = \bar{E}$$

$$\begin{bmatrix} 1 & 3 \\ 2 & 2 \\ 3 & 1 \end{bmatrix} - \begin{bmatrix} 2 & -3 \\ 1 & 4 \\ 2 & -1 \end{bmatrix} = \begin{bmatrix} 1-2 & 3-(-3) \\ 2-1 & 2-4 \\ 3-2 & 1-(-1) \end{bmatrix} = \begin{bmatrix} -1 & 6 \\ 1 & -2 \\ 1 & 2 \end{bmatrix}$$

Matrix multiplication is possible when the number of columns in the first (lead) matrix equals the number of rows in the second (lag) matrix.

$$\bar{A} = \begin{bmatrix} 2 & 3 & 1 \\ 1 & -4 & 1 \\ -3 & 6 & 2 \end{bmatrix}; \qquad \bar{B} = \begin{bmatrix} 4 & -1 & 2 \\ 3 & 1 & -3 \end{bmatrix}; \qquad \bar{C} = \begin{bmatrix} -1 \\ 2 \\ 3 \end{bmatrix}$$

Four of the nine matrix combinations are conformable for multiplication: $\bar{A} \cdot \bar{A}$, $\bar{A} \cdot \bar{C}$, $\bar{B} \cdot \bar{A}$, and $\bar{B} \cdot \bar{C}$. The combination $\bar{A} \cdot \bar{A}$ has 3 columns and 3 rows; $\bar{A} \cdot \bar{C}$ has 3 columns and 3 rows; $\bar{B} \cdot \bar{A}$ has 3 columns and 3 rows; and $\bar{B} \cdot \bar{C}$ has 3 columns and 3 rows.

The rule for matrix multiplication is row elements in the first matrix times the corresponding column elements in the second matrix. The final matrix has the order defined by the number of rows in the first matrix and number of columns in the second matrix. In the four conformable cases preceding, the following matrix orders are produced by multiplication:

$$\bar{A}_{33} \cdot \bar{A}_{33} = \bar{L}_{33}$$

$$\bar{A}_{33} \cdot \bar{C}_{31} = \bar{M}_{31}$$

$$\bar{B}_{23} \cdot \bar{A}_{33} = \bar{N}_{23}$$

$$\bar{B}_{23} \cdot \bar{C}_{31} = \bar{O}_{21}$$

For example,

$$\bar{B}\bar{C} = \bar{O}$$

$$\bar{B} = \begin{bmatrix} 4 & -1 & 2 \\ 3 & 1 & -3 \end{bmatrix}; \quad \bar{C} = \begin{bmatrix} -1 \\ 2 \\ 3 \end{bmatrix}$$

$$\bar{O} = \begin{bmatrix} (4)(-1) + (-1)(2) + (2)(3) \\ (3)(-1) + (1)(2) + (-3)(3) \end{bmatrix} = \begin{bmatrix} 0 \\ -10 \end{bmatrix} = \begin{bmatrix} 0 \\ -10 \end{bmatrix}$$

As a second example,

$$\bar{B}\bar{A} = \bar{N}$$

$$\bar{B} = \begin{bmatrix} 4 & -1 & 2 \\ 3 & 1 & -3 \end{bmatrix}; \quad \bar{A} = \begin{bmatrix} 2 & 3 & 1 \\ 1 & -4 & 1 \\ -3 & 6 & 2 \end{bmatrix}$$

$$\bar{N} = \begin{bmatrix} 8 - 1 - 6 & 12 + 4 + 12 & 4 - 1 + 4 \\ 6 + 1 + 9 & 9 - 4 - 18 & 3 + 1 - 6 \end{bmatrix}$$

$$\bar{N} = \begin{bmatrix} 1 & 28 & 7 \\ 16 & -13 & -2 \end{bmatrix}$$

The general matrix multiplication sequence can be further illustrated by defining and multiplying two matrices, $\bar{A}_{23}$ and $\bar{B}_{32}$, which yield a matrix $\bar{C}_{22}$.

$$\bar{A} = \begin{bmatrix} a_{11} & a_{12} & a_{13} \\ a_{21} & a_{22} & a_{23} \end{bmatrix}; \quad \bar{B} = \begin{bmatrix} b_{11} & b_{12} \\ b_{21} & b_{22} \\ b_{31} & b_{32} \end{bmatrix}$$

$$\bar{C} = \begin{bmatrix} a_{11}b_{11} + a_{12}b_{21} + a_{13}b_{31} & a_{11}b_{12} + a_{12}b_{22} + a_{13}b_{32} \\ a_{21}b_{11} + a_{22}b_{21} + a_{23}b_{31} & a_{21}b_{12} + a_{22}b_{22} + a_{23}b_{32} \end{bmatrix}$$

Finally, matrices may be multiplied by constants which are called *scalars*. Each element in the matrix is multiplied by the scalar.

$$\text{scalar } K = -3; \qquad \bar{A} = \begin{bmatrix} 2 & 1 \\ 4 & 4 \\ 3 & 2 \end{bmatrix}$$

$$K\bar{A} = -3\begin{bmatrix} 2 & 1 \\ 4 & 4 \\ 3 & 2 \end{bmatrix} = \begin{bmatrix} -6 & -3 \\ -12 & -12 \\ -9 & -6 \end{bmatrix}$$

3. THE COMMUTATIVE LAW FOR MATRIX MULTIPLICATION AND INVERSE MULTIPLICATION

One important difference between matrix algebra and the algebra taught in public school concerns the order of placement of matrices in equations. In standard algebra $2 \times 3 \times 4 = 4 \times 3 \times 2 = 24$. In matrix algebra, with one important exception, the placement of matrices is important. Usually, $\bar{B}\bar{A} \neq \bar{A}\bar{B}$. For example,

$$\bar{A} = \begin{bmatrix} 2 \\ 1 \\ 3 \end{bmatrix}; \qquad \bar{B} = [1 \quad 1 \quad 2]$$

$$\bar{A} \cdot \bar{B} = \begin{bmatrix} 2 & 2 & 4 \\ 1 & 1 & 2 \\ 3 & 3 & 6 \end{bmatrix}, \qquad \text{whereas} \quad \bar{B} \cdot \bar{A} = [9]$$

and

$$\bar{C} = \begin{bmatrix} 2 & 1 \\ 2 & 2 \end{bmatrix}; \qquad \bar{D} = \begin{bmatrix} 1 \\ 2 \end{bmatrix}$$

$$\bar{C} \cdot \bar{D} = \begin{bmatrix} 4 \\ 6 \end{bmatrix}; \qquad \text{whereas} \quad \bar{D} \cdot \bar{C} = \text{not conformable}$$

In linear programming, the exception to the commutative law is as important as the general case. The exception is that a matrix $\bar{B}$ may be obtained which, when multiplied by $\bar{A}$, yields the identity matrix $\bar{I}$:

$$\bar{A} \cdot \bar{B} = \bar{I} = \bar{B} \cdot \bar{A}$$

The $\bar{B}$ matrix is called the inverse of $\bar{A}$ and is written $\bar{A}^{-1}$. Hence,

$$\bar{A} \cdot \bar{A}^{-1} = \bar{I} = \bar{A}^{-1} \cdot \bar{A}$$

In standard algebra, the operations of dividing by a whole number and of multiplying by the fractional inverse accomplishes the same result:

$$\tfrac{3}{4} = 3 \times (\tfrac{1}{4})$$

Division is not defined in matrix algebra, and the inverse matrix substitutes for the role of the diviser.

There are numerous ways of finding the inverse of a matrix. These are presented in standard texts.[2] Only *synthetic elimination* will be presented in this volume, because it is the method used in performing matrix operations in linear programming. The matrix $\bar{A}^{-1}$ is found by augmenting $\bar{A}$ on the right by an identity matrix $\bar{I}$ and then transforming $\bar{A}$ into $\bar{I}$. When $\bar{A}$ is transformed into $\bar{I}$, the initial $\bar{I}$ matrix becomes the inverse $\bar{A}^{-1}$. The series of operations is called synthetic elimination.

We start with a matrix $\bar{A}$:

$$\bar{A} = \begin{bmatrix} 3 & 1 & 2 \\ 2 & 1 & 1 \\ 0 & -1 & 2 \end{bmatrix}$$

which is augmented by an identity matrix

$$\begin{bmatrix} 3 & 1 & 2 \\ 2 & 1 & 1 \\ 0 & -1 & 2 \end{bmatrix} \begin{bmatrix} 1 & 0 & 0 \\ 0 & 1 & 0 \\ 0 & 0 & 1 \end{bmatrix}$$

To convert $\bar{A}$ to an identity matrix, the principal diagonal elements (a_{11}, a_{22}, a_{33}) are converted to 1, and all of the other elements in $\bar{A}$ are converted to zeros.

Step 1 Multiply row 1 by $\frac{1}{3}$ to convert a_{11} to 1:

$$\tfrac{1}{3}[3 \quad 1 \quad 2][1 \quad 0 \quad 0] = [1 \quad \tfrac{1}{3} \quad \tfrac{2}{3}][\tfrac{1}{3} \quad 0 \quad 0]$$

Step 2 Multiply the new row 1 by -2 and add it to row 2 to convert a_{21} to 0:

$$-2[1 \quad \tfrac{1}{3} \quad \tfrac{2}{3}][\tfrac{1}{3} \quad 0 \quad 0] = [-2 \quad -\tfrac{2}{3} \quad -\tfrac{4}{3}][-\tfrac{2}{3} \quad 0 \quad 0]$$

transformed row 1	-2	$-\frac{2}{3}$	$-\frac{4}{3}$	$-\frac{2}{3}$	0	0
+row 2	2	1	1	0	1	0
new row 2	0	$\frac{1}{3}$	$-\frac{1}{3}$	$-\frac{2}{3}$	1	0

Since the element a_{31} is already zero, additional transformations of column 1 are not required.

[2] Faddeeva (1959) provides an excellent summary of these methods.

whoah! all the way back to 1959! to cite a matrices book?

At the conclusion of the first column operation, $\bar{A}$ has been transformed into $\bar{A}_1$ and the identity matrix into a matrix $\bar{B}$.

$$\bar{A}_1 = \begin{bmatrix} 1 & \frac{1}{3} & \frac{2}{3} \\ 0 & \frac{1}{3} & -\frac{1}{3} \\ 0 & -1 & 2 \end{bmatrix}; \qquad \bar{B} = \begin{bmatrix} \frac{1}{3} & 0 & 0 \\ -\frac{2}{3} & 1 & 0 \\ 0 & 0 & 1 \end{bmatrix}$$

The third, fourth, and fifth steps are to convert a_{22} in $\bar{A}_1$ into 1, and a_{12} and a_{32} into zeros. The element a_{22} in $\bar{A}_1$ is converted into 1 by multiplication of row 2 by 3.

Step 3

$$\text{row 2} \qquad 3[0 \;\; \tfrac{1}{3} \;\; -\tfrac{1}{3}][-\tfrac{2}{3} \;\; 1 \;\; 0]$$

equals

$$\text{new row 2} \qquad [0 \;\; 1 \;\; -1][-2 \;\; 3 \;\; 0]$$

Step 4 Multiply new row 2 by $-\frac{1}{3}$ and add it to row 1.

$$\text{row 2} \qquad -\tfrac{1}{3}[0 \;\; 1 \;\; -1][-2 \;\; 3 \;\; 0] = [0 \;\; -\tfrac{1}{3} \;\; \tfrac{1}{3}][\tfrac{2}{3} \;\; -1 \;\; 0]$$

transformed row 2 from step 4	0	$-\frac{1}{3}$	$\frac{1}{3}$	$\frac{2}{3}$	-1	0
+ row 1	1	$\frac{1}{3}$	$\frac{2}{3}$	$\frac{1}{3}$	0	0
new row 1	1	0	1	1	-1	0

Step 5 Add new row 2 to row 3.

transformed row 2 from step 3	0	1	-1	-2	3	0
+ row 3	0	-1	2	0	0	1
new row 3	0	0	1	-2	3	1

At the end of the second column operation, $\bar{A}$ has been transformed into $\bar{A}_2$ and the identity matrix into $\bar{B}_1$:

$$\bar{A}_2 = \begin{bmatrix} 1 & 0 & 1 \\ 0 & 1 & -1 \\ 0 & 0 & 1 \end{bmatrix}; \qquad \bar{B}_1 = \begin{bmatrix} 1 & -1 & 0 \\ -2 & 3 & 0 \\ -2 & 3 & 1 \end{bmatrix}$$

Finally, the sixth and seventh steps convert a_{33} into 1 and a_{13} and a_{23} into 0. The element a_{33} is already 1 and does not need further transformation.

Step 6 Multiply row 3 by -1 and add it to row 1.

$$\text{row 3} \quad -1[0 \ \ 0 \ \ 1][-2 \ \ 3 \ \ 1] = [0 \ \ 0 \ \ -1][2 \ \ -3 \ \ -1]$$

transformed row 3	0	0	−1	2	−3	−1
+row 1	1	0	1	1	−1	0
new row 1	1	0	0	3	−4	−1

Step 7 Add row 3 to row 2.

transformed row 3 from step 5	0	0	1	−2	3	1
+row 2	0	1	−1	−2	3	0
new row 2	0	1	0	−4	6	1

The final matrices $\bar{A}_3$ and $\bar{B}_2$ are as follows:

$$\bar{A}_3 = \begin{bmatrix} 1 & 0 & 0 \\ 0 & 1 & 0 \\ 0 & 0 & 1 \end{bmatrix}; \qquad \bar{B}_2 = \begin{bmatrix} 3 & -4 & -1 \\ -4 & 6 & 1 \\ -2 & 3 & 1 \end{bmatrix}$$

Matrix $\bar{A}_3$ is an identity matrix, and $\bar{B}_2$ should be $\bar{A}^{-1}$. If $\bar{B}_2$ is $\bar{A}^{-1}$, then $\bar{A} \cdot \bar{B}_2 = \bar{I}$.

$$\bar{A} = \begin{bmatrix} 3 & 1 & 2 \\ 2 & 1 & 1 \\ 0 & -1 & 2 \end{bmatrix}; \qquad \bar{B}_2 = \begin{bmatrix} 3 & -4 & -1 \\ -4 & 6 & 1 \\ -2 & 3 & 1 \end{bmatrix}$$

$$\bar{A} \cdot \bar{B}_2 = \begin{bmatrix} 1 & 0 & 0 \\ 0 & 1 & 0 \\ 0 & 0 & 1 \end{bmatrix}, \qquad \text{therefore} \quad \bar{B}_2 = \bar{A}^{-1}$$

At this point, synthetic elimination may seem hopelessly tedious and complicated. The only solution for the tedium is to recognize that once you learn the basic procedure, you will turn to the computer algorithm presented in Chapter 4. The complication can be avoided by using three guidelines.

(1) Always begin with the uppermost principal diagonal element a_{11} and move systematically through $a_{22}, a_{33}, \ldots, a_{mn}$. The element that is the focal point for the row transformation is called the *pivot* element. While this

guideline may add 5 or 10 minutes work, it might save several hours of hair-pulling and fingernail-biting frustration caused by a step which was out of order.

(2) Only use the principal diagonal elements to transform the nonprincipal diagonal elements. This guideline will mean that multiplications and additions will be systematic throughout the operation.

(3) Move down column vector 1 first, column vector 2 second, and column vector n last. Specifically, once a_{11} has been converted into 1, then convert a_{21} into 0, then a_{31} into 0, then a_{41}, then ... a_{m1} into 0. Again, deviation from this guideline could save a few minutes work, but it might cause hours of additional work looking for errors.

If you follow these three guidelines, the maximum number of row transformations that are required equals the number of elements in the matrix. In the matrix $\bar{A}$ case presented above, seven transformations were necessary to convert $\bar{A}$ into $\bar{I}$. Two steps below the maximum were possible because in the column 1 operations (between steps 2 and 3), the element a_{31} was 0 in the original $\bar{A}$ matrix. In the column 3 operations, the element a_{33} was the required value 1 after step 5.

As a summary of matrix addition, subtraction, and multiplication, the following equation will be solved:

$$\bar{A}\bar{A} + \bar{A}\bar{B}^{-1} - 2\bar{B}\bar{A}$$

where

$$\bar{A} = \begin{bmatrix} 2 & 1 \\ 0 & 1 \end{bmatrix}; \qquad \bar{B} = \begin{bmatrix} 1 & 2 \\ -1 & 2 \end{bmatrix}$$

To find $\bar{A}\bar{A}$

$$\begin{bmatrix} 2 & 1 \\ 0 & 1 \end{bmatrix} \cdot \begin{bmatrix} 2 & 1 \\ 0 & 1 \end{bmatrix} = \begin{bmatrix} 4 & 3 \\ 0 & 1 \end{bmatrix}$$

To find $\bar{B}^{-1}$

$$\begin{bmatrix} 1 & 2 \\ -1 & 2 \end{bmatrix}\begin{bmatrix} 1 & 0 \\ 0 & 1 \end{bmatrix} = \begin{bmatrix} 1 & 2 \\ 0 & 4 \end{bmatrix}\begin{bmatrix} 1 & 0 \\ 1 & 1 \end{bmatrix}$$

$$= \begin{bmatrix} 1 & 2 \\ 0 & 1 \end{bmatrix}\begin{bmatrix} 1 & 0 \\ \frac{1}{4} & \frac{1}{4} \end{bmatrix} = \begin{bmatrix} 1 & 0 \\ 0 & 1 \end{bmatrix}\begin{bmatrix} \frac{1}{2} & -\frac{1}{2} \\ \frac{1}{4} & \frac{1}{4} \end{bmatrix}$$

To find $\bar{A}\bar{B}^{-1}$

$$\begin{bmatrix} 2 & 1 \\ 0 & 1 \end{bmatrix} \cdot \begin{bmatrix} \frac{1}{2} & -\frac{1}{2} \\ \frac{1}{4} & \frac{1}{4} \end{bmatrix} = \begin{bmatrix} \frac{5}{4} & -\frac{3}{4} \\ \frac{1}{4} & \frac{1}{4} \end{bmatrix}$$

To find $\bar{B}\bar{A}$

$$\begin{bmatrix} 1 & 2 \\ -1 & 2 \end{bmatrix} \cdot \begin{bmatrix} 2 & 1 \\ 0 & 1 \end{bmatrix} = \begin{bmatrix} 2 & 3 \\ -2 & 1 \end{bmatrix}$$

To find $2\bar{B}\bar{A}$

$$2\begin{bmatrix} 2 & 3 \\ -2 & 1 \end{bmatrix} = \begin{bmatrix} 4 & 6 \\ -4 & 2 \end{bmatrix}$$

To find $\bar{A}\bar{A} + \bar{A}\bar{B}^{-1} - 2\bar{B}\bar{A}$

$$\begin{bmatrix} 4 & 3 \\ 0 & 1 \end{bmatrix} + \begin{bmatrix} \frac{5}{4} & -\frac{3}{4} \\ \frac{1}{4} & \frac{1}{4} \end{bmatrix} - \begin{bmatrix} 4 & 6 \\ -4 & 2 \end{bmatrix} = \begin{bmatrix} \frac{5}{4} & -\frac{15}{4} \\ \frac{17}{4} & -\frac{3}{4} \end{bmatrix}$$

4. SOLVING SIMULTANEOUS EQUATIONS

In the previous section, we found that augmenting $\bar{A}$ by $\bar{I}$ on the right side and converting $\bar{A}$ into $\bar{I}$ by synthetic elimination produced $\bar{A}^{-1}$. Synthetic elimination may also be used to solve a set of simultaneous equations such as the following:

$$2X_1 + 3X_2 = 8$$

$$-X_1 + X_2 = 1$$

To derive the unique values for X_1 and X_2, we first transform the problem into a matrix format.

$$\bar{A}\bar{X} = \bar{B}$$

The $\bar{A}$ values are the coefficient values of X_1 and X_2. The X_1 and X_2 values are unknown, and the B values are right-hand sides.

$$\bar{A} = \begin{bmatrix} 2 & 3 \\ -1 & 1 \end{bmatrix}; \quad \bar{X} = \begin{bmatrix} x_1 \\ x_2 \end{bmatrix}; \quad \bar{B} = \begin{bmatrix} 8 \\ 1 \end{bmatrix}$$

Next, we augment $\bar{A}$ with $\bar{B}$.

$$\begin{bmatrix} 2 & 3 \\ -1 & 1 \end{bmatrix} \quad \begin{bmatrix} 8 \\ 1 \end{bmatrix}$$

Third, $\bar{A}$ is transformed into $\bar{I}$ by synthetic elimination and in the process solves for $\bar{X}$.

Step 1 Multiply row 1 by $\frac{1}{2}$ and add the new row 1 to row 2.

$$\text{row 1} \quad \tfrac{1}{2}[2 \quad 3][8] = [1 \quad \tfrac{3}{2}][4]$$

new row 1	1	$\frac{3}{2}$	4
+row 2	−1	1	1;
new row 2	0	$\frac{5}{2}$	5

$$\begin{bmatrix} 1 & \frac{3}{2} \\ 0 & \frac{5}{2} \end{bmatrix}\begin{bmatrix} 4 \\ 5 \end{bmatrix}$$

Step 2 Multiply the new row 2 by $\frac{2}{5}$.

$$\text{row 2} \quad \tfrac{2}{5}[0 \quad \tfrac{5}{2}][5] = [0 \quad 1][2]$$

Step 3 Multiply row 2 produced in step 2 by $-\frac{3}{2}$ and add it to row 1.

$$\text{row 2} \quad -\tfrac{3}{2}[0 \quad 1][2] = [0 \quad -\tfrac{3}{2}][-3]$$

transformed row 2	0	$-\frac{3}{2}$	−3
+row 1	1	$\frac{3}{2}$	4
new row 1	1	0	1

The final values of vector $\bar{B}$ are $\begin{bmatrix} 1 \\ 2 \end{bmatrix}$, which can be verified by substitution back into the original equation:

$$2(1) + 3(2) = 8 \qquad \text{and} \qquad -(1) + (2) = 1$$

Recapitulating, synthetic elimination can be used to obtain an inverse and the unknown values of simultaneous equations.

Usually both the inverse and the unknowns are obtained at the same time. For example, assume the following equation: $\bar{A}\bar{X} = \bar{B}$:

$$x_1 + x_2 - x_3 = 2$$
$$-2x_1 + x_2 + x_3 = 3$$
$$x_1 + x_2 + x_3 = 6$$

where $\bar{B}$ is the total available resources (e.g., land, capital, labor, water), $\bar{A}$ is the per unit resource requirement for production of the $\bar{X}$'s, and $\bar{X}$ are the production choices (e.g., farming, recreation, residence, commerce, industry).

The inverse is

$$\bar{A}^{-1} = \begin{bmatrix} 0 & -\frac{1}{3} & \frac{1}{3} \\ \frac{1}{2} & \frac{1}{3} & \frac{1}{6} \\ -\frac{1}{2} & 0 & \frac{1}{2} \end{bmatrix}$$

and

$$\bar{X} = \begin{bmatrix} 1 \\ 3 \\ 2 \end{bmatrix}$$

Assume that the total resource picture is going to change from

$$\bar{B} = \begin{bmatrix} 2 \\ 3 \\ 6 \end{bmatrix} \quad \text{to} \quad \bar{B}_1 = \begin{bmatrix} 4 \\ -2 \\ 4 \end{bmatrix}$$

One could once again transform the system $\bar{A}\bar{B}$ into $\bar{I}\bar{X}$ through synthetic elimination and find that

$$\bar{X} = \begin{bmatrix} 2 \\ 2 \\ 0 \end{bmatrix}$$

Or $\bar{A}^{-1}$ can be used to obtain the second set of $\bar{X}$ values. Specifically, given the original equation, $\bar{A}\bar{X} = \bar{B}$, we premultiply both sides by $\bar{A}^{-1}$ and get $\bar{A}^{-1}\bar{A}\bar{X} = \bar{A}^{-1}\bar{B}$. Since $\bar{A}^{-1}\bar{A} = \bar{I}$, the expression reduces to $\bar{X} = \bar{A}^{-1}\bar{B}$. In short, if we wish to test alternative values of $\bar{B}$, we can perform these tests with matrix multiplications by $\bar{A}^{-1}$. Indeed, this procedure is the typical method used in input–output analysis. In linear programming the process of estimating the impact of a change in one or more elements of $\bar{B}$ on $\bar{X}$ is called *parametric programming.* Parametric programming will be reviewed in Chapter 3.

5. BASIS VECTORS IN LINEAR PROGRAMMING

The simultaneous equation systems that have been considered in this chapter are of the simplest form: all the matrices are square and the constraints are equalities. As you saw in the previous chapter, linear programming constraints may be equalities or inequalities. Furthermore, the number of activities (X's) that are optimized may be greater than, equal to, or less than the number of constraint equations. These two characteristics produce rectangular matrices and solutions which do not include nonzero values for all activities. Those vectors which are identified in the solution to the equations are *basis* vectors. The value of a basis vector is usually nonzero, but may be zero in the case of degenerate solutions Those vectors which are not defined in the solution are *nonbasis* vectors, and their value is zero.

In the case illustrated below, there are three production choices (X_1, X_2, and X_3) and two constraints:

Maximize: $$Z = E_1X_1 + E_2X_2 + E_3X_3$$

subject to

$$a_{11}X_1 + a_{12}X_2 + a_{13}X_3 \le b_1$$
$$a_{21}X_1 + a_{22}X_2 + a_{23}X_3 \le b_2$$

Unless a third constraint is added, one of the three $\bar{X}$ vector elements will be zero. The two $\bar{X}$ elements which are part of the solution and which are probably nonzero are the basis vectors; the third is the nonbasis vector.

At any step in the synthetic elimination process the basis vectors may be differentiated from the nonbasis vectors by two characteristics: the basis vectors will, with one exception, have nonzero solution (b) values, and a basis vector will appear in the tableau as part of the identity matrix, while the nonbasis vector will not. In the following tableau, vectors X_1 and X_3 are part of the basis and have the values of 2 and 1, respectively. Vector 2 is not part of the basis and has a value of 0.

Basis vectors	Solution values (b)	Vectors 1	2	3
1	2	1	3	0
3	1	0	2	1

In the second tableau, vector X_3 has a value of 4 and vector 2 has a value of 2. Vector 1 has a value of 0.

Basis vectors	Solution values (b)	Vectors 1	2	3
3	4	2	0	1
2	2	6	1	0

The exception to the rule that basis vectors assume nonzero values is the *degenerate* solution. A degenerate solution contains the number of basis vectors equal to the number of constraint equations; however, one or more of the solution values is 0. The causes and implications of a degenerate solution are reviewed in Chapter 3, and a means of recognizing a degenerate solution while executing the simplex process is presented later in this chapter.

6. MATRIX NOTATION FOR LINEAR PROGRAMMING

Matrix notation, especially the flamboyant use of summation signs and subscripts, is a problem for the uninitiated. However, once learned it becomes a precise means of expressing complex relationships.

The Greek letter Σ (capital sigma) is used to signify a summation. When ever $\sum$ appears, all quantities to the right are to be summed. Next, rather than using X, Y, Z, and other symbols as indicators of different variables, a single letter, usually X with subscripts i, j, k is used. Normally, the first value will be symbolized by X_1, the second by X_2, and so on. The first example states that we have five values of $\bar{X}$ and are to add the last three or the summation of all of the X subscript i quantities from $i = 3$ to $i = 5$.

$$\sum_{i=3}^{5} X_i = X_3 + X_4 + X_5$$

$$\bar{X} = [3,4,6,8,9]$$

$$\sum_{i=3}^{5} X_i = 23$$

The more general case is

$$\sum_{i=1}^{n} X_i = X_1 + X_2 + \cdots + X_n$$

The following cases also appear frequently in the literature:

1. $\sum_{i=1}^{n} C_i X_i = C_1 X_1 + C_2 X_2 + \cdots + C_n X_n$

Assuming vector $\bar{X}$ above and vector $\bar{C} = [1,2,3,4,5]$, we have

$$\sum_{i=1}^{n} C_i X_i = (1)(3) + (2)(4) + (3)(6) + (4)(8) + (5)(9)$$

$$= 3 + 8 + 18 + 32 + 45 = 106$$

2. $\sum_{i=1}^{n} CX_i$ or $C \sum_{i=1}^{n} X_i = CX_1 + CX_2 + \cdots + CX_n$

C in this case is a single coefficient (2). Assuming vector $\bar{X}$, we have

$$C \sum_{i=1}^{n} X_i = 2(3) + 2(4) + 2(6) + 2(8) + 2(9)$$

$$= 6 + 8 + 12 + 16 + 18 = 60$$

3. $\sum_{i=1}^{n}(X_i + Y_i) = (X_1 + Y_1) + (X_2 + Y_2) + \cdots + (X_n + Y_n)$

$\overline{Y} = [5,4,3,2,1]$ and $\overline{X}$ as the preceding.

$$\sum_{i=1}^{n}(X_i + Y_i) = (3 + 5) + (4 + 4) + (6 + 3) + (8 + 2) + (9 + 1)$$
$$= 8 + 8 + 9 + 10 + 10 = 45$$

4. $\sum_{i=1}^{n}(X_i - Y_i) = (X_1 - Y_1) + (X_2 - Y_2) + \cdots + (X_n - Y_n)$

Assuming vectors $\overline{X}$ and $\overline{Y}$ as the preceding,

$$\sum_{i=1}^{n}(X_i - Y_i) = (3 - 5) + (4 - 4) + (6 - 3) + (8 - 2) + (9 - 1)$$
$$= -2 + 0 + 3 + 6 + 8 = 15$$

5. $\sum_{i=1}^{n}(X_i - C)$ or $\sum_{i=1}^{n} X_i - nC = (X_1 - C) + (X_2 - C) + \cdots + (X_n - C)$

Assuming $\overline{X}$ and C as the preceding,

$$\sum_{i=1}^{n}(X_i - C) = (3 - 2) + (4 - 2) + (6 - 2) + (8 - 2) + (9 - 2)$$
$$= 1 + 2 + 4 + 6 + 7 = 20$$

In linear programming, complex equations are typically expressed in terms of a double summation over two variables of $X: X_i$ and X_j. Each quantity is written with the double subscript ij. A standard expression is

$$\sum_{i=1}^{n}\sum_{j=1}^{m} X_{ij} = \sum_{i=1}^{n}(X_{i1} + X_{i2} + X_{i3} + \cdots + X_{im})$$
$$= (X_{11} + X_{12} + \cdots + X_{1m}) + (X_{21} + X_{22} + \cdots + X_{2m})$$
$$+ \cdots + (X_{n1} + X_{n2} + \cdots + X_{nm})$$

Assume a matrix

$$X = \begin{bmatrix} 2 & 3 & 4 \\ 1 & 2 & 3 \\ 1 & 5 & 9 \end{bmatrix}$$

$$\sum_{i=1}^{n}\sum_{j=1}^{m} X_{ij} = (2 + 3 + 4) + (1 + 2 + 3) + (1 + 5 + 9)$$
$$= 9 + 6 + 15 = 30$$

7. SUMMARY

This section of the chapter has reviewed the basic operations of matrix algebra. With a few notable exceptions, matrix and standard algebra operations are identical. The most important exception is that division is not defined for matrices. Multiplication by the inverse substitutes for division.

B. MATHEMATICAL SOLUTION OF LINEAR PROGRAMMING PROBLEMS

In this section you will use the algebraic methods reviewed in the previous section to solve linear programming problems. The procedure, known as the *simplex method*, consists of finding an *initial feasible solution* to the problem and then finding other feasible solutions that *improve on the initial solution.* The simplex method is not the most rapid means of solving a linear programming problem by hand calculation. Nevertheless, a conscious decision has been made to present only the simplex method. The simplex is the most useful technique because the simplex process provides a thorough understanding of how each successive iteration moves the solution toward the *optimum.* Other methods of solving linear programming problems are not presented. While the user should be thoroughly familiar with one technique, the computer algorithm is the technique you will use to solve real-world problems. Complete familiarity with computer manuals is more important than knowing two or more methods of solving a simple example by hand.

A second choice that the author has made was to eliminate all mathematical proofs from the chapter. On the one hand, some students who approach linear programming from a mathematics background find theorems, corollaries, and lemmas and their proofs to be necessary to their complete comprehension of the process. On the other hand, the vast majority of students whom I have taught linear programming find mathematical proofs to be extraordinary consumers of study time. They prefer to accept the mathematicians word that the problems have proofs or would prefer to find the proofs in appendices.

Next, my students insist that the presentation of the calculation methods suffers from a lack of clarity in standard texts. Accordingly, I have been urged to present the steps in the solution process in simple flow diagrams.

1. SOLVING LINEAR PROGRAMMING PROBLEMS

All of the steps are presented in Figures 1–3. We will proceed through the process step by step, concentrating first on the single-phase process and then, later in the chapter, on the two-phase process. These terms will be defined during the course of the presentation.

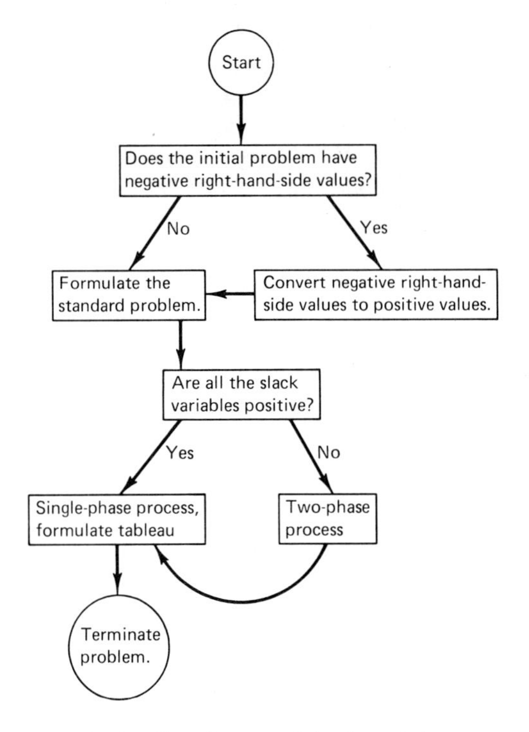

Fig. 1 Flow diagram of simplex solution.

The sample problem will be used to illustrate the process.

Maximize: $$Z = 2X_1 + X_2 \tag{1}$$

subject to

$$X_1 + 2X_2 \leq 10 \tag{2}$$

$$2X_1 + 3X_2 \leq 12 \tag{3}$$

$$3X_1 + X_2 \leq 15 \tag{4}$$

The simplex process consists of converting a system of linear inequality equations to linear equality equations with positive right-hand side values and solving for the X_j values. After each iteration, the X_j values move the solution toward the optimum. If the solution does not improve, either a solution does not exist or the solution has multiple optima.

We begin by converting any negative right-hand-side values to positive values. The sample problem has positive right-hand-side values. If, however, Eq. (4) had been $3X_1 + X_2 \leq -15$, then we would have multiplied Eq. (4) by -1 to obtain the necessary positive right-hand-side value.

$$-3X_1 - X_2 \geq 15$$

All the signs are changed and the inequality sign is reversed.

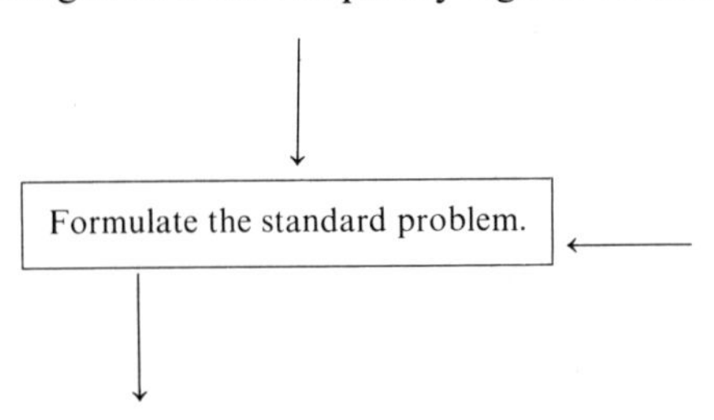

Formulating the standard problem consists of converting the nonnegative inequalities into equalities by defining slack variables. The slack variables convert inequalities to equalities, which is a necessary step for the mathematical solution of the problem. In addition, the slack variables are meaningful indicators of the system's need for additional resources or its supply of excess resources. For example, if a slack variable appears in the solution as $X_j = 10$, this is interpreted to mean that 10 additional units of X_j are available. If $X_j = -10$, then 10 additional units of X_j are required. Finally, the values of the slack variables are 0 because they do not contribute to maximizing the value of the problem. In the preceding sample problem

$$\text{Maximize:} \qquad Z = 2X_1 + X_2 \tag{1}$$

subject to

$$X_1 + 2X_2 \leq 10 \tag{2}$$

$$2X_1 + 3X_2 \leq 12 \tag{3}$$

$$3X_1 + X_2 \leq 15 \tag{4}$$

The standard problem is

$$\text{Maximize:} \qquad Z = 2X_1 + X_2 + 0X_3 + 0X_4 + 0X_5 \tag{1'}$$

subject to

$$X_1 + 2X_2 + X_3 = 10 \tag{2'}$$

$$2X_1 + 3X_2 + X_4 = 12 \tag{3'}$$

$$3X_1 + X_2 + X_5 = 15 \tag{4'}$$

The slack variables are X_3, X_4, X_5. By adding X_3 to Eq. (2), X_4 to Eq. (3), and X_5 to Eq. (4) we have derived an initial feasible solution.

$$(X_1 = 0; \quad X_2 = 0; \quad X_3 = 10; \quad X_4 = 12; \quad X_5 = 15).$$

2. SINGLE-PHASE PROCESS

The process used is called "single phase" because a single objective function is used. Assuming for the moment that all the slack variables are positive and that the initial feasible solution is not optimum, the standard problem is placed in a tableau in which the matrix operations can conveniently be performed. The sample problem is displayed in the accompanying tableau.

Basis vectors (Bv)	Solution or right-hand side-values (b)	Objective function values (C_j)	All vectors (a)				
			V_1	V_2	V_3	V_4	V_5
V_3	10	0	1	2	1	0	0
V_4	12	0	2	3	0	1	0
V_5	15	0	3	1	0	0	1

From left to right, the tableau contains four major pieces of information: (1) the vectors that constitute the feasible solution (V_3,V_4,V_5), (2) the solution or right-hand-side values (10,12,15), (3) the values of the basis vectors to the problem (C_j's: 0,0,0), and (4) the elements of the constraint matrix.

Before embarking upon an iteration, check the column vectors. If any column vector has *only* negative values when the b values are positive, then the problem is unbounded and has no solution. You need additional constraint(s) to bound the problem or you have made an error.

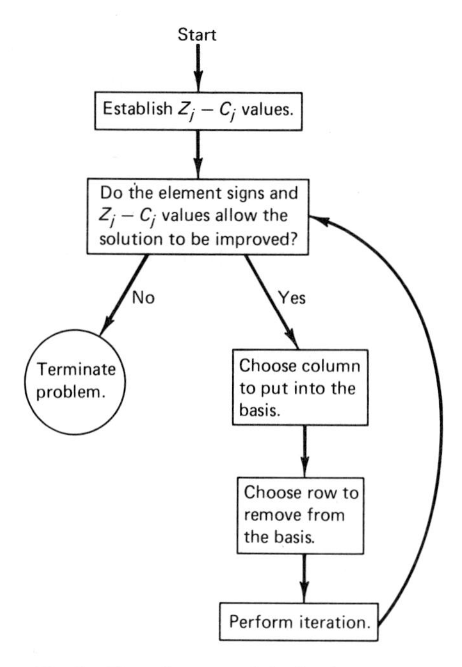

Fig. 2 Flow diagram of single-phase process.

At this point, one vector could be inserted randomly into the basis and another removed from it. One or two undirected iterations at the beginning would probably move the solution toward the optimum. However, unless the user is somehow psychic, trial-and-error iterations lead only slowly, if at all, to an optimum.

Fortunately, the decision of which substitutions to make can be systematically determined with the aid of a row vector called $Z_j - C_j$. The $Z_j - C_j$ values are the per unit profitableness or unprofitableness of including an activity in the solution which is not in the solution. The C_j values are the multipliers attached to each variable in the objective function of the standard problem. In the sample problem, the C_j values are $X_1 = 2$, $X_2 = 1$, $X_3 = 0$, $X_4 = 0$ and $X_5 = 0$. The Z_j values are formed by multiplying the elements in each column by the objective function values (C_j's) of the basis vectors and summing the vector. In the preceding problem for column vector 1, we perform the following calculations to obtain the Z_j value from column vector 1.

BV	C_j	V_1
V_3	0	1
V_4	0	2
V_5	0	3

$$Z_j = \sum C_j V_j$$
$$Z_j \quad \text{for} \quad V_1 = 0$$

$$\begin{aligned} 0 \times 1 &= 0 \\ +0 \times 2 &= 0 \\ +0 \times 3 &= \underline{0} \\ Z &= 0 \end{aligned}$$

For column vector 2, the calculation is

BV	C_j	V_2
V_3	0	2
V_4	0	3
V_5	0	1

$$Z_j \quad \text{for} \quad V_2 = 0$$

$$\begin{aligned} 0 \times 2 &= 0 \\ +0 \times 3 &= 0 \\ +0 \times 1 &= \underline{0} \\ Z &= 0 \end{aligned}$$

Indeed, all the Z_j values in the initial feasible solution are 0 because the slack variables constitute the solution. Calculating $Z_j - C_j$ for each column vector, we get:

		V_1	V_2	V_3	V_4	V_5
($\sum C_j V_j$)	Z_j	0	0	0	0	0
objective function weights	$+C_j$	2	1	0	0	0
	$Z_j - C_j$	−2	−1	0	0	0

The $Z_j - C_j$ row vector gives us two pieces of information. First, it tells us if the present solution is optimal. In a maximization problem, the solution can be improved if one or more of the $Z_j - C_j$ values is negative. In the preceding case, two $Z_j - C_j$ values are negative: $V_1 = -2$ and $V_2 = -1$. If all the $Z_j - C_j$ values are positive or zero, then the problem should be terminated.

In a minimization problem, if one or more of the $Z_j - C_j$ values is positive, the solution can be improved. The V_1 value of -2 means that, for each unit of activity 1 that inserted into the basis, the value of the solution will be increased by 2. Summarizing, for maximization problems look for negative $Z_j - C_j$ values; for minimization problems, look for positive $Z_j - C_j$ values. If the correct signs do not appear, the solution is terminated.

Second, the $Z_j - C_j$ row vector can tell us which nonbasis vector should be inserted into the basis. Specifically, the greatest positive or negative value (depending upon whether we are maximizing or minimizing) should be inserted into the basis. In the problem above, the two negative $Z_j - C_j$ values for the maximization problem are $V_1 = -2$, $V_2 = -1$. V_1 has the higher absolute value in the appropriate negative direction and should be placed in the basis.

To further illustrate the choice, assume the accompanying $Z_j - C_j$ row is in a minimization problem.

	V_1	V_2	V_3	V_4
$Z_j - C_j$	-6	4	0	3

In this case, V_2 should be placed in the basis because it has a higher positive value than the other three vectors. If two column vectors have the same and highest $Z_j - C_j$ values, then either one can be chosen. For example, in the succeeding maximization problem either V_1 or V_3 can be chosen.

	V_1	V_2	V_3	V_4
$Z_j - C_j$	-2	1	-2	3

Summarizing, the $Z_j - C_j$ row vector tells you if the solution can be improved upon and which vector will improve the solution as much as possible. The value of $Z_j - C_j$ is cumbersome to calculate. After the initial calculation of $Z_j - C_j$, you will find that the $Z_j - C_j$ row can be transformed during the iteration as you would any other row vector.

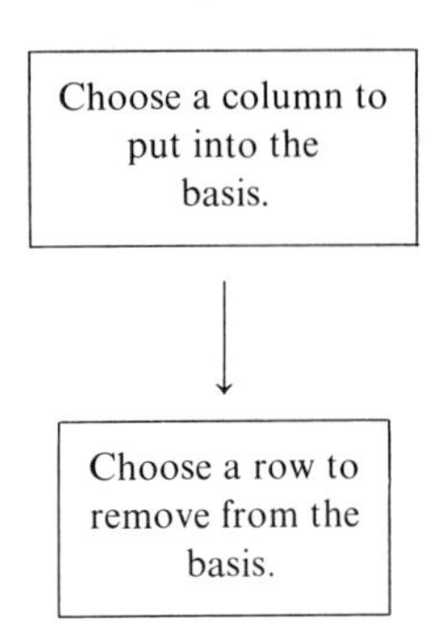

Once the vector to move into the basis has been chosen, we must decide which vector to remove from the basis. The decision is made by dividing the solution values (b's) by the corresponding elements in the column vector which is to be placed in the matrix. The lowest b/a ratio becomes the pivot element.

In the sample problem, vector 1 will be inserted into the problem.

BV	b	V_1	
			For V_3, $b_3/a_{31} = 10/1 = 10$
V_3	10	1	For V_4, $b_4/a_{41} = 12/2 = 6$
V_4	12	2	
V_5	15	3	For V_5, $b_5/a_{51} = 15/3 = 5$

The b/a ratio of the intersection of row vector V_5 and column vector V_1, element a_{51} is 5; it is the smallest ratio. The mechanical procedure has a specific geometric translation. The lowest b/a ratio indicates the most impinging constraint in the column. While you should use the most impinging constraint, do not use a b/a ratio which is negative, or one for which the divisor or the numerator is 0.

Once the decisions on which vector to insert and which to remove have been made, you are ready to perform an iteration. The pivot is made on the intersection of the row vector that is to be removed and the column vector that is to be placed in the basis. The tableau, with the pivot element circled, is as follows:

BV	b	C_j	1	2	3	4	5
3	10	0	1	2	1	0	0
4	12	0	2	3	0	1	0
5	15	0	[3]	1	0	0	1
$Z_j - C_j$	$Z = 0$		-2	-1	0	0	0

The value of the problem is $Z = 0(10) + 0(12) + 0(15) = 0$. All the steps in this problem will be presented and discussed to reinforce the methods presented in the previous section.

First, we convert element a_{51} (3) into 1 by multiplying the whole row by $\frac{1}{3}$.

BV	b	1	2	3	4	5
5	5	1	$\frac{1}{3}$	0	0	$\frac{1}{3}$

Second, we convert a_{31} (1) into 0 by multiplying row 5 by -1 and adding it to row 3 to form a new row 3. Row 5 itself stays unchanged after initial reduction of the pivot element to 1.

BV	b	1	2	3	4	5
3	10	1	2	1	0	0
5	-5	-1	$-\frac{1}{3}$	0	0	$-\frac{1}{3}$
3	5	0	$\frac{5}{3}$	1	0	$-\frac{1}{3}$

Third, we transform a_{41} (2) into 0 by multiplying row 5 by -2 and adding it to row 4 to form a new row 4.

BV	b	1	2	3	4	5
4	12	2	3	0	1	0
5	-10	-2	$-\frac{2}{3}$	0	0	$-\frac{2}{3}$
4	2	0	$\frac{7}{3}$	0	1	$-\frac{2}{3}$

Fourth, we transform the $Z_j - C_j$ element -2 into 0 by multiplying row 5 by 2 and adding it to the $Z_j - C_j$ row.

BV	1	2	3	4	5
$Z_j - C_j$	-2	-1	0	0	0
5	2	$\frac{2}{3}$	0	0	$\frac{2}{3}$
$Z_j - C_j$	0	$-\frac{1}{3}$	0	0	$\frac{2}{3}$

These four steps complete the insertion of row 1 into the basis and the removal of row 5 from the basis.

BV	b	C_J	1	2	3	4	5
3	5	0	0	$\frac{5}{3}$	1	0	$-\frac{1}{3}$
4	2	0	0	$\frac{7}{3}$	0	1	$-\frac{2}{3}$
1	5	2	1	$\frac{1}{3}$	0	0	$\frac{1}{3}$
$Z_j - C_j$	$Z = 10$		0	$-\frac{1}{3}$	0	0	$\frac{2}{3}$

The value of the problem is $Z = 0(5) + 0(2) + 2(5) = 10$. If you want to check the accuracy of calculations, pull each of the elements out of the tableau and insert them back into the standard format equations.

For row 3,

$$5 = \tfrac{5}{3}X_2 + X_3 - \tfrac{1}{3}X_5$$

$$5 = \tfrac{5}{3}(0) + 5 - \tfrac{1}{3}(0)$$

$$5 = 5$$

The $Z_j - C_j$ row can be checked through direct calculation.

For column vector 1,

BV	C_j	V_1
V_3	0	0
V_4	0	0
V_1	2	1

Z_j for $V_1 = 2$

$$\begin{array}{r} 0 \times 0 \\ +0 \times 0 \\ +2 \times 1 \\ \hline Z = \quad 2 \end{array}$$

For column vector 2,

BV	C_j	V_2
V_3	0	$\frac{5}{3}$
V_4	0	$\frac{7}{3}$
V_1	2	$\frac{1}{3}$

Z_j for $V_2 = \frac{2}{3}$

$$\begin{array}{r} 0 \times \frac{5}{3} \\ +0 \times \frac{7}{3} \\ +2 \times \frac{1}{3} \\ \hline Z = \quad \frac{2}{3} \end{array}$$

The remaining $Z_j - C_j$ calculations are listed in the following:

	V_1	V_2	V_3	V_4	V_5
Z_j	2	$\frac{2}{3}$	0	0	$\frac{2}{3}$
C_j	2	1	0	0	0
$Z_j - C_j$	0	$-\frac{1}{3}$	0	0	$\frac{2}{3}$

The first iteration has been completed with the value of the problem at 10. Each subsequent iteration, if any, should increase the value by less than 10.

The second iteration begins by examining the elements of the new tableau matrix and the $Z_j - C_j$ values to see if the problem should be terminated (see Fig. 2).

BV	b	C_j	1	2	3	4	5
3	5	0	0	$\frac{5}{3}$	1	0	$-\frac{1}{3}$
4	2	0	0	$[\frac{7}{3}]$	0	1	$-\frac{2}{3}$
1	5	2	1	$\frac{1}{3}$	0	0	$\frac{1}{3}$
$Z_j - C_j$	$Z = 10$		0	$[-\frac{1}{3}]$	0	0	$\frac{2}{3}$

The elements and the $-\frac{1}{3}$, $Z_j - C_j$ value in column 2 mean that if row 2 is inserted into the problem, the value of the problem will be increased. If X_2 becomes 1, the value of the problem will increase from 10 to $10\frac{1}{3}$.

We calculate the b/a values to determine which basis vector to remove.

For V_3,

$$b_3/a_{32} = 5/\tfrac{5}{3} = 3$$

For V_4,

$$b_4/a_{42} = 2/\tfrac{7}{3} = \tfrac{6}{7}$$

For V_1,

$$b_1/a_{12} = 5/\tfrac{1}{3} = 15$$

Clearly, row 4 should be removed from the basis and vector 2 entered into the basis.

First, we convert row 4, element a_{42} into 1 by multiplication by $\frac{3}{7}$.

BV	b	1	2	3	4	5
4	$\frac{6}{7}$	0	1	0	$\frac{3}{7}$	$-\frac{2}{7}$

Second, we convert a_{12} ($\frac{5}{3}$) into 0 by multiplying row 4 by $-\frac{5}{3}$ and adding it to row 3 to form a new row 3.

BV	b	1	2	3	4	5
3	5	0	$\frac{5}{3}$	1	0	$-\frac{1}{3}$
4	$-\frac{10}{7}$	0	$-\frac{5}{3}$	0	$-\frac{5}{7}$	$\frac{10}{21}$
3	$\frac{25}{7}$	0	0	1	$-\frac{5}{7}$	$\frac{1}{7}$

Third, we convert element a_{12} ($\frac{1}{3}$) into 0 by multiplying row 4 by $-\frac{1}{3}$ and adding it to row 1 to get a new row 1.

BV	b	1	2	3	4	5
1	5	1	$\frac{1}{3}$	0	0	$\frac{1}{3}$
4	$-\frac{2}{7}$	0	$-\frac{1}{3}$	0	$-\frac{1}{7}$	$\frac{2}{21}$
1	$\frac{33}{7}$	1	0	0	$-\frac{1}{7}$	$\frac{9}{21}$

Finally, we convert the $Z_j - C_j$ element $-\frac{1}{3}$ to 0 by multiplying row 4 by $\frac{1}{3}$ and adding it to the $Z_j - C_j$ row.

BV	1	2	3	4	5
$Z_j - C_j$	0	$-\frac{1}{3}$	0	0	$\frac{2}{3}$
4	0	$\frac{1}{3}$	0	$\frac{1}{7}$	$-\frac{2}{21}$
$Z_j - C_j$	0	0	0	$\frac{1}{7}$	$\frac{4}{7}$

The second completed tableau is

BV	b	C_j	1	2	3	4	5
V_3	$\frac{25}{7}$	0	0	0	1	$-\frac{5}{7}$	$\frac{1}{7}$
V_2	$\frac{6}{7}$	1	0	1	0	$\frac{3}{7}$	$-\frac{2}{7}$
V_1	$\frac{33}{7}$	2	1	0	0	$-\frac{1}{7}$	$\frac{9}{21}$
$Z_j - C_j$	$Z = 10\frac{2}{7}$		0	0	0	$\frac{1}{7}$	$\frac{4}{7}$

The solution of the problem is $Z = 0(\frac{25}{7}) + 1(\frac{6}{7}) + 2(\frac{33}{7})$.

The solution is now terminated because there are no negative values in the $Z_j - C_j$ row. Since the value of X_2 was $\frac{6}{7}$, the value of the objective function rose $\frac{2}{7}$. $[(\frac{6}{7})(\frac{1}{3})]$.

The optimal value is $10\frac{2}{7}$ and the constraints are satisfied.

Maximize Z: $$2X_1 + X_2 + 0X_3 + 0X_4 + 0X_5 \qquad (1')$$

$$2(\tfrac{33}{7}) + \tfrac{6}{7} + 0(\tfrac{25}{7}) = \tfrac{66}{7} + \tfrac{6}{7} + 0 = \tfrac{72}{7} = 10\tfrac{2}{7}$$

subject to

$$X_1 + 2X_2 + X_3 = 10 \qquad (2')$$

$$\tfrac{33}{7} + 2(\tfrac{6}{7}) + \tfrac{25}{7} = 10$$

$$10 = 10$$

$$2X_1 + 3X_2 + X_4 = 12 \qquad (3')$$

$$2(\tfrac{33}{7}) + 3(\tfrac{6}{7}) + 0 = 12$$

$$12 = 12$$

$$3X_1 + X_2 = 15 \qquad (4')$$

$$3(\tfrac{33}{7}) + \tfrac{6}{7} = 15$$

$$15 = 15$$

3. THE TWO-PHASE PROCESS

Most linear programming problems do not limit themselves to positive slack variables. If one or more of the slack variables is negative, or if the problem has one or more equality constraints, then we do not have an initial feasible solution because the number of basis vectors must equal the number of constraint equations. A basis vector must be a positive one, a condition not met by a negative slack variable or an equality constraint which does not have a slack variable. When this condition is not met, we turn to the two-phase process.

The two-phase process develops an initial feasible solution during the first phase and returns to the single-phase procedures (Fig. 2) during the second phase. Each of the two phases has a unique objective function. To illustrate the two-phase process, a fourth constraint has been added to the sample problem:

Maximize: $$Z = 2X_1 + X_2 \tag{1}$$

subject to

$$X_1 + 2X_2 \leq 10 \tag{2}$$

$$2X_1 + 3X_2 \leq 12 \tag{3}$$

$$3X_1 + X_2 \leq 15 \tag{4}$$

$$X_1 + X_2 \geq 4 \tag{5}$$

Equation (5), written in standard form, produces a negative slack variable. The problem has four constraints and only three basis vectors (X_3, X_4, and X_5).

$$X_1 + X_2 - X_6 = 4 \tag{5'}$$

We could multiply equation (5) by -1. However, the result would be an unacceptable, negative right-hand side value.

To form an initial feasible solution, an artificial variable is added.

$$X_1 + X_2 - X_6 + \bar{X}_7 = 4 \tag{5''}$$

The artificial variable X_7 produces an initial feasible solution. If two negative slack variables were in the problem, then two artificial variables would be necessary. In short, an artificial variable must be added for every negative slack variable and when the constraints are equalities. The artificial variables are purely statistical and have no real value.

The first phase of the process consists of preparing a standard problem that will ensure the artificial variables can be removed from the basis. For a maximization problem, the removal of an artificial variable is ensured by systematically making the C_j values for the artificial variable negative numbers and all of the other real and slack variables 0.

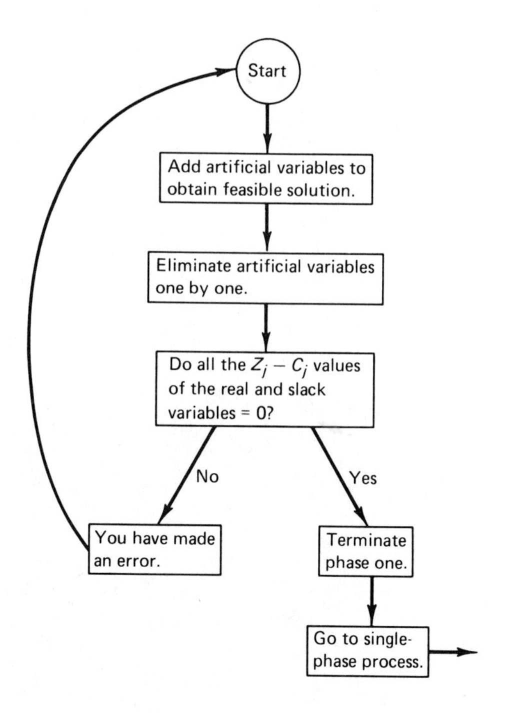

Fig. 3 Flow diagram of two-phase process.

In the sample problem we have:

$$\text{Maximize: } Z = 0X_1 + 0X_2 + 0X_3 + 0X_4 + 0X_5 + 0X_6 - X_7 \qquad (1'')$$

subject to

$$X_1 + 2X + X_3 = 10 \quad (2'')$$

$$2X_1 + 3X_2 + X_4 = 12 \quad (3'')$$

$$3X_1 + X_2 + X_5 = 15 \quad (4'')$$

$$X_1 + X_2 - X_6 + X_7 = 4 \quad (5'')$$

Given the above tableau, we should be able to immediately remove the artificial variable through normal row operations. Once an artificial variable is removed from the basis, you eliminate it from the tableau. Before proceeding to eliminate X_7 from the basis, it should be noted that, had the sample problem called for minimization, we would have made the artificial variables positive numbers.

The accompanying is the initial tableau for the first phase.

BV	b	C_j	1	2	3	4	5	6	7
3	10	0	1	2	1	0	0	0	0
4	12	0	2	3	0	1	0	0	0
5	15	0	3	1	0	0	1	0	0
7	4	−1	[1]	1	0	0	0	−1	1
$Z_j - C_j$	$Z = -4$		−1	−1	0	0	0	1	0

The initial value of the problem is −4. It can be improved because column vectors 1 and 2 have negative $Z_j - C_j$ values. The artificial vector 7 may be eliminated by introducing vectors 1 or 2 into the basis.

We will pivot on a_{71}. The pivot yields the following matrix:

BV	b	C_j	1	2	3	4	5	6	7
3	6	0	0	1	1	0	0	1	−1
4	4	0	0	1	0	1	0	2	−2
5	3	0	0	−2	0	0	1	3	−3
1	4	0	1	1	0	0	0	−1	1
$Z_j - C_j$	$Z = 0$		0	0	0	0	0	0	1

Vector 7 has been eliminated and the first phase has been completed. You can always be sure that the first phase has been completed by two checks. First, the $Z_j - C_j$ row has become a null vector except for the artificial variable element. Second, the number of iterations in the first phase should equal the number of artificial variables. If the problem has four (4) artificial variables, then four (4) iterations are necessary to complete the first phase, and at the end of the fourth iteration the $Z_j - C_j$ row will be null except for the four artificial variables. The number of iterations in the first phase may be less than the number of artificial variables if the problem does not have a solution.

The second phase of the two-phase process is identical to the single-phase process (Fig. 2). Using the tableau of the completed first phase, the standard problem without the artificial variable is prepared and placed in the tableau.

BV	b	C_j	1	2	3	4	5	6
3	6	0	0	1	1	0	0	1
4	4	0	0	1	0	1	0	2
5	3	0	0	−2	0	0	1	3
1	4	2	1	1	0	0	0	−1
$Z_j - C_j$	$Z = 8$		0	1	0	0	0	−2

Four differences between the final tableau of the first phase and the first tableau of the second phase are apparent. First, vector 7 has been eliminated and is not included in the second phase. Second, the C_j values are drawn from the second objective function.

Maximize: $Z = 2X_1 + X_2 + 0X_3 + 0X_4 + 0X_5 + 0X_6$ (1‴)

As a result, the value of the problem is 8. Finally, the $Z_j - C_j$ row has been recomputed to reflect the new objective function.

	1	2	3	4	5	6
Z_j	2	2	0	0	0	−2
C_j	2	1	0	0	0	0
$Z_j - C_j$	0	1	0	0	0	−2

The problem can be improved by inserting vector 6 into the basis. The lowest b/a ratio is vector 5. Accordingly, vector 5 will be removed from the basis and vector 6 inserted. Substitute the values into the objective function and equations as in the first sample problem.

BV	b	C_j	1	2	2	4	5	6
3	5	0	0	$\frac{5}{3}$	1	0	$-\frac{1}{3}$	0
4	2	0	0	$[\frac{7}{3}]$	0	1	$-\frac{2}{3}$	0
6	1	0	0	$-\frac{2}{3}$	0	0	$\frac{1}{3}$	1
1	5	2	1	$\frac{1}{3}$	0	0	$\frac{1}{3}$	0
$Z_j - C_j$	$Z = 10$		0	$-\frac{1}{3}$	0	0	$\frac{2}{3}$	0

Again the problem can be improved by moving vector 2 into the basis and vector 4 out of the basis.

BV	b	C_j	1	2	3	4	5	6
3	$\frac{25}{7}$	0	0	0	1	$-\frac{5}{7}$	$\frac{1}{7}$	0
2	$\frac{6}{7}$	1	0	1	0	$\frac{3}{7}$	$-\frac{2}{7}$	0
6	$\frac{11}{7}$	0	0	0	0	$\frac{2}{7}$	$-\frac{4}{21}$	1
1	$\frac{33}{7}$	2	1	0	0	$-\frac{1}{7}$	$\frac{9}{21}$	0
$Z_j - C_j$	$Z = 10\frac{2}{7}$		0	0	0	$\frac{1}{7}$	$\frac{4}{7}$	0

The problem can no longer be improved and is terminated. The optimum solution is almost the same as the solution of the single-phase problem which had only three constraints. The value is the same ($10\frac{2}{7}$). The only difference is that vector X_6 is $\frac{11}{7}$. Vector 6 was not in the first problem.

4. A MINIMIZATION PROBLEM

Next, a minimization problem will be solved

Minimize: $$Z = -3X_1 + 6X_2 \tag{6}$$

subject to

$$-2X_1 + 2X_2 \leq 6 \tag{7}$$

$$2X_1 + 4X_2 \leq 21 \tag{8}$$

$$-4X_1 + 2X_2 \geq -12 \tag{9}$$

Using Fig. 1 as a guide, we begin by checking for negative right-hand-side values. Equation (9) has a negative right-hand-side value, so it is multiplied by -1.

$$(-1)(-4X_1 + 2X_2 \geq -12) = 4X_1 - 2X_2 \leq 12 \tag{9$'$}$$

The right-hand-side signs are now positive, and the inequality sign for Eq. (9) is reversed.

Second, we convert the nonnegative inequalities into equalities and create the standard problem by defining slack variables.

Minimize: $$Z = -3X_1 + 6X_2 + 0X_3 + 0X_4 + 0X_5 \tag{6$'$}$$

subject to

$$-2X_1 + 2X_2 + X_3 = 6 \tag{7$'$}$$

$$2X_1 + 4X_2 + X_4 = 21 \tag{8$'$}$$

$$4X_1 - 2X_2 + X_5 = 12 \tag{9$''$}$$

A two-phase solution is not necessary because all the slack variables are positive.

Third, we formulate the tableau.

BV	b	C_j	1	2	3	4	5
3	6	0	-2	2	1	0	0
4	21	0	2	4	0	1	0
5	12	0	4	-2	0	0	1

Fourth, the $Z_j - C_j$ values are established.

	1	2	3	4	5
Z_j	0	0	0	0	0
C_j	−3	6	0	0	0
$Z_j - C_j$	3	−6	0	0	0

This is a minimization problem. We are looking for positive $Z_j - C_j$ values. Vector 1 has the only positive value and will be put into the basis.

Fifth, to determine which row to remove from the basis, we look for the most impinging constraint given by the lowest positive b/a ratio:

BV	b	V_1
3	6	−2
4	21	2
5	12	4

For V_3, $b/a = -\frac{6}{2} = -3$

For V_4, $b/a = \frac{21}{2} = 10.5$

For V_5, $b/a = \frac{12}{4} = 3$

The vector V_5 has the lowest positive b/a ratio. We will therefore insert vector 1 and remove vector 5.

BV	b	C_j	1	2	3	4	5
3	6	0	−2	2	1	0	0
4	21	0	2	4	0	1	0
5	12	0	[4]	−2	0	0	1
$Z_j - C_j$	$Z = 0$		3	−6	0	0	0

First, we convert a_{51} into 1 by dividing row 5 by 4.

BV	b	C_j	1	2	3	4	5
3	6	0	−2	2	1	0	0
4	21	0	2	4	0	1	0
5	3	0	1	$-\frac{1}{2}$	0	0	$\frac{1}{4}$
$Z_j - C_j$			3	−6	0	0	0

Second, we convert a_{31} into 0 by multiplying row 5 by 2 and adding it to row 3.

BV	b	1	2	3	4	5
3	6	-2	2	1	0	0
5	6	2	-1	0	0	$\frac{1}{2}$
3	12	0	1	1	0	$\frac{1}{2}$

Third, we transform element a_{41} into 0 by multiplying row 5 by -2 and adding it to row 4.

BV	b	1	2	3	4	5
4	21	2	4	0	1	0
5	-6	-2	1	0	0	$-\frac{1}{2}$
4	15	0	5	0	1	$-\frac{1}{2}$

Fourth, we transform the $Z_j - C_j$ element of vector 1 into 0 by multiplying row 5 by -3 and adding it to the $Z_j - C_j$ row.

BV	1	2	3	4	5
5	-3	$1\frac{1}{2}$	0	0	$-\frac{3}{4}$
$Z_j - C_j$	3	-6	0	0	0
$Z_j - C_j$	0	$-4\frac{1}{2}$	0	0	$-\frac{3}{4}$

These four steps complete the insertion of row 1 into the basis and the removal of row 5 from the basis.

BV	b	C_j	1	2	3	4	5
3	12	0	0	1	1	0	$\frac{1}{2}$
4	15	0	0	5	0	1	$-\frac{1}{2}$
1	3	-3	1	$-\frac{1}{2}$	0	0	$\frac{1}{4}$
$Z_j - C_j$		$Z = -9$	0	$-4\frac{1}{2}$	0	0	$-\frac{3}{4}$

There are no longer any positive $Z_j - C_j$ values, so the solution is terminated. The optimal value is -9 and the constraints are satisfied.

Maximize:
$$
\begin{aligned}
Z &= -3X_1 + 6X_2 + 0X_3 + 0X_4 + 0X_5 \qquad (6') \\
&= -3(3) + 6(0) + 0(12) + 0(15) + 0(0) \\
&= -9
\end{aligned}
$$

subject to

$$-2X_1 + 2X_2 + X_3 = 6 \qquad (7')$$
$$-2(3) + 2(0) + 12 = 6$$
$$-6 + 12 = 6$$
$$2X_1 + 4X_2 + X_4 = 21 \qquad (8')$$
$$2(3) + 4(0) + 15 = 21$$
$$+6 + 15 = 21$$
$$4X_1 - 2X_2 + X_5 = 12 \qquad (9'')$$
$$4(3) - 2(0) + 0 = 12$$
$$12 = 12$$

5. RECOGNIZING UNBOUNDED AND DEGENERATE SOLUTIONS WHILE SOLVING PROBLEMS BY HAND

In the course of the first and second chapters, two problems with obtaining a solution beyond writing contradictory constraints have been mentioned. This section summarizes these problems. One problem is the absence of legitimate leaving variables. This signifies an unbounded solution of one or more constraints. The examples at the end of the chapter contain an unbounded problem that can be solved by changing several constraints.

When two vectors tie for removal from the basis (i.e., their b/a ratios are equal), then a degenerate solution in which one of the basis variables assumes a value of zero follows. Theoretically, the solution process could cycle, repeating solutions rather than moving toward an optimum. The causes of and remedies for degenerate solutions are discussed in Chapter 3.

C. SUMMARY

This chapter has presented the simplex method of solving linear programming problems. All linear programming problems may be solved by the single-phase or two-phase processes.

EXERCISES

Perform feasible calculations on the following:

$$A = \begin{bmatrix} 2 & 1 & -2 \\ 3 & 1 & -4 \\ -1 & 2 & 6 \end{bmatrix}; \quad B = \begin{bmatrix} 5 & 2 \\ 3 & 3 \\ 1 & 0 \end{bmatrix}; \quad C = \begin{bmatrix} 1 \\ -1 \\ 4 \end{bmatrix} \quad D = \begin{bmatrix} -1 & 0 \\ 2 & 4 \\ -1 & 7 \end{bmatrix};$$

$$E = \begin{bmatrix} -3 & 1 \\ 2 & -2 \\ 4 & -1 \end{bmatrix}; \quad F = \begin{bmatrix} 1 & 0 & 2 \\ 0 & 1 & 2 \\ 5 & 3 & 4 \end{bmatrix}; \quad \text{scalar, } a = 3$$

(P1) $D + E$
(P2) $C + D$
(P3) $E - B + D$
(P4) $B \cdot A$
(P5) $A \cdot E$
(P6) $A \cdot C$
(P7) $a \cdot B$
(P8) A^{-1} (Show that this has no inverse.)
(P9) D^{-1}
(P10) $A \cdot F^{-1} - a \cdot F$

Solve the following problems. These same problems should be used at the end of Chapter 3 (for the dual) and for Chapter 4 to test your knowledge of the algorithm.

Maximize:
$$Z = X_1 - X_2 + 2X_3 \quad \text{(P11.1)}$$

subject to

$$2X_1 + 2X_2 - X_3 \geq 1 \quad \text{(P11.2)}$$
$$X_1 - X_2 + 2X_3 \leq 7 \quad \text{(P11.3)}$$
$$X_1 + X_2 + X_3 \leq 9 \quad \text{(P11.4)}$$

The problem requires one artificial variable. It is a multiple-optima problem because the objective function (11.1) parallels constraint (11.3).

Solution: $Z = 7$; $X_1 = 0$; $X_2 = 3.67$; $X_3 = 5.33$; other solutions include, for example, $X_1 = 1.8$; $X_3 = 2.6$. On computer printout, look for symbol A.

Maximize:
$$Z = 2X_1 + 6X_2 - 4X_3 \quad \text{(P12.1)}$$

subject to

$$-4X_1 + 2X_2 - 2X_3 \leq 20 \quad \text{(P12.2)}$$
$$2X_1 + 4X_2 + 4X_3 \leq 24 \quad \text{(P12.3)}$$
$$-2X_1 + 2X_2 + 2X_3 \leq 10 \quad \text{(P12.4)}$$

Use the single-phase process.

Solution: $Z = 35.36$; $X_1 = .67$; $X_2 = 5.67$; $X_3 = 0$.

Maximize:
$$Z = 2X_1 + 3X_2 - X_3 + X_4 \tag{P13.1}$$

subject to

$$X_1 - X_2 + 2X_3 + X_4 \geq 9 \tag{P13.2}$$
$$2X_2 + X_3 - X_4 \leq 5 \tag{P13.3}$$
$$-2X_1 + X_2 - 3X_3 + X_4 \leq -1 \tag{P13.4}$$
$$X_1 + X_3 \geq 3 \tag{P13.5}$$

This is a two-phase problem requiring three artificial variables. The solution is unbounded. Change Eq. (13.2) to $X_1 - X_2 + 2X_3 + X_4 = 9$. Change Eq. (13.4) to $-2X_1 + X_2 - 3X_3 + X_4 = -12$. You will need two artificial variables.

Solution: $Z = 35.6$; $X_1 = 9.8$; $X_2 = 4.2$; $X_4 = 3.4$.

Maximize:
$$Z = 2X_1 + X_2 \tag{P14.1}$$

subject to

$$X_1 - X_2 \leq 2 \tag{P14.2}$$
$$2X_1 + 3X_2 = 3 \tag{P14.3}$$
$$X_1 + X_2 \leq 5 \tag{P14.4}$$

You will need an artificial variable.

Solution: $Z = 3$; $X_1 = \frac{3}{2}$; $X_2 = 0$.

Minimize:
$$Z = 2X_1 + X_2 + X_3 \tag{P15.1}$$

subject to

$$X_1 + 2X_2 - X_3 \geq 2 \tag{P15.2}$$
$$X_1 + 2X_3 \leq 6 \tag{P15.3}$$
$$X_2 + 4X_3 \leq 16 \tag{P15.4}$$

This problem requires one artificial variable.

Solution: $Z = 1$; $X_1 = 0$; $X_2 = 1$; $X_3 = 0$.

Minimize:
$$Z = 2X_1 + 3X_2 - X_3 + X_4 \tag{P16.1}$$

subject to

$$X_1 + 2X_2 - X_3 + 2X_4 \geq 10 \tag{P16.2}$$

$$-2X_1 - 2X_2 + X_3 + X_4 \leq -5 \tag{P16.3}$$

$$X_2 + X_3 + X_4 \leq 0 \tag{P16.4}$$

$$X_1 - X_2 + 4X_4 \geq 5 \tag{P16.5}$$

This problem requires three artificial variables.
Solution: $Z = 20$; $X_1 = 10$; $X_2, X_3, X_4 = 0$.

Minimize: $$Z = -X_1 - X_2 + 2X_3 + X_4 \tag{P17.1}$$

subject to

$$2X_1 - X_2 + X_3 + X_4 \geq 1 \tag{P17.2}$$

$$-X_1 + 2X_2 - X_3 + 2X_4 \geq 5 \tag{P17.3}$$

$$X_1 + X_4 \leq 4 \tag{P17.4}$$

This problem requires two artificial variables.
Solution: $Z = -11$; $X_1 = 4$; $X_2 = 7$; $X_3, X_4 = 0$.

REFERENCES

Almon, C., Jr. (1969). "Matrix Methods in Economics." Addison-Wesley, Reading, Massachusetts.

Faddeeva, V. N. (1959). "Computational Methods of Linear Algebra." Dover, New York.

Hohn, F. E. (1964). "Elementary Matrix Algebra." MacMillan, New York.

Marcus, M., and Minc, H. (1965). "Introduction to Linear Algebra." MacMillan, New York.

School Mathematics Study Group (1960). "Introduction to Matrix Algebra." Yale Univ. Press, New Haven.

Chapter 3

Beyond the Global Optimum

Linear and Nonlinear Programming Options

The purpose of this chapter is to extend your knowledge beyond the optimal solution to a linear programming problem. The first section focuses on extensions of linear programming: sensitivity analysis, parametric programming, spot testing, and the dual problem. The second section is concerned with methods for handling integers and nonlinearities, including linear programming approximations and special algorithms such as branch and bound, backtracking, dynamic programming, and heuristic programming. Section three focuses on mixed linear and integer methods.

A. POSTOPTIMAL ANALYSIS

A linear programming model that has been carefully designed and tested for many months should represent a mine of information. An optimal solution is the most obvious output. But the optimal solution only scratches the surface of the information that can readily be mined. Three other packages of information may be developed: (1) the sensitivity of the optimal solution to changes in the activities, the constraints, and the matrix coefficients, (2) the optimal solution to the problem when one or more of the parameters (objective function weights, constraints, matrix coefficients) is allowed to systematically vary, and (3) spot tests of suboptimal solutions and specific parameters.

1. SENSITIVITY ANALYSIS

Sensitivity analysis determines the stability of the solution to small changes in different model parameters. If the solution is insensitive to changes in its key parameters, then the user can recommend its use with confidence. If, however, small changes in any of the parameters drastically change the solution, then the relationship between the solution and the critical parameters should be investigated in detail.

Clues to the stability of the solution are developed in the process of determining the optimal solution. Specifically, the report of the optimal solution contains two important pieces of information: (1) how much the value of the optimal solution will change if available resources are changed, and (2) how much the optimal value will change if an activity not in the optimal solution is forced into the solution.

Let us use a linear programming problem from Chapter 2 [Eqs. (1′)–(4′)] to illustrate the development and use of this information.

Maximize: $$Z = 2X_1 + X_2 + 0X_3 + 0X_4 + 0X_5 \qquad (1)$$

subject to:

$$\overbrace{\qquad\qquad\qquad\qquad}^{\text{slack variables}}$$

$$X_1 + 2X_2 + X_3 = 10 \qquad (2)$$

$$2X_1 + 3X_2 + X_4 = 12 \qquad (3)$$

$$3X_1 + X_2 + X_5 = 15 \qquad (4)$$

$$X_1 \geq 0 \qquad (5)$$

$$X_2 \geq 0 \qquad (6)$$

Optimal solution: $Z = 10\frac{2}{7}$; $X_1 = \frac{33}{7}$; $X_2 = \frac{6}{7}$; $X_3 = \frac{25}{7}$; $X_4 = 0$; $X_5 = 0$.

Inspection of the graphical solution (Fig. 1) indicates that Eqs. (3) and (4) are the restrictive constraints. If we change the availability of the resource in constraint Eq. (2) from $X_1 + 2X_2 + X_3 = 10$ to $X_1 + 2X_2 + X_3 = 9$, the optimal value of the solution will remain at $10\frac{2}{7}$. The amount of slack from activity X_3 will be reduced. Since the slack variable has no value, the optimal value of the problem is intact.

If, however, we change either constraints (3) or (4), the optimal solution will change. For example, if we change Eq. (4) from $3X_1 + X_2 + X_5 = 15$ to $3X_1 + X_2 + X_5 = 16$, the value of the optimal solution will increase from $10\frac{2}{7}$ to $10\frac{6}{7}$, an increase of $\frac{4}{7}$. If the right-hand-side value goes from 15 to 18, the value of the solution changes from $10\frac{2}{7}$ to 12. In short, for every unit

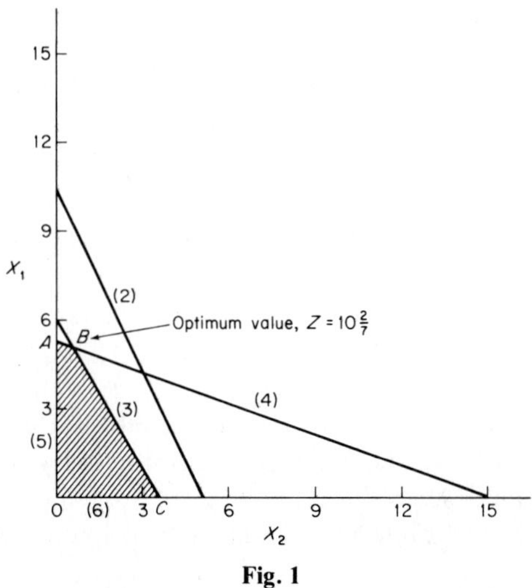

Fig. 1

increase in the Eq. (4) resource, the value of the solution increases $\frac{4}{7}$ of a unit until the solution changes. When the availability of the resource decreases, the value decreases by $\frac{4}{7}$ of a unit. Figure 2 illustrates some of these changes.

The unit change in the value of the objective function per unit change in a resource is known by many names, including *dual activity*, *opportunity cost*,

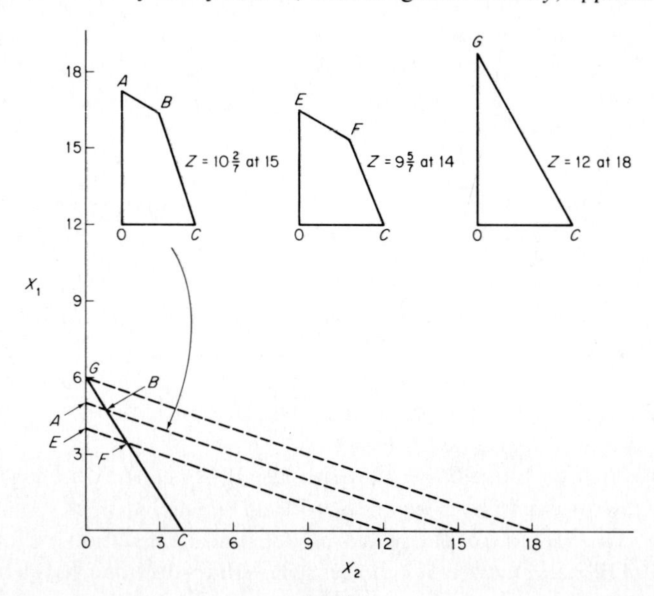

Fig. 2

marginal cost, and *shadow price*. If you examine the $Z_j - C_j$ values in the algebraic solution of this problem in Chapter 2, you will find that the dual activity of constraint (3) is $\frac{1}{7}$ and of constraint (4) is $\frac{4}{7}$. Each time you develop a new $Z_j - C_j$ row you are determining the relationship between the value of the problem and the right-hand-side values. Summarizing, the dual activity tells us the rate of change in the objective function per unit change in the availability of resources.

Another useful piece of information is the penalty associated with not following the mix recommended by the optimal solution. Using the same sample problem, let us review this information.

Maximize: $Z = 2X_1 + X_2 + 0X_3 + 0X_4 + 0X_5$ (1)

subject to

$$X_1 + 2X_2 + X_3 = 10 \quad (2)$$

$$2X_1 + 3X_2 + X_4 = 12 \quad (3)$$

$$3X_1 + X_2 + X_5 = 18 \quad (4')$$

$$X_1 \geq 0 \quad (5)$$

$$X_2 \geq 0 \quad (6)$$

Solution: $Z = 12; X_1 = 6; X_2 = 0; X_3 = 4; X_5 = 0; X_6 = 0$ (see Fig. 3).

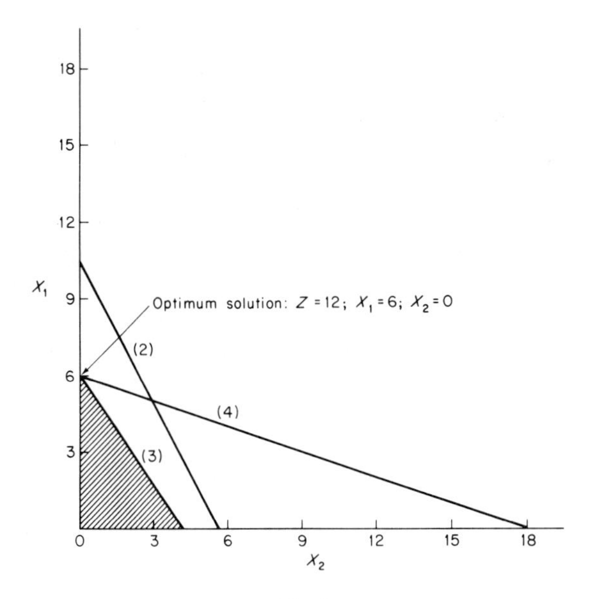

Fig. 3

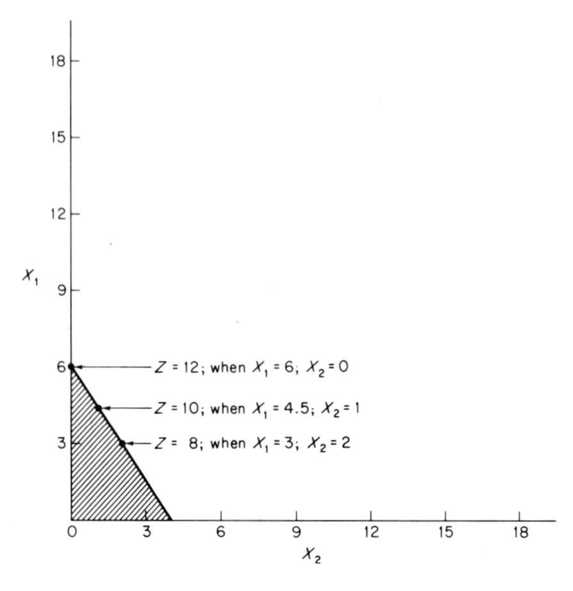

Fig. 4

Assume that the user insists on producing at least one unit of X_2. He will want to know how much of an economic penalty will result. If the value of X_2 is 1, then the value of X_1 becomes $4\frac{1}{2}$. Accordingly, the value of the problem drops to 10 [$Z = 2(\frac{9}{2}) + 1(1) = 10$]. For every unit of X_2 we put into the solution, the value of the problem is reduced by 2. If X_2 is raised to 2, the value of the problem will be reduced to 8. The computer reports this information as reduced costs for the nonbasis activities (see Chapter 4 for sample computer printouts). Figure 4 illustrates the penalty of placing X_2 into the solution.

Dual activity and reduced cost are important indicators of the sensitivity of the solution. However, great care must be taken when using the dual activity and reduced cost indicators because you do not know the range over which the relationships implied by the dual activity and reduced costs can vary. The range within which the optimal basis operates may be determined for the cost coefficient of the activities, the constraints, and the matrix coefficients.

Analysis of the objective function cost coefficients will specify the range over which the weights can vary without changing the optimal solution. Ranges are developed for those activities that are part of the optimal solution. The initial sample problem, with the right-hand side of Eq. (4) equal to 15, will be used to illustrate the cost coefficient range (Fig. 5). The dotted objective function line $Z = 2X_1 + X_2$ passes through the optimum solution combination at point B (line 1). If the C coefficient of activity X_1 is reduced

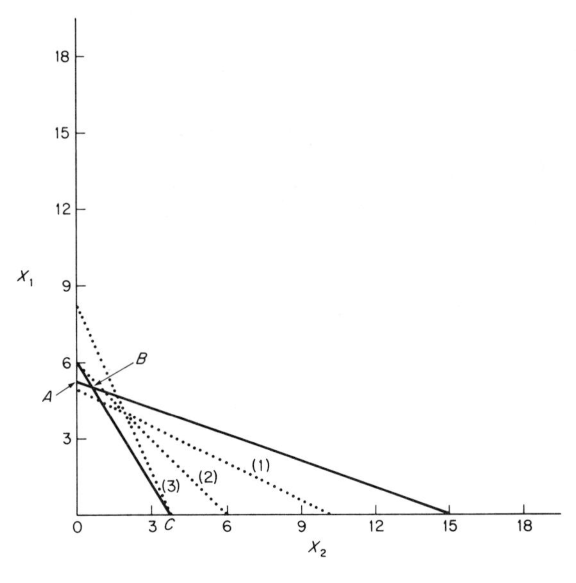

Fig. 5

from 2 to 1, then the slope of the objective function equation becomes a 45° line (line 2). Each time the C weight is reduced by the fractional unit, the angle of the objective function line becomes closer to the perpendicular and the objective function value decreases. In short, a decrease in the value of solution results from a decrease in the value of the C coefficient of activity X_1. A major change in the optimal solution occurs when the weight of activity X_1 becomes less than $\frac{2}{3}$. If Eq. (1) is to maximize: $Z = .667X_1 + X_2$, then X_1 and X_2 are both nonzero activities. If Eq. (1) is to maximize: $Z = .666X_1 + X_2$, then X_1 becomes a zero-valued activity because the optimal solution passes through point $X_2 = 4$; $X_1 = 0$.

The obvious next step is to determine the upper range of the X_1 weight. At what point is X_2 driven out of the solution by increasing the coefficient of X_1? After several iterations we find that X_2 is driven out of the nonzero basis when $X_1 \geq 3$. Summarizing, the cost coefficient ranges indicate the weights to which the model is most sensitive. In the sample case, the optimal solution consisting of X_1, X_2, and X_3 as nonzero values is stable for objective function weights of $X_1 \geq .667$ and < 3.00. If the X_1 coefficient is $< .667$ or > 3.00, the basis changes.

The extent to which right-hand-side values can change without the solution becoming infeasible is typically critical to the decision maker. Graphically, a constraint is moved up and down parallel to itself until the optimal solution is no longer feasible. With the addition of a single card, the MPS 360 computer package automatically finds objective function and right-hand-side ranges (Chapter 4).

Finally, individual coefficients in the resource-use matrix should be subjected to sensitivity analysis if they are likely to seriously change the solution. Unfortunately, most of the computer package programs do not automatically develop ranges for every matrix coefficient. You must change a few coefficients manually. The author, however, has used parametric programming to make the desired matrix coefficient sensitivity tests. This use of parametric programming will be explained in the next section.

2. PARAMETRIC PROGRAMMING

Sometimes sensitivity and range data will not answer all your questions. You may wish to know how the solution changes when one, two, or more objective function, right-hand side, and matrix coefficient parameters change over a particular range. Such questions are answered with parametric programming.

We can illustrate a typical application with the aid of the sample problem.

Maximize:
$$Z = 2X_1 + X_2 \tag{1}$$

subject to

$$X_1 + 2X_2 \leq 10 \tag{2}$$
$$2X_1 + 3X_2 \leq 12 \tag{3}$$
$$3X_1 + X_2 \leq 15 \tag{4}$$
$$X_1 \geq 0 \tag{5}$$
$$X_2 \geq 0 \tag{6}$$

Solution: $Z = 10\frac{2}{7}$; $X_1 = \frac{33}{7}$; $X_2 = \frac{6}{7}$.

The decisionmaker is unhappy with the findings that the optimal solution produces only $\frac{6}{7}$ of a unit of activity X_2. You are instructed to determine how much it will cost if between 0 and 4 units of X_2 are produced in 1-unit increments.

Mathematically, this question can be answered by solving the problem five times. Each time, constraint (6) is incremented by one unit as below:

Run 1:	$X_2 \geq 0$	Run 4:	$X_2 \geq 3$
Run 2:	$X_2 \geq 1$	Run 5:	$X_2 \geq 4$
Run 3:	$X_2 \geq 2$		

Graphically (Fig. 6), we are passing a plane across the feasible solution space from left to right (dotted lines). Each time we move the line to the right,

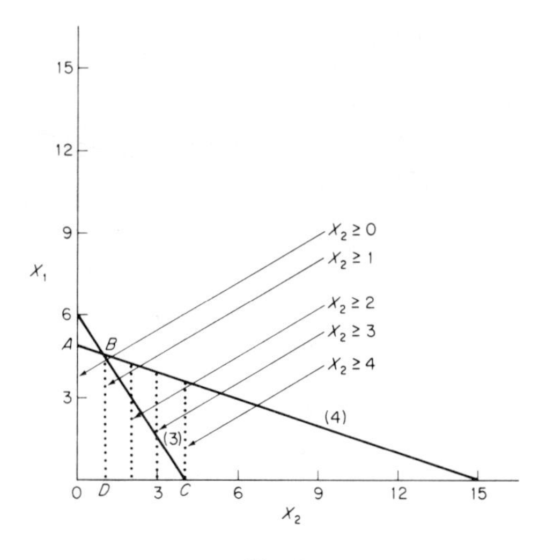

Fig. 6

the feasible solution space shrinks and the value of the optimal solution decreases.

Five computer runs will give us the desired information. Parametric programming will perform the calculations in one run. We define a parameter C and add it to Eq. (6): $X_2 \geq 0 + C$. We instruct the computer to let C vary from 0 to 4. When $C = 0$, the optimal solution $Z = 10\frac{2}{7}$, $X_1 = \frac{33}{7}$, $X_2 = \frac{6}{7}$ results. When $C = 1$, the optimal solution drastically changes. Specifically, if $X_2 \leq \frac{6}{7}$, Eqs. (3) and (4) are limiting. When $X_2 > \frac{6}{7}$, then Eq. (4) is no longer limiting. Instead, Eqs. (3) and (6) are limiting.

Actually, our sample problem is not the best one to illustrate the power of the parametric programming technique, because only Eqs. (3) and (4) are initially binding. The feasible polygon as pictured in Fig. 7 is a better example because five different constraints are limiting at different points. As we move the $X_2 \geq C$ equation across the solution space, the computer will identify changes in the value of the optimal solution, value of the activities, and marginal costs. Summarizing, right-hand-side parametric programming will test the impact of changing a resource on the solution within any range and at any increment specified by the user.

Parametric programming on the objective function values is appropriate when one or more of the weights is expected to change. For example, in our sample problem we might expect the coefficient variable X_2 to change systematically from 1 to 5. We define a new objective function: maximize: $Z = 2X_1 + (1 + C)X_2$ where C ranges from 0 to 5 in 0.5 increments. The computer will develop new solutions as C changes in increments of 0.5.

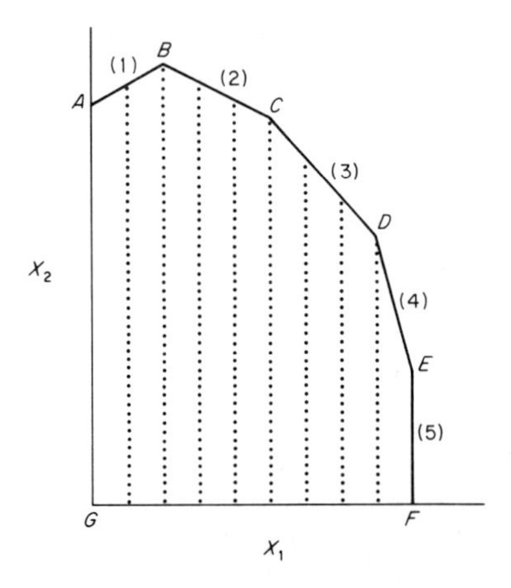

Fig. 7

Parametric programming of specific matrix coefficients is often important. For example, let us assume that the X_1 coefficient in Eq. (3) is likely to change from 2 to 2.5. We want to determine the impact of this change on the solution. As previously indicated most of the package programs do not automatically make this calculation. However, you can perform the test. From a previous run you have found that the optimum solution to activity X_1 is 4.71. Write a new constraint equation which sets $X_1 = 4.71$. Constraint (3) accordingly becomes

$$2(4.71) + 3X_2 + X_4 = 12$$

$$3X_2 + X_4 = 12 - 9.42$$

$$3X_2 + X_4 = 2.58$$

Since $X_4 = 0$ in the optimal solution, $X_2 = 0.86$. If we want the X_1 weight of Eq. (3) to change from 2 to 2.1, we can calculate a new value for X_2 assuming that $X_1 = 4.71$.

$$[2(4.71) + .1(4.71)] + 3X_2 + X_4 = 12$$

$$3X_2 + X_4 = 2.11$$

$$X_2 = .703$$

To make this calculation on the computer, we subtract a constant (C) .471 from Eq. (3). $2(4.71) + 3X_2 + X_4 = 12 - C$. The parametric change coefficient is $-.471$ and is carried out from $C = 0$ to $C = [(.5)(4.71) = 2.35]$.

Summarizing, two changes are required to systematically test the matrix coefficients: (1) a constraint that fixes the value of the activity of the coefficient being tested at the optimal value, and (2) a right-hand-side change coefficient.

3. SPOT TESTING

Do not use the automatic sensitivity, range, and parametric programming capabilities of computer codes as a substitute for common sense and ingenuity. Testing the full range of solutions, parameters, and matrix coefficients may not be necessary. Decisionmakers or a group may recommend the selection of a few critical parameters. These parameters should then be systematically tested by postoptimal techniques.

In general, the author can distinguish three applications of postoptimal analyses. One uses parametric programming to systematically relax all the constraints and objective function coefficients by a constant percentage. A second application involves spot testing a few parameters by a specific amount at a specific number of times. And a third tests those parameters which are subject to unpredictable changes or to the control of the decisionmaker, or both.

4. THE DUAL PROBLEM

An economic interpretation of the dual variables was presented in the previous section. Recapitulating, the dual variables indicate the value of scarce resources. However, these values are valid only over the range of the optimal solution. This section elaborates upon the previous treatment by presenting the dual problem.

Every linear programming problem with n activity variables and m constraints (not counting nonnegativity restrictions) has an associated complementary problem with m activity variables and n constraints. The initial problem is called the primal and the associated problem the dual. The weights of the activities (columns) of the primal problem become the constraints of the dual problem. The constraints of the primal problem become the activities of the dual. The constraint values of the primal become the activity weights of the dual. The rows of the primal problem become the columns of the dual. If the primal is a maximization problem, the dual is a minimization and vice versa. If the primal has $\geq$ row signs, the dual has $\leq$ row signs. Both problems have precisely the same solution value.

These symmetrical problems are described in mathematical symbols that follow.

Primal

Maximize: $Z_p = C_1X_1 + C_2X_2 + \cdots + C_nX_n$

subject to

$$a_{11}X_1 + a_{12}X_1 + \cdots + a_{1n}X_n \leq b_1$$
$$a_{21}X_1 + a_{22}X_2 + \cdots + a_{2n}X_n \leq b_2$$
$$\vdots$$
$$a_{m1}X_1 + a_{m2}X_2 + \cdots + a_{mn}X_n \leq b_m$$
$$X_1; \quad X_2; \quad \cdots; \quad X_n \geq 0$$

Dual

Minimize: $Z_d = b_1Y_1 + b_2Y_2 + \cdots + b_mY_m$

subject to

$$a_{11}Y_1 + a_{21}Y_2 + \cdots + a_{m1}Y_m \geq C_1$$
$$a_{12}Y_1 + a_{22}Y_2 + \cdots + a_{m2}Y_m \geq C_2$$
$$\vdots$$
$$a_{1n}Y_1 + a_{2n}Y_2 + \cdots + a_{mn}Y_m \geq C_n$$
$$Y_1; \quad Y_2; \quad \cdots; \quad Y_m \geq 0$$

At this point, the reader may believe that a dual problem exists, but he may wonder why its existence is worth highlighting. Two properties of the dual are important: (1) the computation of the dual may be simpler than that of the primal, and (2) the solutions to the dual often provide insights about the primal that might have been overlooked.

While the high-speed computer has removed some of the tedium from developing a linear programming solution, the fact remains that it usually, though not always, takes more time to solve a problem with a few variables and many constraints than a problem with many activities and few rows. If the primal problem has many constraints, the dual will have few constraints. The computer may automatically solve the dual problem to save time.

While the computational advantages of the dual problem are often used, most of the interest in it stems from the potentially useful insights it provides. On the one hand, some economists and regional scientists have attributed a good deal of theoretical value to the dual solution. For example, the literature of the 1970s contains statements such as, "The dual solution was also shown to provide a decision rule for controlling urban growth" [Wellar and LaCava (1971)]. On the other hand, the dual has two important limitations.

First, the marginal costs implied by the dual variables are restricted to the range of the optimal solution. If the optimal solution is unstable and the solution suddenly changes, policies based on the interpretation of the dual solution will be dubious. Second, it is difficult to interpret the dual variables of noneconomic activities. For example, suppose you maximized the value of land for residential development subject to a requirement for low-income housing. The dual would indicate the worth of each low-income housing unit. Clearly, the issue of low-income housing in middle- and upper-income areas is far more complex than the values derived from the dual. In short, the dual may yield useful information and it may yield misleading and oversimplified working rules.[1]

The author has infrequently used the dual. One of these infrequent uses will be reviewed to exemplify the strengths and weaknesses of the dual. The author and his associates developed, tested, and ran a linear programming model to determine the economic cost of meeting projected potable water demands for seventy-two agency service areas in the northeastern portion of New Jersey for the years 1975–2000. Tests were performed to determine the tradeoffs among increasing the supply, controlling the demand, and increasing the capacity for interagency water transfers.

The model (Chapter 7) was structured as follows:

Minimize:
$$Z_p = \sum_i \sum_j C_{ij} X_{ij} \tag{7}$$

subject to

$$\sum_i X_{ij} = D_j \tag{8}$$

$$\sum_j X_{ij} \leq S_i \tag{9}$$

$$X_{ij} \leq T_{ij} \tag{10}$$

$$X_{ij} \geq 0 \tag{11}$$

where

Z_p is the average total daily cost of supplying water;
X_{ij} is the average daily quantity of water shipped from purveyor i to demand area j in million gallons per day (mgd);
C_{ij} is the retail cost of water for supplying water from purveyor i to demand area j in dollars per mgd;
D_j is the average daily demand for water at area j in mgd;
S_i is the average daily supply of water at source i in mgd;
T_{ij} is the maximum daily amount of water that the system can transfer from purveyor i to purveyor j in mgd.

[1] See Dorfman *et al.* (1958) and Baumol (1965) for good interpretations of the dual.

Equation (7) represents the *objective function* Z, i.e., the average total daily cost of supplying water. The model minimizes Z by calculating the least-cost solution for each unknown X_{ij}. The *demand constraints* [Eq. (8)] require that the amount of water delivered to any demand area j be equal to the demand for water in that area (D_j). This constraint requires the model to distribute all of the available supply of water until the demand is satisfied. The *supply constraints* [(Eq. (9)] require that the amount of water diverted from any source i be less than or equal to the amount of water available at that source (S_i). The *transfer capacity constraints* [Eq. (10)] limit the amount of water shipped from any source i to any demand area j to the capacity of the pipe network and pumps between that source and that area (T_{ij}). The non-negativity constraints [Eq. (11)] prohibit negative water allocations.

Three applications of the model were as follows:

(1) The first application of the model was to determine the best allocation of the available public potable water given the existing system.

(2) The second application of the model was to determine the average total daily cost of meeting forecasted water demands and the optimal allocation of the water given forecasted supplies, transfer capacity, and retail costs.

(3) The third application of the model was to perform sensitivity analyses on different components of the system. Three situations were considered. First, the impact of increasing and decreasing demand D_j on cost was determined. Second, the model was used to measure the cost implications of increasing and decreasing the supply S_i. And third, the linear programming model was used to estimate the cost of different transfer capacities.

The dual of the problem is the following:

Maximize:
$$Z_d = \sum_j D_j Y_j - \sum_i S_i V_i - \sum_i \sum_j T_{ij} W_{ij} \tag{12}$$

subject to

$$Y_j - V_i - W_{ij} \leq C_{ij} \tag{13}$$

$$Y_j \geq 0 \tag{14}$$

$$V_i \geq 0 \tag{15}$$

$$W_{ij} \geq 0 \tag{16}$$

where

Z_d is the average daily benefit of supplying water;
Y_j is the shadow price (or marginal cost) of the water at the sink j in dollars per mgd;
V_i is the shadow price of the water at the source i in dollars per mgd;
W_{ij} is the shadow price of transferring water from purveyor i to demand area j in dollars per mgd.

Equation (12) represents the *objective function* Z_d. The dual of the problem maximizes the benefits that can be expected in supplying the water.

Equation (13) expresses the cost of water supply as constraints: the cost of supplying water from source i to sink j (C_{ij}) must be greater than or equal to the shadow price of water at sink j minus the shadow price of water at source i and minus the shadow price of transferring water from i to j. Equations (14)–(16) are the nonnegativity constraints that serve to prohibit negative shadow prices.

The dual was most useful in the selection of an optimal pricing system. Given a set of costs C_{ij} (wholesale or retail costs), the dual determines the marginal costs of supplying water in each demand area and at each source and the marginal cost of transferring water in the case of an optimal allocation of water. The comparison of the marginal costs with the actual costs indicate how the pricing system could be modified to encourage an optimal allocation of water.

Marginal cost information was important but had to be considered in the context of the realities of managing a public utility. The water supply system in the region we analyzed has a long list of political restrictions on optimal water movements arising from home rule. The marginal cost information helped us locate communities where the price structure was most inequitable to the citizen. Frankly, however, the inequities will be difficult to correct because of the complex political climate. While the dual solution helped us draw additional economic information, economic considerations are only one of the important considerations in this case.

We cannot leave the dual without reviewing its role in multiple-optima and degenerate problems. As you will recall from Chapter 1, if a problem has more than one solution, then it has an infinite number of solutions. Multiple optima need not be a problem and indeed may be useful to the decisionmaker. A degenerate solution is one in which the number of nonzero activities in the solution is less than the number of rows (e.g., five constraints, four nonzero variables, four constraints, two nonzero variables). Degeneracy occurs when more than two of the limiting constraints form a corner point.

The following problem produces a degenerate solution:

Maximize:
$$Z = X_1 + X_2 \tag{17}$$

subject to

$$3X_1 + X_2 \le 7 \tag{18}$$

$$X_1 + 2X_2 \le 2 \tag{19}$$

$$X_1 - X_2 \le 2 \tag{20}$$

$$X_1 \ge 0 \tag{21}$$

$$X_2 \ge 0 \tag{22}$$

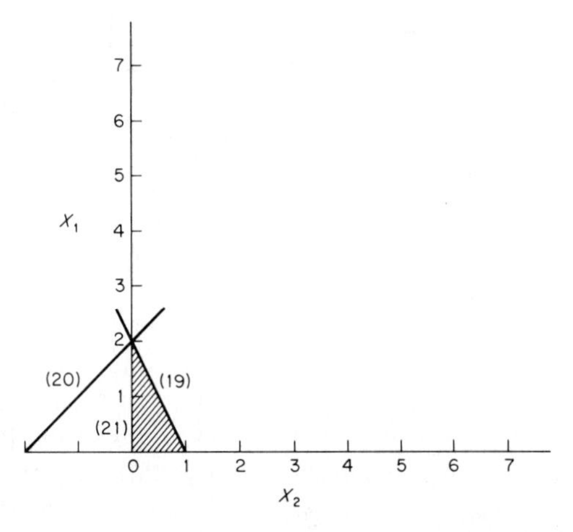

Fig. 8

Three constraints, (19), (20), and (21), pass through $X_1 = 2$, $X_2 = 0$, which also is the optimal point. One is redundant. (See Fig. 8.)

If you solve this problem by hand with the single-phase method you will find that it produces the same value $Z = 2$ with different slack variables in the solution as zero.

$Z = 2$	$Z = 2$
$X_1 = 2$	$X_1 = 2$
$X_3 = 1$	$X_3 = 1$
$X_5 = 0$	$X_4 = 0$

This problem has a degenerate optimal solution. Its dual is a multiple-optima solution. Indeed, if a problem has a degenerate optimal solution, then its dual will have multiple-optima solution and vice versa.[2]

The multiple-optima and degenerate optima solution combination can cause a mathematical problem and a problem of interpretation. If you try to solve a problem which has a degenerate solution, you will find the $Z_j - C_j$ marginal values call for another iteration, and yet the value of the solution is not improved! If cycling and recycling of answers that do not improve the solution occur, you may not get an optimal solution. This mathematical problem had to be solved by computer programmers to prevent a degenerate problem from infinitely recycling and therefore not producing an optimum. The mathematical problem has been solved in most cases. Cycling can occur in large problems that have a great number of zero right-hand-side values.

[2] See Miller and Dossani (1967) for an interesting discussion of this characteristic and other possibilities.

The second problem is, if a problem has an infinite number of solutions, how does one interpret the dual variables? If a degenerate solution occurs or if you wish to avoid one, the author's advice is to drop only completely redundant constraints. If you drop a constraint that is not completely redundant, you lose information. Rather than drop constraints, slightly change the value of the constraint (e.g., from 5 to 5.01). This change will also prevent cycling.

Summarizing, the dual variables are the implied economic costs of scarce resources. The dual problem determines the total value of all of the available resources. Chapters 9 and 10 present several examples of the application of the dual. In the next section the dual will be illustrated drawing upon game theory.

5. AN EXAMPLE OF THE DUAL: LINEAR PROGRAMMING AND GAME THEORY

Game theory was developed by von Neumann and Morgenstern (1944) as a means of understanding economic behavior. The theory is an attractive means of explaining how conflicts arise and are solved. The uses of game theory as a formal mathematical tool have been relatively rare and largely in the area of warfare because of the great complexity introduced by multi-party, cooperative decisions. Nevertheless, the relationship of game theory and linear programming will be briefly explained in this section because of game theory's historical importance as an initial effort to model and resolve conflicts and because game theory problems are a special case of linear programming.[3] The first part of this section will introduce the terminology of game theory. Then the mathematical relationship of game theory to linear programming will be demonstrated with a simple example.

Linear programming maximizes or minimizes the goal of one person or a group of persons who agree upon goals. When resources are limited and objections are raised by other parties, these problems become constraints. Game theory explicitly recognizes the different goals of each party in conflicts that range from total opposition to slight disagreement.

In the social and environmental sciences, conflicts are rarely fought solely by two diametrically opposed parties. Rather, a series of interest groups are involved at different levels. Some focus only on one aspect of a conflict, others in nearly all the aspects. Some have little money and power and participate with the hope of winning through persuasion and delay. Other parties are well funded and represent powerful interests. While conflict strategies may be guessed, predicting the strategies and outcomes of conflicts

[3] See Vajda (1967), particularly Chapter 3, section 4, and Chapter 9. Also see Karlin (1959).

is extraordinarily difficult. Only the simplest cases can be modeled without great simplification. The simplest of the cases is the *two-person, zero-sum game*. The two persons represent the opposing two individuals or groups of persons behaving as two individuals for the conflict. Zero sum refers to their goals relative to one another. Zero sum means that every gain by one party is a loss to the other party. Quick agreements in such a game are rare.

Each party defines its goals. These goals are translated into a *payoff matrix* which indicates the value of the potential activities for the two persons given specific activities by each. While this section will discuss only two-person, zero-sum games, the theory is generalized to nonzero sum games and to n-person games. These vastly more intricate games permit cooperative coalitions and noncooperative strategies among multiple opponents [see Rapoport (1970)]. An example two-person, zero-sum game follows.

A citizen's environmental group and a development-oriented group are campaigning to win public support. In this case they are in conflict about different uses of a strip of land along a river. The development-oriented group has four alternatives that it can support in a referendum:

(1) build a major road along the river,
(2) build a dock for barges,
(3) build an amusement part, and
(4) build oil storage tanks.

The environmental group's choices are two:

(1) preserve the land and
(2) build a park that contains a zoo.

Overall, while there are six possibilities, only four can draw support from the developers and two from the environmental group.

Voter reaction to the interacting strategies of the two parties is at stake. A total of eight voter outcomes to the intersecting strategies is possible. The value of each possible outcome is measured in the voters' estimated appeal for each strategy. The values from the perspective of the developers are shown in Table I.

The developers want the highest possible positive score: they are the maximizing player. The environmentalists are the minimizing player. Both groups are hoping to build their image and credibility by demonstrating a responsible image and a willingness to bargain. If the development-oriented group argues for a major road when the environmental group calls for a park containing a zoo, then the developers will look bad (payoff of -15). On the other hand, if the developers suggest an amusement park when the environmental group suggests the park and zoo, then the developers will seem to be more reasonable (payoff of 10). Which strategies should

Table I Payoff Matrix

Strategy	Environmentalists: 1. Build park and zoo	2. Preserve the land
Developers		
1. Build road	−15	−5
2. Build barge dock	−8	3
3. Build amusement park	10	−5
4. Build oil storage tanks	−30	−20

each party stress in its releases, given the fact that neither party knows which strategy the opponent will stress?

We begin answering the question by looking for strategies that one party should not use. Oil storage tanks are a poor strategy for the developers. Similarly, though not quite as obvious, building the road is also a poor choice for the developers. Both these unfavorable strategies are dominated by other strategies. For the general case, one strategy *dominates* another when the result of the dominant strategy is at least as good or better than the outcome of the dominated strategy. Neither of the environmental group's strategies is dominant. If the developers are thinking logically, the choices open to both parties are reduced to four (see Table II).

If, by eliminating illogical choices, we were able to arrive at a mutually favorable strategy for both parties, we would have found a *saddle point*. Formally, a saddle point is an intersection of strategies which includes the lowest value in the row of the maximizing player and the highest value in the intersecting column of the minimizing player.

Many games, however, do not contain saddle points. In these games, linear programming may be used to solve the problem. The primal represents one player's strategies, the dual problem yields the opponent's solution.

The problem from the perspective of the environmentalists is to minimize the positive vote that the developers can attract.

Minimize: $$Z = V_{\text{primal}} \tag{23}$$

Table II Reduced Payoff Matrix

Strategy	Environmentalists: 1. Build park and zoo	2. Preserve the land
Developers		
2. Build barge dock	−8	3
3. Build amusement park	10	−5

The constraints are that they must choose a strategy or a combination of strategies.

$$S_1 + S_2 = 1 \tag{24}$$

where S_1 is the choice of the park and zoo strategy and S_2 is the preservation choice.

Next, the opponents' choices act as constraints. All the environmentalists can assume is that their opponents will use some best strategy and try to maximize its value. Both of the opponents' logical strategies must be considered. The first opponent strategy will yield an average payoff of

$$-8S_1 + 3S_2 \leq V \quad \text{or} \quad -8S_1 + 3S_2 - V \leq 0 \tag{25}$$

and the second strategy yields

$$10S_1 - 5S_2 \leq V \quad \text{or} \quad 10S_1 - 5S_2 - V \leq 0 \tag{26}$$

Finally, negative strategies are not permitted

$$S_1, S_2 \geq 0 \tag{27}$$

The problem can be solved by the simplex method: $V = -.384$, $S_1 = .308$, $S_2 = .692$. The environmental group should put about 70% of their efforts in the preservation activity and about 30% into the park and zoo concept.

From the perspective of the developers, their goals are to maximize positive voter reaction. Their problem is the dual of the environmental group's choices:

Maximize:
$$Z = V_{\text{dual}} \tag{28}$$

subject to

$$C_1 + C_2 = 1 \tag{29}$$

$$-8C_1 + 10C_2 - V \geq 0 \tag{30}$$

$$3C_1 - 5C_2 - V \geq 0 \tag{31}$$

$$C_1, C_2 \geq 0 \tag{32}$$

The solution is $V = -.384$, $C_1 = .577$, $C_2 = .423$. The developers should put the majority of their efforts into pressing for the barge dock.

Summarizing, game theory and the dual problem of linear programming share some common threads. Two-person, zero-sum, and other simplified games may be solved by linear programming. Few problems in the economic and environmental sciences are amenable to such simplification.

B. DEALING WITH INTEGERS AND NONLINEARITIES

The purpose of this section is to explore some mathematical programming problems which are closely related to linear programming. Indeed, linear programming may be used to solve some of these problems and should, if possible, be used because of the simplicity and low cost of the linear programming optimization process. The section is divided into three parts: integer requirements and linear programming solutions, nonlinear objective functions or constraints and linear programming solutions, and special algorithms for solving integer and nonlinear programming problems (branch and bound, backtrack, dynamic programming, and heuristics).

1. INTEGER REQUIREMENTS AND LINEAR PROGRAMMING SOLUTIONS

One of the assumptions of linear programming is that activities may assume continuous values (e.g., 3.16 million gallons of water; 85.2 hours of labor; 6.3 acres of land). Some decisions require integer solutions (e.g., 3.0 not 3.2 schools, 5 not 5.4 trips). If all or the vast majority of the activities must be integers, then linear programming cannot be applied. If only a few integers are required, then linear programming will probably suffice. Three ways of solving mixed linear and integer programming problems with linear programming will be presented: (1) rounding off of linear programming solutions, (2) cutting plane, and (3) spot testing.

The simplest type of mixed problem which can use the mathematics of linear programming is one in which the rounding of fractional answers is reasonable. For example, let us assume that a minimize hospital management cost problem yields a solution of 2113.12 nurses, 862.4 doctors, and 1765.84 nurse's aides. Given the large number of activities in each of the solution variables, rounding the activities to 2113 nurses, 862 doctors, and 1766 nurse's aides should have little impact on the solution. Indeed, given the softness of much of the data that are used in establishing parameters, the author would recommend rounding of large numbers.

If the solution to the hypothetical problem had included the opening of 4.7 hospitals and of 7.2 ambulance dispatch centers, then rounding would be inappropriate because of the large cost difference between 4 and 5 hospitals and 7 and 8 ambulance dispatch centers. A series of linear programming runs may be used to transform the 4.7 hospitals and 7.2. dispatch centers to integers. First, using the linear programming optimal solution as your first solution, choose the activity which is the furthest from the integer value (4.7 hospitals in this case). Write a constraint which sets the number of hospitals equal to 4. Then run the problem and obtain a solution. If the solution is feasible, record the activities and value of the problem. If the solution is infeasible, then the lower-valued solution will not be used. Next, write a constraint which sets the number of hospitals equal to 5. Run the problem. If you obtain feasible solutions when the number of hospitals is set at either 4 or 5, then choose the number of hospitals which minimizes the value of the solution. If the problem was a maximization problem, you would select the solution which produced a higher value.

In the hospital management case, let us assume that the problem was feasible and minimized when the 4-hospital constraint was inserted. This solution is to be your second working solution.

Once the first integer has been fixed at 4, then use the second working solution as a starting point for working out the best integer value for the ambulance dispatch activities. The procedure is identical to the previous process. Write an equality constraint for the lower integer value of ambulance dispatch centers. Obtain a solution. Write an equality constraint for the higher integer value and obtain a second solution. If both are feasible, choose the one which minimizes the value of the problem.

Recapitulating, the cutting plane method allows you to solve mixed-integer and linear programming problems by systematically adding new integer constraints to an initial optimal linear programming solution. The name cutting plane method refers to the fact that you are inserting planes into feasible solution space when you write the integer constraints.[4] Providing you had an infinite amount of patience, any number of integer values could be chosen by the cutting plane method. If too many integer values are decided by this process, the solution may not converge.

Spot testing is a third alternative that may be used to solve mixed linear and integer problems with linear programming. If a few integer variables are involved, select a range of appropriate integer solutions and then use linear programming to test the linear activity alternatives for each of the integer solutions. For example, in the hospital management case, an experienced user is likely to be able to estimate the number of hospitals that are economically and politically viable as between 3 and 6. By pre-screening on the basis of minimum entry thresholds, political acceptability, and other criteria, a narrow range of feasible integer solutions may be set. The integer variables are taken as givens, and linear programming is used to develop a solution for the linear variables such as nurses, doctors, beds, etc. When more than one activity must be an integer, runs representing combinations of these integers can be tried. The author has successfully worked combinations drawn from three integer variables.

[4] See Land and Doig (1960) for an illustrated and early example of the cutting plane method.

Some mathematical programming problems with mixed integer and linear activities cannot be conveniently solved by rounding, cutting plane, and spot testing procedures because the number of integers is too great. Other problems have only integer activities. A variety of methods are available to solve integer and mixed programming problems. They will be briefly reviewed later in this chapter.

2. NONLINEAR OBJECTIVE FUNCTIONS OR CONSTRAINTS AND LINEAR PROGRAMMING SOLUTIONS

If one or more constraints are nonlinear or the objective function is nonlinear, the linear programming simplex method will probably not work. The reason the simplex methodology will not work is because its method for searching out the optimal solution assumes that the feasible solution space is without holes and dents. The simplex procedure finds a feasible solution and then tests adjacent corner points. It moves from corner to corner with the instruction that, as soon as none of the adjacent corner points yields a better solution, the optimal solution has been found. The corner point hopping procedure may not work if a constraint or the objective function is nonlinear.

Figure 9 will illustrate the problem. Figure 9a is a typical feasible region polygon. The optimum point is at A. Figure 9b contains one nonlinear curve. If the initial search for an optimum solution began at point C, it would find A, then stop because the value of the problem declines in either direction from A. The simplex procedure will not go up the hill to find point B. A local optimum has been found, but the global optimum has been missed.

The dent between points A and B is one problem with nonlinear constraints. The second problem is a hole in the feasible region, which can wreak havoc with solutions to integer problems that seek a solution within the feasible polygon region.

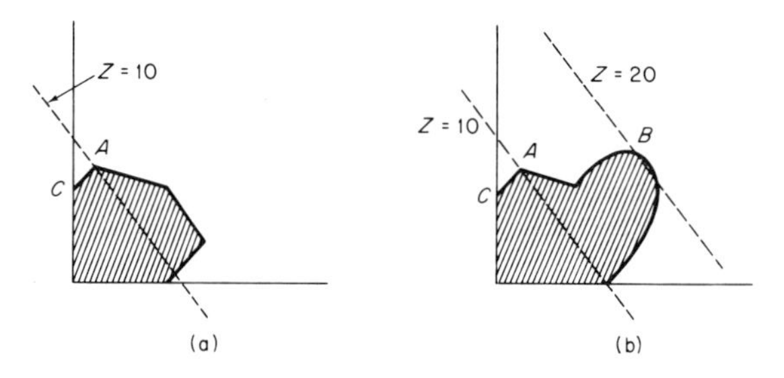

Fig. 9 (a) Linear constraints, (b) one nonlinear constraint.

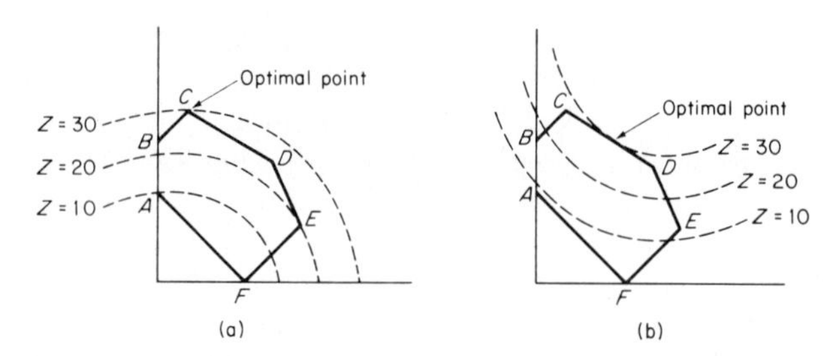

Fig. 10 (a) Concave objective function with maximization problem. (b) Convex objective function with maximization problem.

Nonlinear objective functions pose problems which depend on the shape of the objective function and the goal of the problem (maximization or minimization). If the objective function is concave (an arc connecting two points falls above a line connecting the two points as in Fig. 10a) a maximization problem can probably be solved. If the objective function is convex (an arc connecting two points falls below a line connecting the two points as in Fig. 10b), a maximization problem can only be solved with luck because moving up the curve of the convex objective function may not get us to an improved solution. The optimal is unlikely to be at a corner point. A concave objective function with a minimization problem is not amenable to the simplex method, and a convex objective function with a minimization problem is likely to work because the simplex process will slide down the curve to pick the lowest unit cost.

A linear programming formulation of a nonlinear programming problem may work if you can make linear approximations of a nonlinear constraint, a maximization-concave, or a minimization-convex objective function problem. Figure 11 illustrates two of the typical approaches. One approach (Fig. 11a) is to estimate the answer and try a linear objective function which

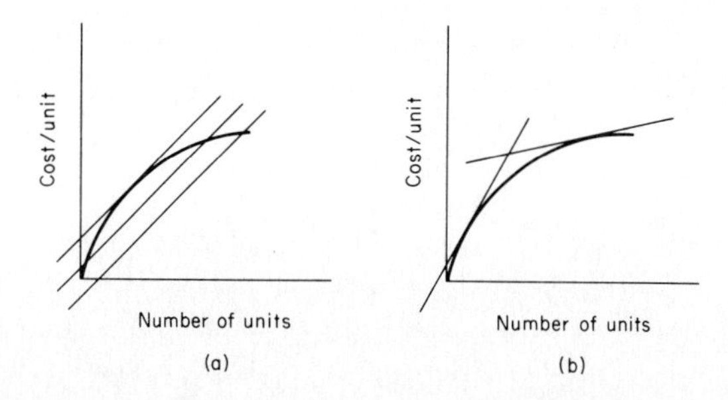

Fig. 11 (a) Linear approximation of nonlinear conditions. (b) Piecewise approximation of nonlinear conditions.

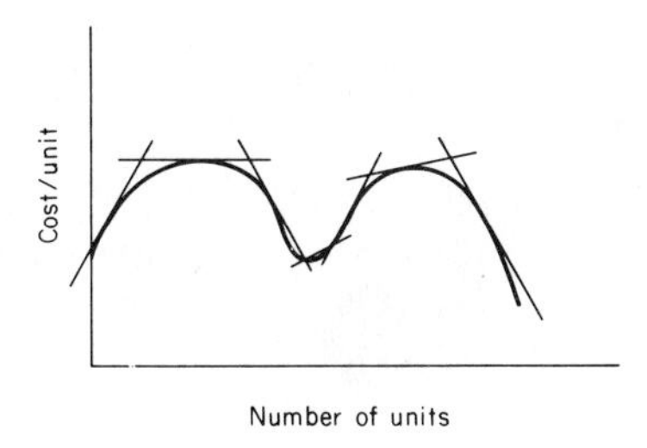

Fig. 12

comes closest to the estimated solution. Check the estimated answer against the linear programming solution. Given the information gained from the initial run, try a second linear approximation. Then try a third, a fourth, or as many as are necessary to closely represent a nonlinear response pattern. A clever computer programmer can write a program that will automatically call linear programming as a subroutine.

A second approach (Fig. 11b) is to piece together linear lines to describe the nonlinear line. The advantage of the piecewise approximation approach is that a representation of the entire curve is included in every run. The piecewise approach has two disadvantages. Each piecewise segment adds to the size of the problem. The linear approximations are difficult to construct without first seeing a feasible solution without the nonlinear conditions included. A good deal of time is therefore required to carefully piece together and test a linear approximation.

If many piecewise segments are required to approximate a complex curve, then separable programming can be tried. Separable programming uses piecewise linear approximations of nonlinear functions such as the objective function shown in Fig. 12.

The mathematical process makes sure that the linear approximations of the curve are followed. Separable programming has been successfully applied to industrial and business problems. Optimal solutions are not necessarily to be found [Beale (1968)].

3. SPECIAL ALGORITHMS FOR INTEGER AND NONLINEAR PROGRAMMING PROBLEMS

If a problem has too many integers, or if the nonlinear constraints and objective function cannot be approximated by linear lines, then specialized processes such as separable programming are used. Four types of special algorithms will be presented: branch and bound; backtrack; dynamic programming; and heuristics. The mathematical formulations of these techniques are beyond the scope of this volume. They are presented in a large literature.[5] The searching approaches and advantages and disadvantages of these algorithms will be reviewed. Then the branch and bound and the backtrack approaches will be presented with an example.

The branch and bound algorithm looks for an optimal solution by arbitrarily defining initial upper (U) and lower (L) values of the objective function and then, through a series of iterations, by pressing in toward the optimal value (Z). The search procedure begins by establishing a set of solutions chosen from the objective function. These solutions are tested to determine if any are feasible. If none are feasible the procedure ends. From among the feasible solutions the best solution is made the upper value of the problem, U. All the other feasible solutions are matched against the best upper value to try to find a better solution. Any feasible or infeasible solutions which produce a Z value higher than U are deleted from the solution because further branching from them would not lead to a better solution. With the best feasible solution set as U and with solutions for which $Z > U$ deleted, the algorithm determines the lowest value of Z from among the remaining solutions and sets it equal to the lower value L. The value L may be infeasible. The lower-bound L is then used as the initial point from which new solutions are sought by branching. The process continues until L equals U.

The branch and bound process has been used for solving moderately sized integer and nonlinear problems. The advantage of the branch and bound algorithm is that normally it is able to find a solution after searching only a fraction of the possible solution combinations. The disadvantage of the process is that it requires enormous computer storage capabilities. Accordingly, a problem with many possible solutions may be too large for the branch and bound method.

Backtracking algorithms are more suitable when computer time is available but core storage is scarce. An arbitrary upper-bound U is initially set. The algorithm searches through possible solutions in a clockwise direction. Starting with single-activity solutions, it tests each solution against the value of U. If $Z < U$, the solution is tested for feasibility. If it is feasible, it replaces the existing U. When $Z > U$, that solution and all combinations branching from it are eliminated. The algorithm then backtracks to the next possible solution to test for $Z < U$ and for feasibility.

Backtracking algorithms keep only one solution in memory. They can handle problems when computer storage is limited. However, this advantage is traded for the disadvantage of increased calculations and therefore for increased machine time.

[5] See Fiacco and McCormick (1968), Whittle (1971), Beale (1968), Hadley (1964), Nemhauser (1966), Howard (1960). For those not inclined toward a mathematical treatment, Scott (1971) presents the models with tree diagrams.

Dynamic programming is, in and of itself, a discrete field of mathematical programming. It has been developed primarily to deal with decisions involving events that occur in a sequence. For example, the solid-waste planner would seek an optimum path of waste through a resource recovery facility. The dynamic programming algorithm requires the user to clearly define the stages. The algorithm optimizes the inputs to the final stage. Since the inputs of the final stage are the outputs of the next to the last stage, the next to the last stage is optimized, and so on. Less frequently, a dynamic programming problem is solved from the beginning to the end, rather than from the end to the beginning. Whether beginning at the front or back end, the procedure searches through the branches formed at each stage to define the global optimum for all of the stages.

In comparison to branch and bound and backtrack algorithms, dynamic programming is a specialized procedure that is efficient for a particular class of problems. The successful use of all three algorithms depends on the size of the problem and the experience of the user. If the user can initially suggest a reasonable upper bound and/or group the activities so that the computer is likely to find the optimal solution early in branching process, the branch and bound and the backtrack approaches will be efficient. One interesting approach is to use both the branch and bound and the backtracking approaches to solve the same problem. Branch and bound is used when storage is available, and backtracking when storage becomes a problem and calculating time is available. Switching back and forth between the two approaches is possible.

When the number of activities exceeds one hundred, the special optimizing algorithms will rarely work. Large problems and some complex, small problems often can only be analyzed for suboptimal solutions. The development of models which may yield solutions, but which will not necessarily yield optimal solutions is called heuristic programming. Heuristic processes are not easily characterized. Some are rigidly defined for a particular problem; others are flexible enough to treat many different problems. A heuristic model may produce an optimal solution; the user, however, will not know that the solution is optimal without time-consuming testing.

The working principle of heuristic programming is that the searching out of a local optimum at every decision point will eventually lead to a global optimum. Beyond this generalization, the methods can be dichotomized. One approach is analogous to linear programming insofar as it begins with an arbitrary feasible solution and works toward optimality. The second approach begins with an arbitrary, frequently infeasible solution and proceeds through local optimization to move toward a global optimum. The author's experience with heuristic programming algorithms suggests that the experienced user can rapidly develop a satisfactory solution if the initial solution is close to the optimum and if the model can be designed to avoid obviously undesirable solutions.

As you read through the application Chapters 6–10 you will find a few examples of nonlinear programming and a fair number of mixed-integer and linear programming examples. Some of the mixed-integer and linear programming problems have deliberately been presented in their linear forms and in their integer forms. Several heuristic programming models are also presented.

C. BRANCH AND BOUND AND BACKTRACK SEARCH PROCEDURES FOR SOLVING MORE COMPLEX PROGRAMMING PROBLEMS

This chapter has used graphs to explain the great complexity of finding solutions to nonlinear problems and has provided suggestions for transforming nonlinear models into linear models. Many nonlinear programming problems should be converted into linear programming problems. Sometimes a nonlinear to linear simplification is unsatisfactory. In the social and environmental sciences, the problem that most often cannot be converted has integer and linear activities. Integer and mixed-integer and linear programming programs are common enough to warrant more detail than has already been provided in the chapter.

This section will discuss branch and bound and backtrack procedures because they are commonly used and can be illustrated graphically and with the algebra of the simplex method. The intent of the overview is to provide the reader with enough background to enable him or her to consult the more difficult nonlinear programming literature and to consult with experienced users.

It is not the intent of this section to provide the reader with enough information to go directly to one of the commercially vended integer and mixed-integer and linear programming packages. While some readers may be able to make this jump, it might turn out to be an expensive leap because problems containing integers often take 5 to 20 times as much in computer time to solve as do strictly linear programming problems.

Briefly, before considering an application, the author suggests reading the literature and/or consulting with people who have used the packages to learn time-saving methods of structuring integers and mixed-integer and linear problems.

Branch and bound and backtrack methods can be applied to more than just the problems of interest in this chapter. In essence, they can be used to solve the general class of problem in which a discrete rather than a continuous solution is sought. As indicated previously in the chapter, branch and bound and backtrack methods are only two methods of solving integer problems. While we will review only branch and bound and backtrack methods, an understanding of these approaches should help the reader understand the other methods, which are generally similar.

1. BRANCH AND BOUND AND BACKTRACK METHODS AND DECISION TREES

A decision tree is a mathematical device which enables its user to systematically evaluate a discrete set of choices. Beginning at the root of the tree, the user examines simple and then increasingly more complex solution combinations. Recently, decision trees have gained notoriety as a means of evaluating the safety of electric generating facilities that use nuclear fuel. In the nuclear power case, the goal of the M.I.T. team led by Professor Norman Rasmussen was to find the probability of accidents occurring because of a combination of mechanical failures and personnel errors.

Decision trees underlie branch and bound and backtrack methods. Initially, decision trees will be explained with an example. Then decision trees and algebra will be used to solve an extremely simple mixed-integer and linear programming problem.

2. WATER SUPPLY PURCHASING EXAMPLE[6]

The Smith public potable water company is developing a two-year plan for acquiring a supply. The company has been plagued for years by high May-through-October demand caused by seasonal residents. The company has its own supply and may purchase water from a neighboring company. On one hand, the neighboring company insists on a minimum average daily purchase contract for a period of at least six months. On the other hand, the neighboring company offers a reduced rate if the service is continuous. The economics of the Smith Company's decision are summarized in Table III.

The company pays $300 per million gallons for its own water during the low-demand period and $400 during the high-demand period. If the Smith Company purchases water from its neighboring company without a previous contract, it will pay $420 per million gallons. Each consecutive six-month period for which the contract is extended, irrespective of demand, will be rewarded with a $10,000 reduction in cost. If the Smith Company purchases water for the four time periods, it will pay $220 ($241 per million gallons)

[6] The decision tree example was modeled after a similar example in van de Panne (1971).

Table III Smith Water Company Economic Data

	First-year demand		Second-year demand	
	High-demand period 1	Low-demand period 2	High-demand period 3	Low-demand period 4
Requirements for six months, mg	912.5	547.5	985.5	584.0
Average daily, mgd	5.0	3.0	5.4	3.2
Costs when self-supplied, $ (rounded to nearest 1000)	365	164	394	175
Costs when purchased				
No contract in previous season				
1	230	230	230	230
2		220	220	220
3			210	210
4				200

for the second six months, and $210 ($360 per million gallons) for the third period, and $200 ($203 per million gallons) for the fourth period. During which six-month periods should the Smith Water Company purchase the water?

While the goal of this problem is a least-cost optimal decision, it cannot be solved by linear programming because it involves discrete choices which must be made in a sequence.

The first season decision tree appears in Fig. 13. Smith can either purchase for $230,000 or draw upon its own supply for $365,000. The second six-month period increases the total choices to four (see Fig. 14).

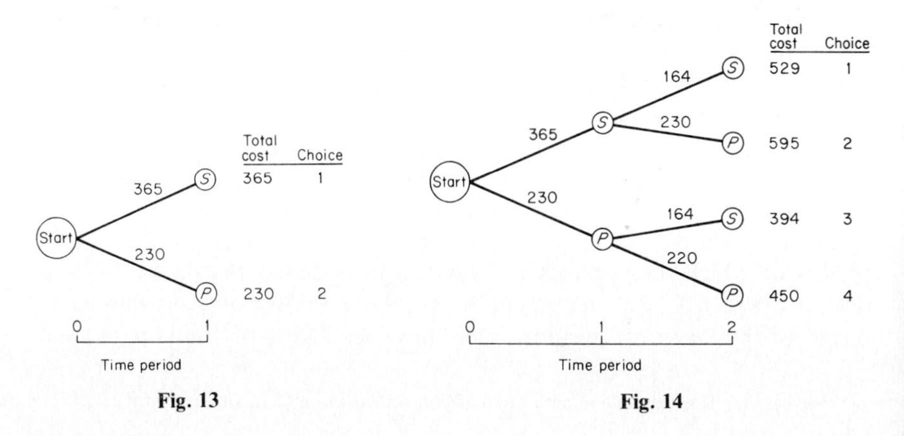

Fig. 13

Fig. 14

The least-cost solution is to draw on the Smith supply for the second six months (for \$164,000) and purchase for the first six months (\$230,000), for a total cost of \$394,000 (choice 3).

Eighteen months produces eight paths, and four periods produces sixteen paths (see Fig. 15).

The total cost favors the decision of self-supplying during the low-demand periods and purchasing during the high-demand periods (choice 11 for \$799,000). The second-lowest cost is to purchase during the first three periods and self-supply for the last period (choice 15 for \$835,000).

Enumerating all 16 decisions required the review of 30 decision points $(2 + 4 + 8 + 16 = 30)$ on the tree.

Branch and bound and backtrack methods use decision trees in a more efficient manner. Two search procedures will be described. First, we will use

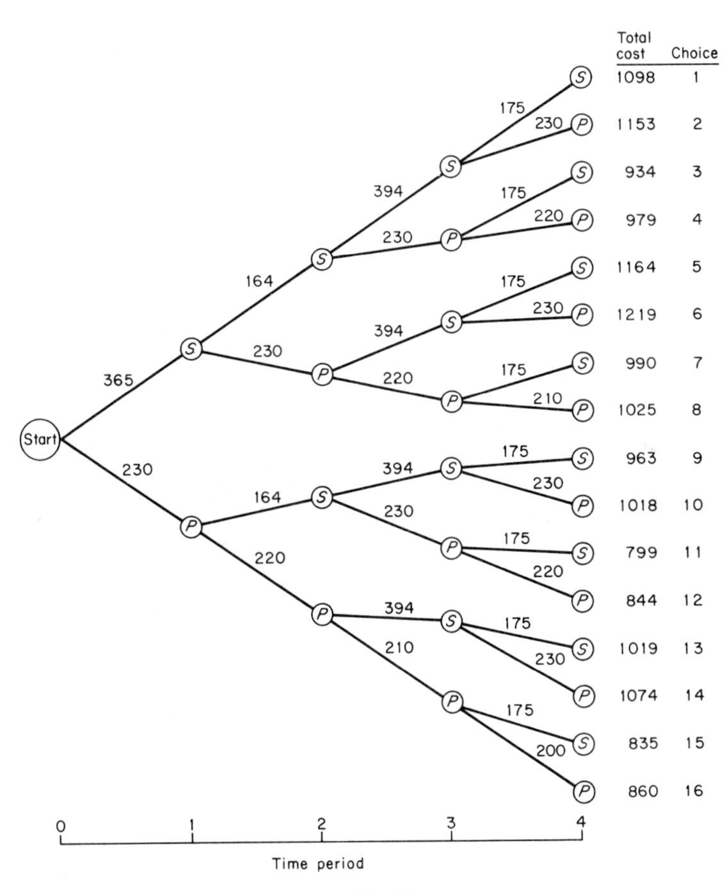

Fig. 15

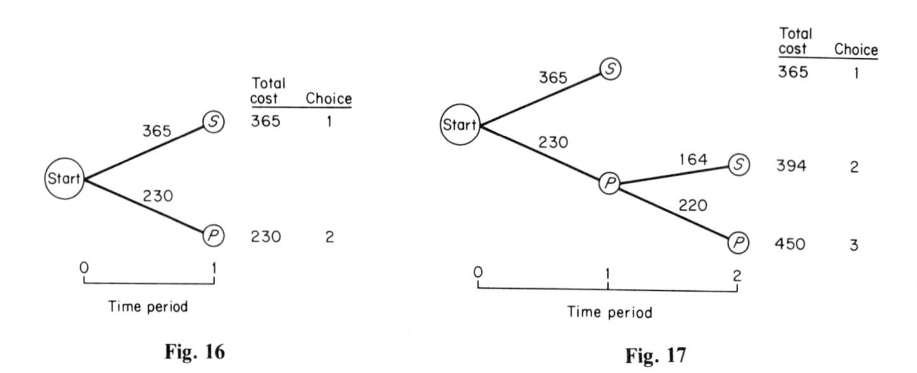

Fig. 16

Fig. 17

the Smith Water Company example and then use a flow diagram to describe the first search procedure. We begin with the initial tree (Fig. 16). The lowest possible cost is \$230,000. We branch out from this lowest cost node (Fig. 17). The lowest cost is now \$365,000. We branch out from that node (Fig. 18). Now \$394,000 (choice 3) is the lowest cost choice, so we branch from that node (Fig. 19). The least-cost node of \$450,000 is the next branching node, then \$529,000, then \$595,000 (Fig. 20). At this point, because all eight branches had to be constructed, the reader may wonder why we used this method. It has not saved any work. The final step demonstrates the utility of the method. Choice 6 is the lowest cost. We branch from choice 6 to the fourth branches (Fig. 21). Choice 6 is the lowest of the two final costs that have been computed—\$799,000. Looking at the eight other alternatives, we find that no further branching from choices 1, 3, 4, and 8 is required since their cost after three steps already exceeds the cost of choice 6 after all four steps have been completed. While choices 2, 5, and 9 would be expanded through the four steps, overall 22 decision nodes have to be calculated to arrive at the best choice of \$799,000. In comparison, complete enumeration required all 30 nodes.

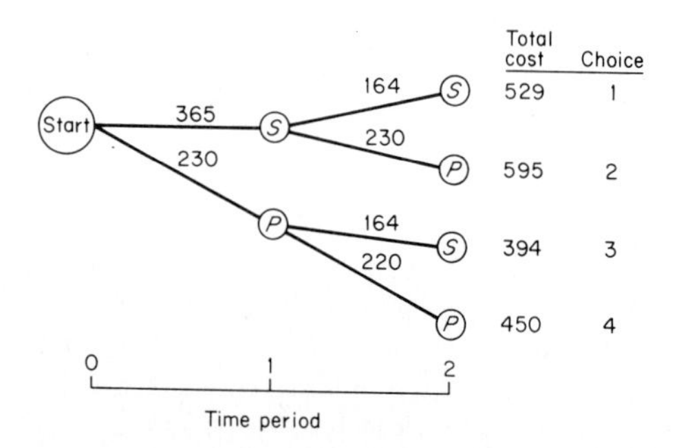

Fig. 18

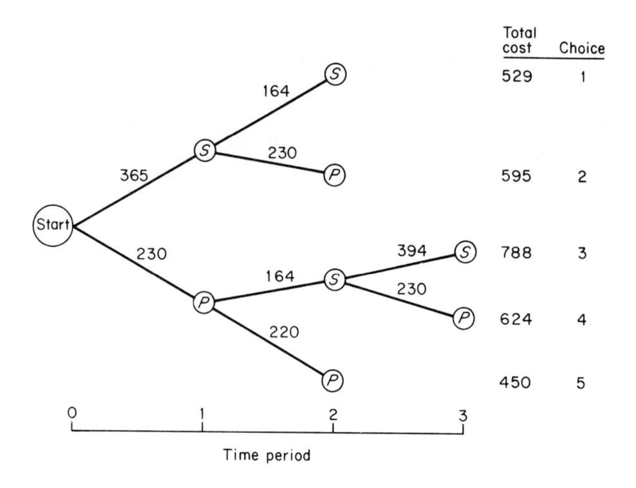
Total cost
Choice
529 1
595 2
788 3
624 4
450 5
Start
365
230
164
230
164
220
394
230
S
P
0
1
2
3
Time period

Fig. 19

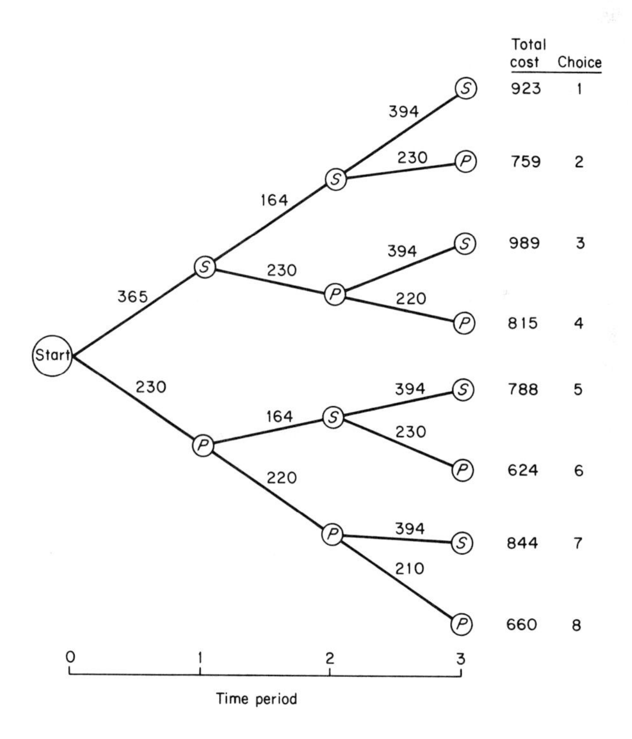
Total cost
Choice
923 1
759 2
989 3
815 4
788 5
624 6
844 7
660 8
Start
365
230
164
230
164
220
394
230
394
220
394
230
394
210
S
P
0
1
2
3
Time period

Fig. 20

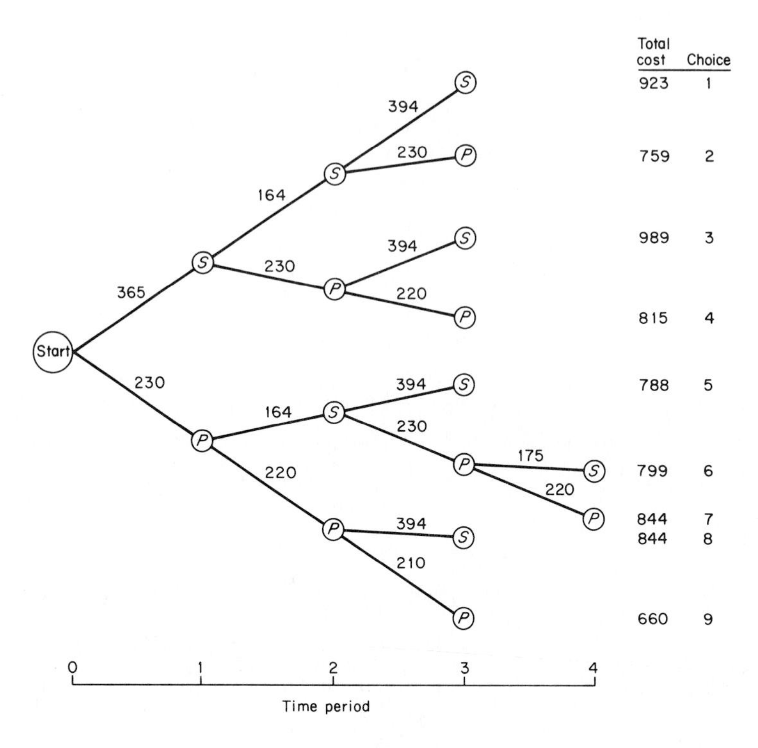

Fig. 21

Figure 22 is a flow diagram of the branch and bound process that has just been used.

The second search process quickly finds an initial solution and checks back for a better solution. We begin once more with the initial branches (Fig. 23). The lowest cost is \$230, so we branch from that node (Fig. 24). The lowest cost after two time periods is \$394,000 so we branch from that point (Fig. 25). The lowest cost after three time periods is \$474, so the branch is made from that decision node (Fig. 26).

In this case, choice 3 is the best solution, and from our previous efforts it is the optimal solution. Of course we would not know this to be the case. Therefore, from the \$799 node, we backtrack to look for a better solution.

From choice 3 at time period 4, we backtrack to the purchase node at time period 3. We have already found that branching from the purchase node at time period 3 to the purchase node at time period 4 gives a total cost of \$844 (choice 4).

Next we backtrack to the self-supplied decision node at time period 2. From there we advance to the self-supply point at time period 3. The total

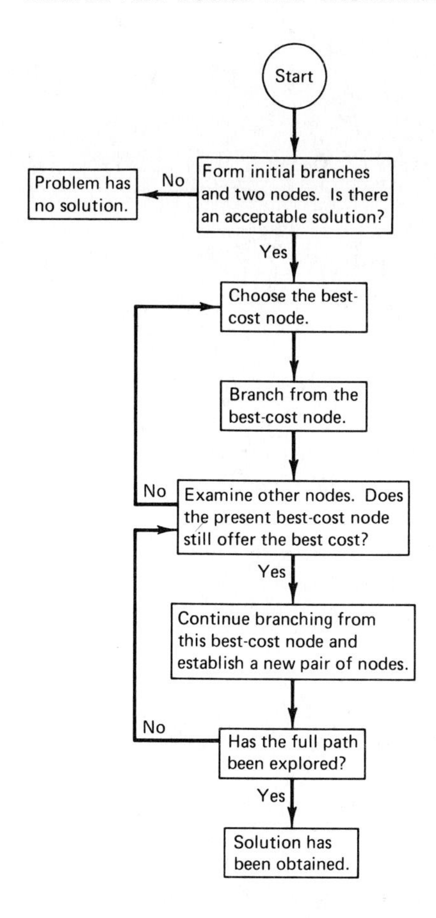

Fig. 22 Flow diagram of one branch and bound search procedure.

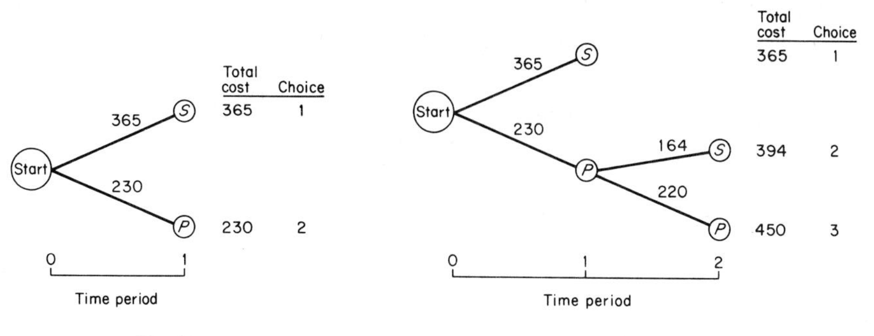

Fig. 23

Fig. 24

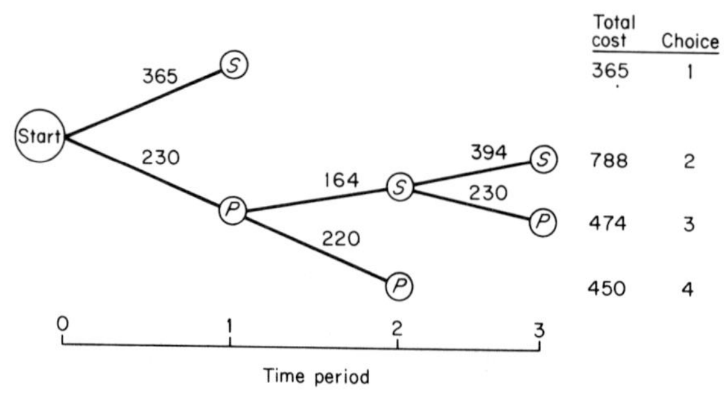

Fig. 25

cost is $788, which is less than the best total cost of $799. Therefore, this tree must be followed out to the four paths.

After exploring this path, we backtrack back to the first time period to the purchase *p* decision point. From there we would branch out to the second time period, to the third, and to the fourth, if necessary. Then, we would backtrack to the origin and explore the branches arising from the initial self-supplied *s* decision point. Figure 27 is a flow diagram of this search process.

In this case, we would have to explore the same 22 decision points. Therefore, this time neither search procedure has an advantage. However, each approach has advantages with some types of problems.

3. INTEGER AND MIXED LINEAR PROGRAMMING EXAMPLE

A simple example will illustrate the use of branch and bound and backtrack methods in conjunction with the simplex algebra method. The town of Tomsville wishes to maximize the return from building on vacant land.

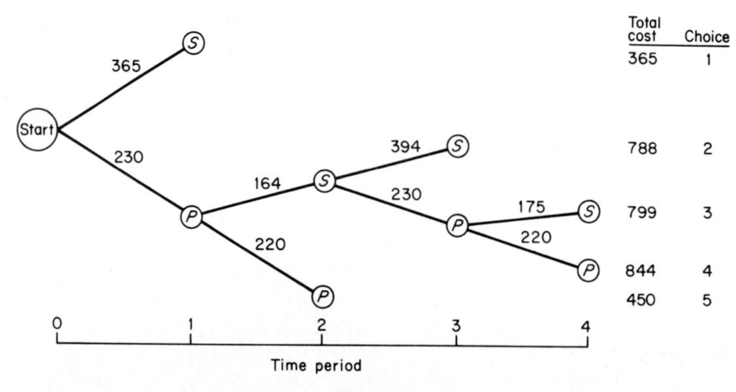

Fig. 26

The value of the land depends upon the number of units to be constructed and value of the units constructed. However, constructing and operating the units implies a fixed capital facility construction charge for services.

The eight following equations describe the problem. Land-use type X_1 implies a gain of $10,000 per acre but means a fixed charge of $3000 per acre. Land-use type X_2 yields $6000 per acre and a fixed charge of $2000 per acre. The first constraint is a building constraint such as labor. Land use X_1 uses 2 units of labor per unit in operation while activity X_2 only uses 1 unit. The construction of the services each uses 1 labor unit. The second constraint is an environmental constraint such as water. Most of the water is used in generating the units.

$$\text{Maximize:} \quad Z = 10X_1 + 6X_2 - 3I_1 - 2I_2 \tag{33}$$

subject to

$$2X_1 + X_2 + I_1 + I_2 \leq 15 \tag{34}$$

$$30X_1 + 24X_2 + 100I_1 + 70I_2 \leq 600 \tag{35}$$

$$X_1, X_2 \geq 0; \quad I_1, I_2 = 1 \text{ or } 0 \tag{36}$$

$$X_1 \geq 0 \quad \text{if } I_1 = 1 \tag{37}$$

$$X_1 = 0 \quad \text{if } I_1 = 0 \tag{38}$$

$$X_2 \geq 0 \quad \text{if } I_2 = 1 \tag{39}$$

$$X_2 = 0 \quad \text{if } I_2 = 0 \tag{40}$$

Equations (36)–(40) are the fixed-charge constraints.

An initial solution is obtained by using the simplex process. The problem can be solved without artificial variables. The initial standard problem and tableau are created.

$$\text{Maximize:} \quad Z = 10X_1 + 6X_2 - 3I_1 - 2I_2 + 0X_3 + 0X_4$$

subject to

$$2X_1 + X_2 + I_1 + I_2 + X_3 = 15$$

$$30X_1 + 24X_2 + 100I_1 + 70I_2 + X_4 = 600$$

BV	b	C_j	X_1	X_2	I_1	I_2	X_3	X_4
3	15	0	[2]	1	1	1	1	0
4	600	0	30	24	100	70	0	1
$Z_j - C_j$	$Z = 0$		-10	-6	3	2	0	0

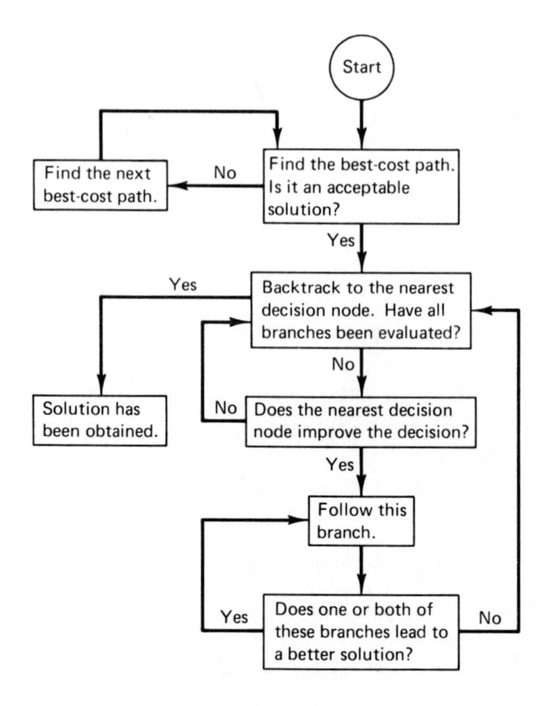

Fig. 27 Flow diagram of the second search procedure.

Next we pivot on element a_{31}. The variable X_1 is introduced into the basis and X_3 is removed.

Row 3 is multiplied by $\frac{1}{2}$ to produce row 1.

	b_j	X_1	X_2	I_1	I_2	X_3	X_4
$(\frac{1}{2})V_3$	$\frac{15}{2}$	1	$\frac{1}{2}$	$\frac{1}{2}$	$\frac{1}{2}$	$\frac{1}{2}$	0

Then multiply row 1 by -30 and add it to row 4.

	b_j	X_1	X_2	I_1	I_2	X_3	X_4
1	-225	-30	-15	-15	-15	-15	0
4	600	30	24	100	70	0	1
4	375	0	9	85	55	-15	1

Then multiply row 1 by 10 and add it to the $Z_j - C_j$ row.

	X_1	X_2	I_1	I_2	X_3	X_4
1	10	5	5	5	5	0
$Z_j - C_j$	−10	−6	3	2	0	0
$Z_j - C_j$	0	−1	8	7	5	0

The second iteration calls for X_2 to be put into the basis and X_1 removed. Element a_{12} is the pivot element.

BV	b	C_j	X_1	X_2	I_1	I_2	X_3	X_4
1	$\frac{15}{2}$	10	1	$[\frac{1}{2}]$	$\frac{1}{2}$	$\frac{1}{2}$	$\frac{1}{2}$	0
4	375	0	0	9	85	55	−15	1
$Z_j - C_j$	$Z = 75$		0	−1	8	7	5	0

The calculations and resulting tableau follow:

BV	b	X_1	X_2	I_1	I_2	X_3	X_4
$2 \times V_1$	15	2	1	1	1	1	0

	b	X_1	X_2	I_1	I_2	X_3	X_4
$-9 \times V_2$	−135	−18	−9	−9	−9	−9	0
4	375	0	9	85	55	−15	1
4	240	−18	0	76	46	−24	1

	X_1	X_2	I_1	I_2	X_3	X_4
V_2	2	1	1	1	1	0
$Z_j - C_j$	0	−1	8	7	5	0
$Z_j - C_j$	2	0	9	8	6	0

BV	b	C_j	X_1	X_2	I_1	I_2	X_3	X_4
2	15	6	2	1	1	1	1	0
4	240	0	−18	0	76	46	−24	1
$Z_j - C_j$	$Z = 90$		2	0	9	8	6	0

The solution is optimum with respect to the noninteger constraints. The integer constraints (37), (38), and (40) are satisfied, but Eq. (39) is not satisfied because $X_2 \geq 0$ while $I_2 = 0$. The integer constraint must be met. If $I_2 = 1$, then X_2 must be less than 15 because of constraint equations [Eqs. (34) and (35)]. We can find the values which satisfy Eqs. (34), (35), and (39) by subtracting column I_2 from the b vector and Z scalar. The results are the b^* and Z^* values.

BV	$b - I_2 = b^*$
2	$15 - 1 = 14$
4	$240 - 46 = 194$
$Z_j - C_j$	$Z - I_2 = Z_j^*$
	$90 - 8 = 82$

Have we an optimum solution? We cannot tell because we have tested only one combination.

$$X_1 = 0, I_1 = 0 \qquad \text{and} \qquad X_2 \geq 0, I_2 = 1$$

Alternatives that were not tested are (1) $X_1 \geq 0$, $I_1 = 1$ and $X_2 = 0$, $I_2 = 0$, (2) the combination of $X_1 \geq 0$, $I_1 = 1$ and $X_2 \geq 0$, $I_2 = 1$, and (3) $X_1 = 0$, $I_1 = 0$ and $X_2 = 0$, $I_2 = 0$. The last combination would certainly satisfy all the constraints but not produce an optimal solution. It can, in essence, be ruled out a priori. The computer algorithm will reject it.

A decision tree summary of the steps to this point is given in Fig. 28.

We can enumerate the other solutions or branch from the optimum and backtrack. Given the size of this problem it makes little difference. However,

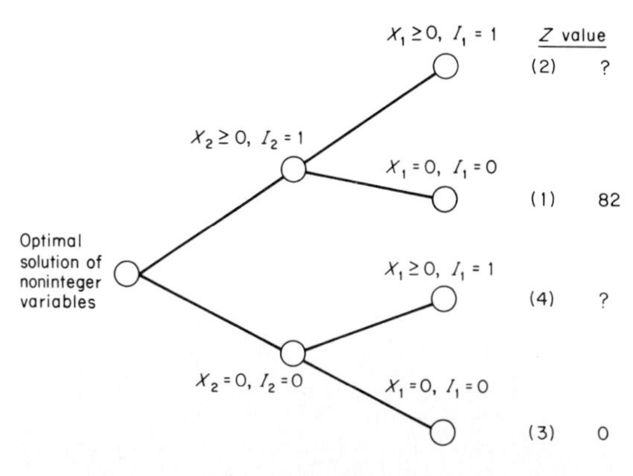

Fig. 28

a multiple-integer problem would be infeasible to solve without one of the search methods. Therefore, the backtrack method will be used.

First, we backtrack from $X_2 \geq 0$, $I_2 = 1$ and $X_1 = 0$, $I_1 = 0$ to the $X_2 \geq 0, I_2 = 1$ decision node. Next, we solve for $X_2 \geq 0, I_2 = 1$ and $X_1 \geq 0$ and $I_1 \geq 1$.

The final tableau for $X_2 \geq 0$, $I_2 = 1$ and $X_1 = 0$, $I_1 = 0$ informs us that such a combination will not improve the solution:

BV	b	C_j	X_1	X_2	I_1	I_2	X_3	X_4
2	15	6	2	1	1	1	1	0
4	240	0	−18	0	76	46	−24	1
$Z_j - C_j$	$Z = 90$		2	0	9	8	6	0

The a_{41} element is -18; therefore, we do not want to pivot X_1 in and X_4 out. Our curiosity for an $X_1 \geq 0$, $X_1 = 1$ and $X_2 \geq 0$, $X_2 = 1$ solution can be satisfied, however, by arbitrarily modifying the existing best solution. Subtract 1 from X_2 and assign it to X_1. The solution becomes:

$$X_1 = 1; \qquad I_1 = 1$$
$$X_2 = 13; \qquad I_2 = 1$$

Checking the constraints we find:

$$\begin{aligned} 2X_1 + \quad X_2 + I_1 + I_2 &\leq 15 \\ 2(1) + 1(13) + \ 1 + \ 1 &\leq 15 \\ 2 + \quad 13 + \ 1 + \ 1 &\nleq 15 \end{aligned} \tag{34}$$

Obviously, an arbitrary assignment will not work because of Eq. (34). Because of Eq. (34), 3 units have to be subtracted from X_2 to add one unit to X_1. Thus the solution is as follows:

$$X_1 = 1; \qquad I_1 = 1$$
$$X_2 = 11; \qquad I_2 = 1$$
$$\begin{aligned} 2(1) + 1(11) + 1 + 1 &\leq 15 \\ 15 &\leq 15 \end{aligned}$$

For Eq. (35),

$$\begin{aligned} 30X_1 + \ 24X_2 + \ 100I_1 + \ 70I_2 &\leq 600 \\ 30(1) + 24(11) + 100(1) + 70(1) &\leq 600 \\ 464 &\leq 600 \end{aligned}$$

The value of the solution is 71 which is less than 82.

Maximize:
$$
\begin{aligned}
Z &= 10X_1 + 6X_2 - 3I_1 - 2I_2 \\
&= 10(1) + 6(11) - 3(1) - 2(1) \\
&= 10 + 66 - 3 - 2 \\
&= 71
\end{aligned} \tag{33}
$$

Indeed, as the value of X_1 increases relative to X_2, the value of the problem decreases until X_2 approaches 0 and X_1 approaches 7.0 as follows:

X_1	X_2	I_1	I_2	Z
0	14	0	1	82
.1	12.8	1	1	72.8
6.45	.1	1	1	60.1

The backtrack from the best solution and test of $X_2 \geq 0$, $I_2 = 1$ and $X_1 \geq 0$, $I_1 = 1$ does not improve the solution (see Fig. 29).

The final backtrack takes us back to the origin and to $X_2 = 0$, $I_2 = 0$, from where we branch to $X_2 = 0$, $I_2 = 0$ and $X_1 \geq 0$ and $X_1 = 1$. Pivoting on element a_{21} of the optimal matrix, we derive the matrix previously obtained. The best solution for $X_1 \geq 0$, $I_1 = 1$ and $X_2 = 0$, $I_2 = 0$ is $X_1 = 7.0$, $I_1 = 1$, $Z = 67$.

BV	b	C_j	X_1	X_2	I_1	I_2	X_3	X_4
1	$\frac{15}{2}$	10	1	$\frac{1}{2}$	$\frac{1}{2}$	$\frac{1}{2}$	$\frac{1}{2}$	0
4	375	0	0	9	85	55	-15	1
$Z_j - C_j$	$Z = 75$		0	-1	8	7	5	0

$Z^* = Z - I_1 \qquad Z^* = 75 - 8 = 67$

The optimal solution is $Z = 82$ for $X_2 = 14$, $I_2 = 1$, $X_1 = 0$, $I_1 = 0$. This simple example vastly understates the complexity of the search for the best combination of continuous and integer variables. Had the problem contained even one more integer and noninteger combination, the number of combinations would have increased to 8. And the selection for the best place to branch would have greatly increased in complexity because the initial optimum solution is unlikely to ultimately be the optimum solution.

Summarizing, branch and bound and backtracking methods are used to solve problems that contain discontinuities in the solution space. This section has presented an example of mixed-integer and linear programming problem.

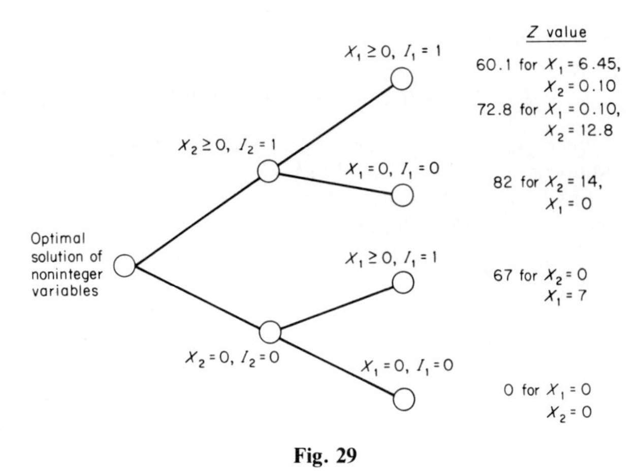

Fig. 29

The branch and bound and backtrack methods are also commonly used to solve problems in which all the variables are 0 or 1 integers and other problems which contain 0, 1, and other integer choices.

EXERCISES

To be sure that you understand the dual, formulate and solve some of the duals of the exercises at the end of Chapter 2. There are no exercises for the nonlinear methods due to their complexity. Postoptimal analysis output is presented in the next chapter. If you wish to test the cutting method for solving mixed-integer and linear programming problems with linear programming, use some of the exercise problems in Chapter 2 which yield mixed-integer and linear solutions. However, preparing the dual for and/or using the cutting plane method on the problems in Chapters 6–10 will be a far more fruitful effort than solving the Chapter 2 problems because the application chapter problems are more meaningful with respect to a real-world context.

REFERENCES

Baumol, W. (1965). "Economic Theory and Operations Analysis," Chapter 6, pp. 103–128. Prentice-Hall, Englewood Cliffs, New Jersey.

Beale, E. M. L. (1968). "Mathematical Programming in Practice," Chapter 14, pp. 124–135. Wiley, New York.

Dorfman, R., Samuelson, P., and Solow, R. (1958). "Linear Programming and Economic Analysis," Chapter 3, pp. 39–63. McGraw-Hill, New York.

Fiacco, A. V., and McCormick, G. (1968). "Nonlinear Programming: Sequential Unconstrained Minimization Techniques." Wiley, New York.
Hadley, G. (1964). "Non-Linear and Dynamic Programming." Addison-Wesley, Reading, Massachusetts.
Howard, R. A. (1960). "Dynamic Programming and Markov Processes." MIT Press, Cambridge, Massachusetts.
Karlin, S. (1959). "Mathematical Methods and Theory in Games, Programming, and Economics," Vol. 1, Addison-Wesley, Reading, Massachusetts.
Land, A. H., and Doig, A. G. (1960). An automatic method of solving discrete programming problems. *Econometrica* **28**(3), 497–520.
Miller, C. R., and Dossani, N. (July 1967). "A Note on Degeneracy and Multiple Optima in Linear Programs." Regional Science Research Institute Discussion Paper No. 17, Philadelphia.
Nemhauser, G. L. (1966). "Introduction to Dynamic Programming." Wiley, New York.
Rapoport, A. (1970). "*N*-Person Game Theory." Univ. of Michigan Press, Ann Arbor.
Scott, A. J. (1971). "Combinatorial Programming, Spatial Analysis, and Planning." Methuen, London.
Vajda, S. (1967). "The Theory of Games and Linear Programming." Science Paperbacks, London.
van de Panne, C. (1971). "Linear Programming and Related Techniques," Chapter 15, pp. 313–334. Amer. Elsevier, New York.
von Neumann, J., and Morgenstern, O. (1944). "Theory of Games and Economic Behaviour." Princeton Univ. Press, Princeton, New Jersey.
Wellar, B., and LaCava, G. (1971). A note on duality and parametric programming in urban planning. *J. Regional Sci.* **11**(2), 267–270.
Whittle, P. (1971). "Optimization Under Constraints." Wiley, New York.

Chapter 4

Computer Algorithms

The purpose of this chapter is to provide the reader with enough background to use the IBM MPS/360 and/or MPSX/370 programming codes for solving most linear programming problems. The author assumes that the reader can use the packages without a thorough knowledge of the codes. Detailed instructions on the codes are available in the following publications from IBM:

MPS/360 Control Language (H20-0290)
MPS/360 Linear and Separable Programming (H20-0476)
MPS/360 Read Communications Format (READCOMM) (H20-0372)
MPS/360 Report Generator, Program Description Manual (H20-0560)
MARVEL/360 Primer (H20-0505)
MARVEL/360 Program Description Manual (H20-0496)
MPSX/370 Primer (GH19-1091-0) and newsletters

A. INTRODUCTION TO MPS

The MPS/360 and MPSX/370 codes are two of several packages available to today's user. Univac and Control Data have developed analogous packages, and other less sophisticated packages exist. The MPS/360 and MPSX/370 seem to be the most useful codes for small problems. They can solve linear programming problems with up to 4000 rows and theoretically with an unlimited number of variables. The MPSX/370 is a more powerful code. However, it is not available everywhere. Therefore, the MPS/360 code is also presented.

A sample problem drawn from the solid-waste chapter is used throughout this chapter. The system is described by the linear model:

$$\text{Minimize:} \quad Z = 7.00X_1 + 4.00X_2 + 4.20X_3 + 7.50X_4 + 6.80X_5 + 5.50X_6 + 6.20X_7 + 1.00X_8 + 0X_9 \tag{1}$$

subject to

$$X_1 + X_2 = 4000 \tag{2}$$

$$X_3 + X_4 + X_5 = 9000 \tag{3}$$

$$X_6 + X_7 = 10{,}000 \tag{4}$$

$$X_2 + X_3 + X_6 \leq 5000 \tag{5}$$

$$X_1 + X_4 + X_8 \leq 10{,}000 \tag{6}$$

$$X_5 + X_7 \leq 10{,}000 \tag{7}$$

$$X_9 \leq 4000 \tag{8}$$

$$.6(X_2 + X_3 + X_6) - X_9 = 0 \tag{9}$$

$$.4(X_2 + X_3 + X_6) - X_8 = 0 \tag{10}$$

where $X_1 \ldots X_9$ are the management choices; the C_j are the cost of each management strategy; the b_i are the constraints: the waste sources [Eqs. (2)–(4)], the processing plant [Eq. (5)], the landfills [Eqs. (6)–(7)], the generating station [Eq. (8)], and the transfer of the waste through the system [Eqs. (9)–(10)]; a_{ij} are constraint set coefficients: .6 for Eq. (9) and .4 for Eq. (10) and Z is the value of the objective function.

An MPS/360 and MPSX/370 program for solving a linear programming problem of this form requires three card types:

1. JCL/360 and 370 Job Control Language Cards (JCL) These cards provide job control information to the operating system. These job cards vary depending on the installation used. Those described in this paper are for Rutgers University, New Brunswick, New Jersey.
2. MPS/360 and 370 Program Control Cards These cards outline the sequence used by the MPS/360 and the MPSX/370 to solve the linear programming problem.
3. Input Data Cards These cards provide the input data to describe the linear programming problem to be solved.

B. CARD LAYOUT

The card layout for the MPS/360 and 370 program for solving a linear programming problem is given in Table I. No blank cards are permitted. All control language statements begin in column 10 of the card. The symbol ȼ indicates that the column should be left blank. Where the MPS/360 and MPSX/370 codes are different, they are indicated as MPS/360 only, MPSX/370 only.

The MPS control program cards for the solution are summarized below. The notation MPSX in the margin signifies a new card for the MPSX/370 system. The notation NC identifies a new card for another purpose.

MPS Control Program for Beginning Linear Programming Analysis

Required MPS Job Control Cards

```
        PROGRAM
        INITIALZ
MPSX    TITLE('MPSCLȼPROGRAMȼEXAMPLE')
        MOVE(XDATA,'22–29 data name')
        MOVE(XPBNAME,'24–31 problem file name')
        CONVERT('CHECK','SUMMARY')
        BCDOUT
        SETUP('MAX') —alternatively, could be SETUP('MIN')
        MOVE(XOBJ,'21–28 objective function name')
        MOVE(XRHS,'21–28 right-hand-side name')
        PICTURE
        PRIMAL
        SOLUTION
        EXIT
        PEND
```

C. PARAMETRIC PROGRAMMING AND SENSITIVITY ANALYSIS

Once you have become accustomed to the MPS system and have begun working with realistic problems, you will want to take advantage of the postoptimal analysis options. These include the ability to perform parametric programming on the objective function and sensitivity analysis on all or any constraints. In many cases, these computer system capabilities may obviate the need to perform additional runs in which the optimization model is, in essence, used as a simulation model.

Table I

Card no.	Type	Columns	Content	Comments
1	JCL	1–2	//	Gives job information to operating system.
		3–10	job name	Up to eight alphameric characters, first character alphabetic, no imbedded blanks.
		11	ƀ	
		12–14	JOB	
		15	ƀ	
		16	(	
		17–27	account number	Unique for each user.
		28	,	
		29–32	bin number	The bin where user desires to collect the printed output.
		33	)	
		34	,	
		35–45	your name	May be longer or shorter than ten characters.
		46	,	Follows directly after last letter in name.
		47–56	MSGLEVEL = 1	Follows directly after the comma. Causes JCL/360 job control cards to be printed out.
2	JCL	1–2	//	
		3–8	JOBLIB	
		9	ƀ	
		10–11	DD	
		12	ƀ	
		13–35	DSN = ECN · MPS360, DISP = SHR	
3	JCL	1–2	//	This card indicates to operating system that MPS job is to be executed. The step name is optional, and if not used, the words EXEC and MPSCG must be moved over accordingly.
		3–10	step name	
		11	ƀ	
		12–15	EXEC	
		16	ƀ	

4	JCL	1–10	//LP.SYSIN	Signals to operating system that the MPS program follows.
		11	b̸	
		12–13	DD	
		14	b̸	
		15	*	
5	MPS	10–16	PROGRAM	Indicates beginning of MPS program.
6	MPS	10–17	INITIALZ	Establishes internal system parameters for errors and time limits.
6.1	MPSX/370	10–39	TITLE ('MPSCL.b̸PROGRAMb̸EXAMPLE')	Names the output that will appear on each page of listing.
7	MPS	10–20	MOVE(XDATA,	Moves name chosen for data into XDATA cell.
		21	'	
		22–29	data name	Up to eight alphameric characters, no imbedded blanks. This card instructs MPS to analyze the problem named in columns 22–29
		30–31	')	
8	MPS	10–22	MOVE(XPBNAME,	Moves name chosen for problem file into XPBNAME
		23	'	
		24–31	problem file name	Up to eight alphameric characters, no imbedded blanks. This card instructs MPS to store the input matrix on a tape file or disk that you have named in columns 24–31.
		32–33	')	
9	MPS	10–35	CONVERT('CHECK','SUMMARY')	This card reads the input data, summarizes and checks the input data, and converts it to problem file in a binary format. Errors are identified.
10	MPS	10–15	BCDOUT	This optional card causes printing of input data, a wise request on the part of the user.

(Continued)

Table I (Continued)

Card no.	Type	Columns	Content	Comments
11	MPS	10–15	SETUP (	Scans statistics of problem and initiates the solution.
		16	'	
		17–19	MAX (or MIN)	Specifies a maximization or minimization problem. If you do not specify MAX or MIN after SETUP, the program will minimize.
		20–21	')	
12	MPS	10–19	MOVE(XOBJ,	Moves name chosen for objective function row into XOBJ cell.
		20	'	
		21–28	objective function row name	Up to eight alphameric characters; no imbedded blanks may be used.
		29–30	')	
13	MPS	10–19	MOVE(XRHS,	Moves name chosen for right-hand side into XRHS cell.
		20	'	
		21–28	right-hand-side name	Up to eight alphameric characters, no imbedded blanks.
		29–30	')	
14	MPS	10–16	PICTURE	This optional card prints a picture of the matrix (Some people find it helpful; others consider it a waste of paper.)
15	MPS	10–15	PRIMAL	An initial feasible solution and, if possible, an optimal solution is found. The DUAL statement may be used instead of or in conjunction with PRIMAL.
16	MPS	10–17	SOLUTION	The solution is printed out in tabular form.
17	MPS	10–13	EXIT	Returns control to operating system.
18	MPS	10–13	PEND	Designates end of MPS program.

19	JCL	1–2	/*	Defines end of MPS program to operating system.
20	JCL	1–10	//GO.SYSIN	Signals to operating system that MPS data follows.
		11	ƀ	
		12–13	DD	
		14	ƀ	
		15	*	
21	DATA	1–4	NAME	Gives name to data check.
		5–14	ƀƀƀƀƀƀƀƀƀƀ	
		15–22	date name	Up to eight alphameric characters, no imbedded blanks. Must be identical to name given on card 7.
22	DATA	1–4	ROWS	Indicates that the rows follow.
23.1–*m*	DATA	1	ƀ	One card for each row.
		2	N,G,L, or E	Row type: N specifies the objective row; G, a ≥ constraint; L, a ≤ constraint; and E, an = constraint.
		3–4	ƀƀ	
		5–12	row name	Up to eight alphameric characters, no imbedded blanks, left justified.
24	DATA	1–7	COLUMNS	This card indicates that column information follows.
25.1–*m*	DATA	1–4	ƀƀƀƀ	
		5–12	column name	Up to two matrix element per card in the column order. Up to eight alphameric characters, no imbedded blanks, left justified.
		13–14	ƀƀ	

(continued)

Table I (Continued)

Card no.	Type	Columns	Content	Comments
		15–22	row name	Use the row name corresponding to matrix element for which the value is to be given, left justified.
		23–24	b̸b̸	
		25–36	matrix element value	Use a sign and a decimal point if required.
		37–39	b̸b̸b̸	All decimal points in same column is desirable, though not required. Numerics may be used. A sign may be indicated. If a sign is not specified, a plus (+) sign is assumed. The position of the decimal and sign are not fixed. Zero values need *not* be entered.
		40–47	row name	Instructions as in 15–22.
		48–49	b̸b̸	
		50–61	matrix element value	Instructions as in 25–36. Note: listing two matrix elements on a card is optional.
26	DATA	1–3	RHS	This card indicates the right-hand-side information follows.
27.1–*m*	DATA		same as cards 25.1–*m*	Up to two right-hand-side values per card. Format same as for column cards except for right-hand-side name (card 13) in place of column name.
28	DATA	1–6	ENDATA	Indicates end of data deck.
29	JCL	1–2	/*	Defines end of MPS data to operating system.
30	JCL	1–3	//*	Defines end of job to operating system.

The control program cards for the sensitivity analysis and parametric programming will now be reviewed. The cards are presented in groups rather than card by card because the card sequence is inappropriate to explain the program instructions. Begin punching all cards in column 10.

1. SENSITIVITY ANALYSIS

RANGE This card in the MPS control program determines the range of input costs for which the optimal basis remains the same. For constraints, it determines the range of the right-hand-side values for which the optimal basis remains unchanged. It outputs these ranges together with information on the basis changes when the ranges are exceeded.

MOVE(XBOUND,'23–27') This card in the control program is an MPSX/370 option. It sets bounds for values of the activities. If no bounds are set, then the activities are assumed to have no bounds. Columns 23–27 are used to name the bounded activity. If we wanted to bound Activity 1 in the solid-waste case we might use LINK as the name.

MOVE(XRANGE,'23–27') is an MPSX/370 option. It sets ranges for the constraints. Columns 23–27 are used to name the range. If we wanted to set a range for source 1 in the solid-waste case, we might call it SOURC. In addition to the control program cards, the RANGES and BOUNDS MPSX/370 cards require data cards. In the BOUNDS section, one data card is used to indicate a bound for a specific activity. For example, if we wanted to restrict link 1 in the solid-waste problem to between 2000 and 4000 tons, and link 2 to an upper limit of 10,000 tons, the following four data cards would be needed.

BOUNDS Columns 1–6; tells the program that you are using this option. This card follows *immediately after* the last RHS data card.

b̸UPb̸LINK1b̸b̸b̸b̸b̸X1b̸b̸b̸b̸b̸b̸b̸4000
b̸LOb̸LINK1b̸b̸b̸b̸b̸X1b̸b̸b̸b̸b̸b̸b̸2000
b̸UPb̸LINK2b̸b̸b̸b̸b̸X2b̸b̸b̸b̸b̸b̸b̸10000

The preceding three cards do the following:
Field 1 (columns 2–3) specifies the type of bound: UP—upper, LO—lower. Field 2 (columns 5–9) are the general names you gave to the activities in the control program cards. Field 3 (columns 15–22) are the names of the specific activities: X_1 and X_2. Field 4 (columns 25–36) are the values of the BOUND.

RANGES Columns 1–6; tells the program that you are using this option. This card also comes after the RHS values. If we wanted to restrict source from 1 to 5000 tons, we would need the following card, which is similar to, though not the same as, the COLUMNS, RHS, and BOUNDS.

```
bbbbSOURCbbbbbS1bbbbbbbb5000
```

This card does the following:
Field 1 (columns 5–9) names the constraints SOURC as given in the MOVE (XRANGE. . .) card. Field 2 (columns 15–22) names the specific constraint. Field 3 (columns 25–36) are the values of the range.

Control Program for Performing Sensitivity Analysis

```
        PROGRAM
        INITIALZ
MPSX    TITLE('MPSCLbPROGRAMbEXAMPLE')
        MOVE(XDATA,'22–29')
        MOVE(XPBNAME,'24–31')
        CONVERT('CHECK','SUMMARY')
MPSX    MOVE(XBOUND,'23–27')
MPSX    MOVE(XRANGE,'23–27')
        BCDOUT
        SETUP('MAX')
        MOVE(XOBJ,'21–28')
        MOVE(XRHS,'21–28')
        PICTURE
        PRIMAL
        SOLUTION
NC      RANGE
        EXIT
        PEND
```

2. PARAMETRIC PROGRAMMING ON THE OBJECTIVE FUNCTION

Seven cards are needed for parametric programming on the objective function. The objective is replaced with the original objective function plus a multiple (XPARAM) of a "change row" (XCHROW). The multiple XPARAM is gradually increased, while the solution is kept optimal and feasible for the current value of XPARAM by changing the basis when necessary. First, after the SOLUTION card add MOVE (XCHROW,'CHROW')

Three parameter values must then be indicated:

XPARAM The initial value of the parameter which is set at zero.

XPARMAX The required maximum value of the parameter which must be a positive number and contain a decimal point. When XPARAM reaches this value, the PARAOBJ procedure will terminate, if not earlier.

XPARDELT The parameter interval at which the solution is to be printed. It should be positive and contain a decimal point. For every XPARDELT that XPARAM changes, a solution will be outputted.

Finally, add PARAOBJ.

The example PARAOBJ statements require that an additional row be added in the input data. This is the change row called CHROW, and it must be a nonconstrained row. *Two additional data cards are needed.*

1. Under ROWS insert ƀNƀƀCHROW after your last row name card 23.1 format.

2. Under columns insert another card which names the column, the parameter (CHROW), and the value of c. Format card 25.1.

Control Program for Objective Function Parametric Analysis

```
        PROGRAM
        INITIALZ
        MOVE(XDATA,'22–29')
        MOVE(XPBNAME,'24–31')
        CONVERT('CHECK','SUMMARY')
        BCDOUT
        SETUP('MAX')
        MOVE(XOBJ,'21–28')
        MOVE(XRHS,'21–28')
        PICTURE
        PRIMAL
        SOLUTION
NC      MOVE(XCHROW,'CHROW')
NC      XPARAM=0.
NC      XPARMAX=5.
NC      XPARDELT=1.
NC      PARAOBJ
        SOLUTION
        EXIT
        PEND
```

3. PARAMETRIC PROGRAMMING ON THE RIGHT-HAND SIDE

Seven cards are required for parametric programming of the RHS of the matrix. The RHS is replaced with the original RHS plus a multiple (XPARAM) of a change column (XCHCOL). As in PARAOBJ, XPARAM is gradually increased, and XPARAM, XPARMAX, and XPARDELT are the same as previously defined.

The example PARARHS statements require that a change column called CHCOL be included in the input data in the RHS section. *Two additional data cards are needed.*

1. Under ROWS insert ȸNȸȸCHROW after your last row name; card 23.1 format.
2. Under RHS, insert another card which names the parameter CHCOL (columns 5–12); the constraint row which is to be changed (columns 15–22); and the value of c (columns 25–36); card 27.1 format.

Control Program for Right Hand Side Parametric Analysis

```
        PROGRAM
        INITIALZ
        MOVE(XDATA,'22–29')
        MOVE(XPBNAME,'24–31')
        CONVERT('CHECK','SUMMARY')
        BCDOUT
        SETUP('MAX')
        MOVE(XOBJ,'21–28')
        MOVE(XRHS,'21–28')
        PICTURE
        PRIMAL
        SOLUTION
NC      MOVE(XCHCOL,'CHCOL')
NC      XPARAM=0.
NC      XPARMAX=5.
NC      XPARDELT=1.
NC      PARARHS
        SOLUTION
        EXIT
        PEND
```

4. OTHER OPTIONS

Other alternatives include parametric programming on a particular row, column, or simultaneously on the objective function and the right-hand side.

PARAROW Performs parametric variations on a specified row of the matrix. The procedure is the same as for PARAOBJ.

PARACOL Performs parametric programming on a specified column of the matrix. The procedure is the same as for PARARHS.

PARARIM Performs parametric analysis on both the objective function and the right-hand side simultaneously as functions of one parameter (XPARAM).

Finally, you can run several postoptimal analyses at the same time.

Control Program for Sensitivity Analysis, RHS Parametric Analysis, and Objective Function Parametric Analysis

```
PROGRAM
INITIALZ
MOVE(XDATA,'22–29')
MOVE(XPBNAME,'24–31')
CONVERT('CHECK','SUMMARY')
SETUP('MAX')
MOVE(XOBJ, '21–28')
MOVE(XRHS, '21–28')
MOVE(XROW,'row name')
MOVE(XCOLUMN,'column name')
PICTURE
PRIMAL
SAVE('NAME','OPTIMAL')
SOLUTION
RANGE
RESTORE('NAME','OPTIMAL')
MOVE(XCHROW,'CHROW')
XPARAM=0.
XPARMAX=5.
XPARDELT=1.
PARAOBJ
SOLUTION
RESTORE('NAME','OPTIMAL')
MOVE(XCHCOL,'CHCOL')
XPARAM=0.
XPARMAX=5.
XPARDELT=1.
PARARHS
SOLUTION
EXIT
PEND
```

Note

SAVE RESTORE	The SAVE and RESTORE cards are required because each postoptimal analysis must begin with the optimal basis.
PARARIM	Put this card between XPARDELT = 1 and SOLUTION for simultaneous changes of rows and columns.

D. INTERPRETATION OF THE PRINTOUT

The simplest output contains a wealth of information, of which the following are the most important:

1. The job control language.
2. The MPS/360 control program.
3. The check and summary statistics.
4. A listing of the input data.
5. A list of the iterations.
6. The value of the problem.
7. The row output.
8. The column output.

1. JOB CONTROL LANGUAGE

See cards 1–4, 19–20, 29–30.

2. MPS/360 CONTROL PROGRAM

The control program cards for the solid waste problem for MPS/360 follow.

```
0001   PROGRAM
0002   INITIALZ
0065   MOVE(XDATA,'MINIMIZE')
0066   MOVE(XPBNAME,'PROBFILE')
0067   CONVERT('CHECK','SUMMARY')
0068   BCDOUT
0069   SETUP('MIN')
0070   MOVE(XOBJ,'COST')
0071   MOVE(XRHS,'CAPACITY')
0072   PRIMAL
0073   DUAL
0074   SOLUTION
0076   EXIT
0077   PEND
```

3. CHECK AND SUMMARY STATISTICS

If no errors have been made, it should appear as follows:

```
    CHECK
    SUMMARY
1—  ROWS SECTION
    0 MINOR ERROR(S) - 0 MAJOR ERROR(S).
2—  COLUMNS SECTION.
    0 MINOR ERROR(S) - 0 MAJOR ERROR(S).
3—  RHS'S SECTION.
    REQUIRED
    0 MINOR ERROR(S) - 0 MAJOR ERROR(S).
```

If errors have been made, they will be identified.[1] New editions and updates are available in a newsletter. The program will run with minor errors. A typical minor error which is corrected is imbedded blanks. The printout will appear as follows:

```
2—COLUMNS SECTION.
IMBEDDED BLANK(S) IN THE NAME'    COST' - CORRECTED WITH 'COST
IMBEDDED BLANK(S) IN THE NAME'    COST' - CORRECTED WITH 'COST
IMBEDDED BLANK(S) IN THE NAME'    COST' - CORRECTED WITH 'COST
```

In this case, 110 imbedded blank errors were made with the following message:

```
110  MINOR ERROR(S) - 0 MAJOR ERROR(S).
```

4. A LISTING OF THE INPUT DATA

The printout includes the ROWS, COLUMNS, and RHS values. As the problem below illustrates for X_9, zero-valued C_{ij} values are not printed. In this case, the cost of transfering along link X_9 to the power plant is zero.

[1] For major errors, see Mathematical Programming System/360 (360A-CO-14X), Message Manual, IBM, White Plains, New York. 1969.

There is therefore no X_9 cost printed.

```
NAME               MINIMIZE
ROWS
 N  COST
 E  S1
 E  S2
 E  S3
 L  I
 L  D1
 L  D2
 L  P
 E  BALANCED
 E  BALANCEP
COLUMNS
  X1  COST        7.00000  S1           1.00000
  X1  D1          1.00000
  X2  COST        4.00000  S1           1.00000
  X2  I           1.00000  BALANCED      .40000
  X2  BALANCEP     .60000
  X3  COST        4.20000  S2           1.00000
  X3  I           1.00000  BALANCED      .40000
  X3  BALANCEP     .60000
  X4  COST        7.50000  S2           1.00000
  X4  D1          1.00000
  X5  COST        6.80000  S2           1.00000
  X5  D2          1.00000
  X6  COST        5.50000  S3           1.00000
  X6  I           1.00000  BALANCED      .40000
  X6  BALANCEP     .60000
  X7  COST        6.20000  S3           1.00000
  X7  D2          1.00000
  X8  COST        1.00000  D1           1.00000
  X8  BALANCED - 1.00000
  X9  P           1.00000  BALANCEP - 1.00000
RHS
  CAPACITY S1   4000.00000  S2   9000.00000
  CAPACITY S3  10000.00000  I    5000.00000
  CAPACITY D1  10000.00000  D2  10000.00000
  CAPACITY P    4000.00000
ENDATA
```

5. A LIST OF THE ITERATIONS

For those who are interested in following the simplex solution, the iterations are printed out. The last feasible step of an infeasible solution is indicated. The solid-waste problem took seven steps to arrive at a feasible solution and an additional two steps for an optimal solution. The symbol *M* indicates that multiple choices were open to the computer for choosing a column to input and a row to output.

ITER NUMBER	NUMBER INFEAS	VECTOR OUT	VECTOR IN	REDUCED COST	SUM INFEAS
M 1	2	7	17	1.00000-	13000.0
2		3	14	1.00000-	4000.00
3		6	11	1.00000-	3000.00
4		4	15	1.00000-	3000.00
5		10	16	1.00000-	3000.00
M 6	1	9	19	1.66667-	3000.00
M 7	0	2	18	1.50000-	.
FEASIBLE SOLUTION					
M 8	4	15	13	1.90000-	143000.
9		5	6	4.83333-	143000.
OPTIMAL SOLUTION					

6. THE VALUE OF THE PROBLEM

The value of the solution is indicated even if it is infeasible.

NAME	ACTIVITY	DEFINED AS
FUNCTIONAL	143000.00000	COST
RESTRAINTS		CAPACITY

7. THE ROW OUTPUT

The heart of the solution output is the row and column output. The ROWS section includes the following information:

NUMBER — The internal number of the row. The sample problem (Table II) has 10 rows.

ROW — Row names: the objective function (COST), three sources (S_1, S_2, S_3), the waste processing center (I), the two landfilling sites (D_1 and D_2), the generating station (P), and the two balance equations linking the sources and facilities.

Table II Rows

Number	Row	At	Activity	Slack Activity	Lower limit	Upper limit	Dual activity
1	COST	BS	143000.00000	143000.00000-	NONE	NONE	1.00000
2	S1	EQ	4000.00000	.	4000.00000	4000.00000	7.00000-
3	S2	EQ	9000.00000	.	9000.00000	9000.00000	7.50000-
4	S3	EQ	10000.00000	.	10000.00000	10000.00000	8.80000-
5	I	UL	5000.00000	.	NONE	5000.00000	2.90000
6	D1	BS	10000.00000	.	NONE	10000.00000	.
7	D2	UL	10000.00000	.	NONE	10000.00000	2.60000
8	P	BS	3000.00000	1000.00000	NONE	4000.00000	.
9	BALANCED	EQ	.	.	.	.	1.00000
A 10	BALANCEP	EQ	.	.	.	.	.

AT — Status of the row and its associated slack variable.
- BS: in basis and feasible
- **: in basis and infeasible
- FR: nonbasis free
- EQ: nonbasis artificial or fixed
- UL: nonbasis, activity at upper limit
- LL: nonbasis, activity at lower limit

ACTIVITY — The right-hand-side value of the solution. For example, the S_3 source constraint is 10,000 and the D_2 landfill capacity constraint is 10,000 tons per day. Mathematically: $\sum_j a_{ij} X_j$.

SLACK ACTIVITY — The slack for a particular row. In the sample case, only the generating station P has any slack; 1000 tons per day. Mathematically: $b_i - \sum_j a_{ij} X_j$.

LOWER LIMIT and UPPER LIMIT — The lower and upper limits placed on a row by the constraint. The objective function has no limits, less-than-or-equal-to constraints have no lower limits, greater-than-or-equal-to constraints have no upper limits, and equality constraints are limited to a single number.

DUAL ACTIVITY — This column indicates the change in the objective function for a unit change in the right-hand-side value. For example, disposal site D_2 has a dual activity of 2.6, which means that every ton increase in the capacity of the landfill will save \$2.60. Every ton increase in solid waste from source 1 will add \$7.00 in cost. Mathematically: $-dZ/db_i$. Multiple optima are indicated by the appearance in the optimal solution of nonbasis variables with zero reduced costs. These variables are indicated by an A in front of the appropriate row.

8. THE COLUMN OUTPUT

The column headings are identical to the row headings, with three exceptions.

NUMBER — The internal number of the row. The sample problem (Table III) has nine columns.

INPUT COST — The C_j values from the objective function—in this case, the dollar cost per ton.

REDUCED COST — The $Z_j - C_j$ value for the specific variable. Specifically, the rate of increase in the objective function for a unit increase in the column activity. Mathematically: dZ/dX_j.

ACTIVITY — Mathematically: X_j.

SLACK ACTIVITY — Mathematically: C_j.

Table III Columns

Number	Column	At	Activity	Input cost	Lower limit	Upper limit	Reduced cost
11	X1	BS	4000.00000	7.00000	.	NONE	.
12	X2	LL	.	4.00000	.	NONE	.3000
13	X3	BS	5000.00000	4.20000	.	NONE	.
14	X4	BS	4000.00000	7.50000	.	NONE	.
15	X5	LL	.	6.80000	.	NONE	1.90000
16	X6	BS	.	5.50000	.	NONE	.
17	X7	BS	10000.00000	6.20000	.	NONE	.
18	X8	BS	2000.00000	1.00000	.	NONE	.
19	X9	BS	3000.00000	.	.	NONE	.

E. POSTOPTIMAL ANALYSIS

The RANGE statement is the most commonly used postoptimal option. The results are presented in four sections.

1. ROWS AT LIMIT LEVEL

This section presents data on those rows with slack variables not in the basis (see Table IV). The columns NUMBER, ROW, ACTIVITY, SLACK ACTIVITY, LOWER LIMIT, and UPPER LIMIT are identical to previous sections. Five statements are different.

LOWER ACTIVITY — This column indicates the level to which the slack activity may be decreased without changing the basis. In the sample case, the tonnages coming from sources 1 and 2 could be considerably reduced without changing the basis.

UPPER ACTIVITY — The level to which the slack activity may be increased without changing the basis. In the sample case, the size of the resource recovery facility may be raised from 5000 to 6666 tons per day.

UNIT COST — This piece of information is the change in the objective function per unit change in the row. In the sample case, each additional ton from source 1 costs $7.00. The upper row is the per unit change for a decrease in row activity; the bottom row is the cost per unit increase in row activity.

UPPER COST and LOWER COST	These are not used.
LIMITING PROCESS	The upper row is the row or variable that would change its status if the activity level of the row is decreased below LOWER ACTIVITY. In the sample case, variable X_4 leaves or enters the basis if row 3 is decreased below 5000 tons per day. The lower row indicates the row or column that changes its status if row activity is increased beyond the UPPER ACTIVITY. In the sample case, row D_1 changes if the S_2 constraint is increased to above 9000 tons per day.
AT	Indicates the status associated with the limiting process. LL indicates that the row or column enters or leaves the basis at the lower limit. UL means that the variable or row leaves or enters the basis at the upper limit.

2. COLUMNS AT LIMIT LEVEL

This section presents data on variables not in the basis (see Table V). The indicators NUMBER, COLUMN, AT, ACTIVITY, INPUT COST, LOWER LIMIT, and UPPER LIMIT are identical to information in output from the optimum solution.

LOWER ACTIVITY	The activity level of a variable that results from a cost coefficient increase from INPUT COST to UPPER COST. In the sample cases, UPPER COST(S) of infinity mean that X_2 and X_5 will remain at their current level.
UPPER ACTIVITY	The level of activity that would result from a cost coefficient decrease from INPUT COST to LOWER COST. In the X_2 case, if the cost changes from 4.0 to 3.7, the level of activity becomes 3999.9, a meaningful change from 0.
UNIT COST	This is the change in the objective function per unit decrease (upper line) or increase (lower row) in the activity level. Within the limits set by UPPER ACTIVITY and LOWER ACTIVITY, this cost per unit change is used to determine the cost implication of a change in the activity levels. In the column X_2 case, the UNIT COSTS are not important because the UPPER ACTIVITY and LOWER ACTIVITY are tightly bounded.

Table IV Rows at Limit Level

Number	Row	At	Activity	Slack activity	Lower limit Upper limit	Lower activity Upper activity	Unit cost Unit cost	Upper cost Lower cost	Limiting process	At At
2	S1	EQ	4000.00000	.	4000.00000	.	7.00000-		X1	LL
					4000.00000	4000.00000	7.00000		D1	UL
3	S2	EQ	9000.00000	.	9000.00000	5000.00000	7.50000-		X4	LL
					9000.00000	9000.00000	7.50000		D1	UL
4	S3	EQ	10000.00000	.	10000.00000	10000.00000	8.80000-		X6	LL
					10000.00000	10000.00000	8.80000		D1	UL
5	I	UL	5000.00000	.	NONE	5000.00000	2.90000		D1	UL
					5000.00000	6666.66675	2.90000-		P	UL
7	D2	UL	10000.00000	.	NONE	10000.00000	2.60000		D1	UL
					10000.00000	10000.00000	2.60000-		X6	LL
9	BALANCED	EQ	.	.	.	.	1.00000		D1	UL
					.	1999.99976	1.00000-		X8	LL
10	BALANCEP	EQ	.	.	.	1000.00000-	.		P	UL
					.	2999.99976	.		X9	LL

Table V Columns at Limit Level

Number	Column	At	Activity	Input cost	Lower limit Upper limit	Lower activity Upper activity	Unit cost Unit cost	Upper cost Lower cost	Limiting process	At At
12	X2	LL	.	4.00000	.	4000.00000-	.30000-	INFINITY	X4	LL
					NONE	3999.99976	.30000	3.70000	X1	LL
15	X5	LL	.	6.80000	.	.	1.90000-	INFINITY	X6	LL
					NONE	4999.99609	1.90000	4.90000	X3	LL

Table VI Rows at Intermediate Level

Number	Row	At	Activity	Slack activity	Lower limit Upper limit	Lower activity Upper Activity	Unit cost Unit cost	Upper cost Lower cost	Limiting Process	At At
6	D1	BS	10000.00000	.	NONE	10000.00000	INFINITY		NONE	
					10000.00000	14999.99219	2.60000		D2	UL
8	P	BS	3000.00000	1000.00000	NONE	3000.00000	4.83333		I	UL
					4000.00000	3000.00000	INFINITY		NONE	

UPPER COST	This is the highest cost coefficient at which the variable will be maintained at its ACTIVITY. If the cost coefficient increased beyond UPPER COST, the activity level would decrease to lower activity.
LOWER COST	The lowest cost coefficient at which the variables will remain at its activity. If the coefficient value is decreased below this limit, then activity will increase to upper activity.

3. ROWS AT INTERMEDIATE LEVEL

This printout (Table VI) provides data on rows with slack variables in the basis. The interpretation is identical to Table IV.

4. COLUMNS AT INTERMEDIATE LEVEL

This section reviews the basic variables (see Table VII). The interpretation is identical to Table V.

F. PARAMETRIC PROGRAMMING OUTPUT

During the parametric programming operations, solution printouts will occur at the frequency determined by the XPARDELT value. The format is identical to the standard SOLUTION printout, with several important differences. First, your printout will contain the new rows, or rows that have been added.

```
      NAME
      ROWS
        N   COST
        .
        .
        .
        E   BALANCEP
NC      N   CHROW
      COLUMNS
        X1       COST     7.000     S1          1.000
NC      X1       D1       1.000     CHROW        .250
```

(This data indicates that you plan to increase the price by $0.25 at each iteration. If you wanted to decrease the price, punch in −.250.)

Table VII Columns at Intermediate Level

Number	Column	At	Activity	Input cost	Lower limit Upper limit	Lower activity Upper activity	Unit cost Unit cost	Upper cost Lower cost	Limiting Process	At At
11	X1	BS	3999.99976	7.00000	.	999.99243-	.30000	7.30000	X2	LL
					NONE	3999.99976	INFINITY	INFINITY-	NONE	
13	X3	BS	4999.99609	4.20000	.	999.99658	.30000	4.50000	X2	LL
					NONE	4999.99609	INFINITY	INFINITY-	NONE	
14	X4	BS	4000.00000	7.50000	.	4000.00000	INFINITY	INFINITY	NONE	
					NONE	7999.99951	.30000	7.20000	X2	LL
16	X6	BS	.	5.50000	.	.	INFINITY	INFINITY	NONE	
					NONE	4999.99219	1.90000	3.60000	X5	LL
17	X7	BS	9999.99609	6.20000	.	5000.00391	1.90000	8.10000	X5	LL
					NONE	9999.99609	INFINITY	INFINITY-	NONE	
18	X8	BS	1999.99976	1.00000	.	1999.99976	7.25000	8.25000	I	UL
					NONE	1999.99976	INFINITY	INFINITY-	NONE	
19	X9	BS	2999.99976	.	.	2999.99976	4.83333	4.83333	I	UL
					NONE	2999.99976	INFINITY	INFINITY	NONE	

```
RHS
    CAPACITY S1      4000.00     S2           9000.00
    .............    ........    ..........   ........
NC  CAPACITY P       4000.00     CHCOL S1     500.00
```

(This card indicates that you plan to increase the waste from S_1 by 500 tons at each iteration.)

The first solution will be the optimal solution with $c = 0$. Additional printouts follow. Each time, instead of getting

```
   NAME            ACTIVITY       DEFINED AS
FUNCTIONAL                        COST
RESTRAINTS       143000.000       CAPACITY
```

you will get, for example

```
FUNCTIONAL       143500.000       COST
RESTRAINTS                        CAPACITY   + 1.0 CHROW
```

The new cost is due to the first iteration. After three iterations you will get a new ACTIVITY value and 3.0 CHROW. If you have changed a constraint, the printouts will include information about CHCOL. When the problem reaches a solution beyond which an infeasibility will result, the computer will indicate that you have reached a premature maximum or minimum and will print out the solution.

For example, we will illustrate parametric programming on the intermediate facility (I in the right-hand-side values). The capacity of I is initially 5000. It will be increased five times by increments of 500 tons. The control program is the following:

```
    0001            PROGRAM
    0002            INITIALZ
    0065            MOVE(XDATA,'MINIMIZE')
    0066            MOVE(XPBNAME,'PAPERS')
    0067            CONVERT('CHECK','SUMMARY')
    0068            BCDOUT
    0069            SETUP('MIN')
    0070            MOVE(XOBJ,'COST')
    0071            MOVE(XRHS,'CAPACITY')
    0072            PICTURE
    0073            PRIMAL
    0074            SOLUTION
NC  0075            MOVE(XCHCOL,'CHCOL')
```

```
NC  0076          XPARAM=0.
NC  0077          XPARMAX=5.
NC  0078          XPARDELT=1.
NC  0079          PARARHS
    0080          SOLUTION
    0081          RANGE
    0082          EXIT
    0083          PEND
```

The data have two changes.

```
    NAME          MINIMIZE
    ROWS
     N  COST
     E  S1
     E  S2
     E  S3
     L  I
     L  D1
     L  D2
     L  P
     E  BALANCED
     E  BALANCEP
NC   N  CHROW
    COLUMNS
        X1        COST        7.00000   S1          1.00000
        X1        D1          1.00000
        X2        COST        4.00000   S1          1.00000
        X2        I           1.00000   BALANCED     .40000
        X2        BALANCEP     .60000
        X3        COST        4.20000   S2          1.00000
        X3        I           1.00000   BALANCED     .40000
        X3        BALANCEP     .60000
        X4        COST        7.50000   S2          1.00000
        X4        D1          1.00000
        X5        COST        6.80000   S2          1.00000
        X5        D2          1.00000
        X6        COST        5.50000   S3          1.00000
        X6        I           1.00000   BALANCED     .40000
        X6        BALANCEP     .60000
        X7        COST        6.20000   S3          1.00000
        X7        D2          1.00000
        X8        COST        1.00000   D1          1.00000
```

```
          X8          BALANCED - 1.00000
          X9          P          1.00000   BALANCEP - 1.00000
   RHS
NC        CHCOL    I           500.00000
          CAPACITY S1         4000.00000   S2          9000.00000
          CAPACITY S3        10000.00000   I           5000.00000
          CAPACITY D1        10000.00000   D2         10000.00000
          CAPACITY P          4000.00000
   ENDATA
```

The initial solution is the optimal solution of $143,000.00. The solution after the first 500-ton increment is the following:

NAME	ACTIVITY	DEFINED AS
FUNCTIONAL	141550.00000	COST
RESTRAINTS		CAPACITY + 1.00000 CHCOL

SECTION 1 - ROWS

	NUMBER	ROW	AT	ACTIVITY	SLACK ACTIVITY
	1	COST	BS	141550.00000	141550.00000-
	2	S1	EQ	4000.00000	.
	3	S2	EQ	9000.00000	.
	4	S3	EQ	10000.00000	.
	5	I	UL	5500.00000	.
	6	D1	BS	9700.00000	300.00000
	7	D2	UL	10000.00000	.
	8	P	BS	3300.00000	700.00000
	9	BALANCED	EQ	.	.
A	10	BALANCEP	EQ	.	.
	11	CHROW	BS	.	.

SECTION 2 - COLUMNS

NUMBER	COLUMN	AT	ACTIVITY	INPUT COST
12	X1	BS	4000.00000	7.00000
13	X2	LL	.	4.00000
14	X3	BS	5500.00000	4.20000
15	X4	BS	3500.00000	7.50000
16	X5	LL	.	6.80000
17	X6	BS	.	5.50000
18	X7	BS	10000.00000	6.20000
19	X8	BS	2200.00000	1.00000
20	X9	BS	3300.00000	.

The addition of 500 tons to the recovery facility has diverted waste from landfills to resource recovery.

After five parametric additions, the results are the following:

NAME	ACTIVITY	DEFINED AS
FUNCTIONAL	138166.66667	COST
RESTRAINTS		CAPACITY + 5.00000 CHCOL

SECTION 1 - ROWS

NUMBER	ROW	AT	ACTIVITY
1	COST	BS	138166.66667
2	S1	EQ	4000.00000
3	S2	EQ	9000.00000
4	S3	EQ	10000.00000
5	I	BS	6666.66667
6	D1	BS	9000.00000
7	D2	UL	10000.00000
8	P	UL	4000.00000
9	BALANCED	EQ	.
10	BALANCEP	EQ	.
11	CHROW	BS	.

SECTION 2 - COLUMNS

NUMBER	COLUMN	AT	ACTIVITY	INPUT COST
12	X1	BS	4000.00000	7.00000
13	X2	LL	.	4.00000
14	X3	BS	6666.66667	4.20000
15	X4	BS	2333.33333	7.50000
16	X5	LL	.	6.80000
17	X6	BS	.	5.50000
18	X7	BS	10000.00000	6.20000
19	X8	BS	2666.66667	1.00000
20	X9	BS	4000.00000	.

G. SUMMARY

This chapter has set forth the codes necessary to derive an optimal and some of the simplest postoptimal outputs. The MPSX/370 code has additional options, mixed integer and linear, and separable programming options. Please venture forth into these options with care. Over the years, many of my students have caught bad cases of information overload because they tried to use every option in the first runs.

PART TWO

APPLICATIONS

Chapter 5

Mathematical Models for Policy Analysis

Programming and Other Options

The purpose of this chapter is to provide a perspective for the five application chapters that follow. To provide a perspective about the applications of linear programming, three reviews are included. First, a brief review is made of different types of models. Second, mathematical models used in the following five parts of the decisionmaking process are reviewed:

(1) estimating the supply of and demand for serivices;
(2) transporting and scheduling goods and services;
(3) choosing sites, technologies, and scales;
(4) evaluating the impact of activities, and
(5) evaluating the efficiency of alternative policies.

Coverage of applications is by no means exhaustive. [Gass and Sisson (1975) provide excellent coverage.] The third part of the chapter is a brief set of guidelines aimed at helping the potential user avoid the many pitfalls in linear programming.

A. AN INTRODUCTION TO MODELS

A successful model is an abstraction of reality that expresses the important relationships in a process. Mathematical modeling in the socioeconomic and environmental planning sciences is part science and part art. The science part consists of isolating key information from among many variables, constants, and relationships and expressing this important information as mathematical relationships. The art part involves the juggling of budgets, staffs, and schedules to develop data and a model, and to communicate the results.

Mathematical models may not be used for three reasons. First, the science and/or the art are not present where and when they are needed. Second, the problem may not be sufficiently complex in the eyes of the decisionmakers to warrant mathematical treatment. The decisionmakers may choose to develop a verbal model of the decision process. Third, a physical model of the process may be more appropriate.

Verbal process models have the advantages of low cost and easy communication to nonexpert audiences. They have the disadvantage of being difficult, if not impossible, to replicate. Most decisions are make with verbal abstractions of reality.

Physical models are miniature replicas of the real world. They are assumed to capture the physical essence and therefore the real-world cause-and-effect relationships. Physical models are frequently used in the socioeconomic and environmental sciences. For example, physical models of rivers and estuaries are used to model the impact of dams, effluent spills, and other changes in factors affecting water quantity or quality. Prototype 5-ton per day resource recovery facilities are constructed and operated to test the system before a 1000-ton per day full-scale facility is constructed. Physical models of new communities are constructed to provide interested parties with needed visual and operating perspectives.

The advantages of physical models are that they readily communicate messages to the nonexpert, the results they produce can be replicated, and they can be used to model unique relationships that cannot be verbalized and are too complex to model efficiently with mathematics. The flow properties of estuaries and lakes are examples of the physical representation of systems that are difficult, though not necessarily impossible, to express mathematically. The disadvantages of physical models are that they are extremely expensive to develop and are unique to a specific case.

Mathematical models use symbols to express variables, constants, and relationships. When decision situations are too complex for verbal models and cannot be inexpensively represented by physical models, mathematical models are appropriate. Mathematical models have the major advantage of being readily replicated and can often be modified to be used in more than one situation. The costs of mathematical models vary. Some quantitative models can be quickly and inexpensively prepared with the aid of a calculator; other models take months, even years to calibrate, and require many months of testing before they can be used for decisionmaking. Some are expensive edifices that have never been used for decisionmaking. A major disadvantage of mathematical models is that they have been difficult to explain to non-experts.

B. MATHEMATICAL MODELS USED IN THE SOCIOECONOMIC AND ENVIRONMENTAL SCIENCES

Thousands of mathematical models have been developed for use in the socioeconomic and environmental sciences. These applications may be grouped by discipline (as in Chapters 6–10), by the resources they use (time, personnel, mathematical expertise), by relative frequency of use, by size of the geographical unit to which they have been applied, and by other criteria. The following review of mathematical models has been organized topically to emphasize to the discipline-oriented reader the similarities among many of the disciplines reviewed in Chapters 6–10 and to place programming methods in their appropriate perspective. The five topical areas were selected after considering groupings of more than a dozen and as small as three. The five topics are by no means independent of one another and tend to occupy different parts of the decisionmaking process. To lend consistency to the review, a regional solid-waste planner's decisions will be used as examples in several places.

1. ESTIMATING THE SUPPLY OF AND DEMAND FOR SERVICES

Most decisions for or against expanding or contracting a service are made after considering the present and projected demand for the service and supply of the service. Initially, the present demand and supply are estimated; then the future is forecasted, presupposing a variety of conditions.

Simple statistical models are used to summarize the present demands for service. For example, per capita solid-waste generation may be estimated by using data on tonnages landfilled within and outside the region. If 900 tons per day are landfilled and the region contains .5 million residents, then the average daily per capita generation is 3.6 pounds [(900 × (2000/500,000)]. If the goal of the solid-waste manager's analysis is to determine if a resource recovery facility is required, then other central tendency indicators should be calculated (mode, median) to reduce the influence of atypical days that could lead the manager to contract to sell more waste-derived products than are available. Measures of dispersion are desirable to provide additional information about potential sales and to indicate minimum and peak requirements for equipment (the standard deviation, high and low inputs, the 10–90 range, quartiles, etc.).

Present demands are important in the other disciplines reviewed in this book. The demands for potable, fresh, brackish, and other categories of water are required for nearly every water resources study. The quantity or quality of water effluents is prepared for environmental impact studies. Air quality studies include statistics about the average quality of the air and the frequency of violations. Dpeending on government standards, the arithmetic mean, geometric mean, or other measures of central tendency indicators of dispersion around the central tendency may be required.

Health, education, and law enforcement agencies, respectively, require summary statistics about the number of patients, students, and criminals being served. These data may be disaggregated by categories such as age, race, and sex. Further disaggregation by geographical location and by different time periods is typical.

Finally, regional and city planners have to obtain many of the above summary statistics and additional information such as trip generation on weekdays and holidays between zones, average and seasonal unemployment rates, and family income. Summarizing, simple mathematical models of central tendency and dispersion are used to model the current demand for the service.

Actions are usually predicated on projected needs. These needs are estimated with models that vary greatly in sophistication and cost. A simple projection is achieved by assuming that the base period consumption will be continued into the future. This simple procedure is rarely used because historical data almost always indicate an increase in demand. As an alternative to assuming the continuation of present demands into the future, time series and causal models are developed. A time series projection is made by extending trends, if any, drawn from historical data into the future. For example, in the solid-waste management case, the data in the accompanying tabulation are typical. Give or take a few pounds, the data show that between 1960 and 1975, the average daily solid-waste load has been increasing about 10 pounds per year. Therefore, the manager could project that the 1980 load is likely to be about 950 pounds per day.

Year	Average tons per day	Year	Average tons per day
1961	755	1970	845
1965	810	1975	900

If the manager had annual data for 15 or more years, he might choose to fit the data with a linear or nonlinear mathematical equation as illustrated in Fig. 1. Equation A is a linear line, equation B a geometric line, and equation C a modified exponential representation of historical data. If extrapolated

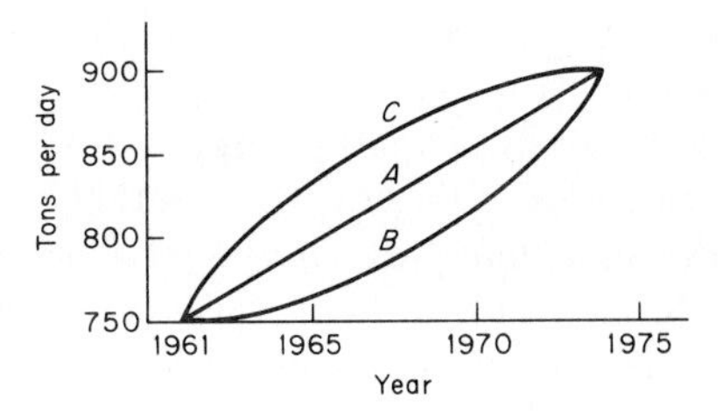

Fig. 1

beyond the historical record, lines *A*, *B*, and *C* forecast quite different solid-waste generation loads. Line *B* would lead to extremely rapid increases in waste load, line *C* to a stable load, and line *A* to a steady increase. Time series analysis has been performed on all of the base period data previously mentioned in this chapter. Sometimes the data are used without modification. At other times, the raw data are transformed to logarithms to obtain a better fit of a mathematical equation. And at other times moving averages are computed to reduce the significance of extreme values.

Developing forecasts by fitting equations to historical data has the major disadvantage of requiring the assumption that future demands are governed by historical relationships. Causal interactions are assumed implicit in the time series. Causal relationships, however, change rapidly in the socioeconomic and environmental sciences. Accordingly, many analysts develop equations that make the causal variables explicit. In the solid-waste planning case, increasing population size and per capita waste generation have been responsible for increasing total waste generation. Population change is caused by changes in fertility, mortality, and migration rates. Interregional migration rates are related to the availability of jobs and amenities. Per capita solid-waste generation is said to be related to affluence, density, and other factors. In short, if they can be computed, causal models specify factors causing changes in demand. If a change in demand is both caused by and influences another component of the system, then more complex, special mathematical models are used that allow feedback between demand and the other components.

Causal models are rapidly replacing time series models in the socioeconomic and environmental sciences. They have not become the vogue in solid-waste planning largely because precisely measured tonnage estimates have been rare [see Hale (1972)]. Causal models have become standard in water resources planning. For example, Linaweaver *et al.* (1967) have developed equations that relate residential water demand to the following: climate, affluence, the presence or absence of sewers and water metering, and price. In health, education, and law enforcement planning, staff and equipment requirements are related to projected population and to specific characteristics of the population. In education, Jaffee (1969) presents numerous methods for forecasting the size of public school populations. Mathematica, Inc. (1971), estimated the demand for financial aid requirements for students in higher education through a complex series of models. The study for the U.S. Office of Education began with estimates of the number of high school graduates grouped by income and sex. Then probability models were used to estimate when and if high school graduates of different ages and sexes would enroll in college. Third, financial aid requirements were estimated by subtracting the projected cost of undergraduate education from expected tuition. Similar forecasting models may be used to estimate the demand for staff, buildings, equipment, and other education resource needs.

Land-use and transportation planners work out complicated relationships between development, transportation, and the provision of water, sewer, fire, and other public services with simple and complex models. Simple equations are used to forecast the generation of trips between zones. Much more complex models with numerous feedback loops have been developed to forecast the need for housing, shopping centers, highways, and other serices.[1]

The most complex demand forecasting models are used to study national and regional economic development problems. Input–output models use mathematical equations to express interindustry sales (e.g., steel manufactures to auto manufactures, leather manufactures to shoe manufactures) and sales by industries to consumers (e.g., autos to people, shoes to people). For example, the construction of a chemical plant requires inputs from the steel industry for beams, the wood industry for walls, the furniture industry and many others. Once the plant begins to operate, it will purchase chemicals from the agricultural and the petroleum industries, among others, and it will sell some of its products directly to the public. Other outputs will be purchased by other chemical industries for further processing. If the public demand for pharmaceuticals increases, an input–output model may be used to project interindustry demands for resources, equipment, and personnel to meet these demands.

While principally used in projecting the demand for interindustry sales [see Miernyk (1965)], input–output models have also been used in other disciplines. For example, Greenberg *et al.* (1971) have used an input–output model to study the flow of public potable water between agencies responsible for meeting water demands. Isard *et al.* (1972) studied the ecology of the demand for natural resources in a New England bay with an input–output

[1] See Ohls and Hutchinson (1975) and Brown *et al.* (1972) for excellent overviews of the large-scale models.

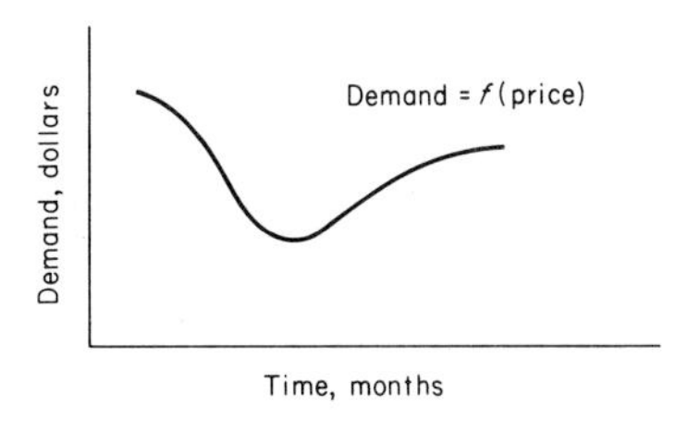

Fig. 2

model. Greater detail about input–output models in general and their relationship to linear programming will be provided in Chapter 10.

Econometric models are highly structured sets of equations that express key relationships between driving forces of the economy. Equations are fitted to historical data and express the relationships between price and demand, wages and prices, investment and demand, supply and demand, and many other indicators of economic change [see Theil (1971)]. Figure 2 illustrates a simple, short-term econometric relationship between price and demand. Initially, a rapid increase in price substantially reduces demand, then demand slowly recovers.

Input–output and econometric models are expensive. Hundreds of thousands—even millions of dollars—may go into the most complex versions of these models. Indeed, the process of constructing and updating these models is in itself a discipline. As other sections of this chapter will demonstrate, input–output and econometric models are used to forecast not only demand, but also many types of impacts of actions.

Summarizing, present demand is normally estimated by using simple statistical models of central tendency and dispersion. Projected demand is estimated by deriving equations from historical data, from causal relationships between demand and factors influencing demand, and by using complex models that include the forecasting of demand among their capabilities.

Estimating the supply of resources and services is normally not as complex as estimating the demand. Agencies and companies periodically inventory their supply of spare parts, trucks, surplus products, personnel and other supplies for which they have made expenditures. The solid-waste manager should have an inventory of available landfills, trucks, and other equipment. The water resource manager will know approximately how many gallons of water will be available under given precipitation conditions. Police, education, and health administrators should have a good idea about their personnel and equipment supplies. And local, regional, state, and national planners will have records and maps of land use and of available services.

Overall, supply is usually the known ingredient and demand the missing ingredient. Such a conclusion would, however, do an injustice to some extremely interesting studies of the variability of supply. Briefly, doctors, nurses, and educators are sometimes ill and take vacations; landfills are sometimes closed because of fire, strikes, and so forth. Water resources is the discipline that has made the most extensive analyses of the availability of supply. Because rainfall is so variable and variability leads to drought-related crop damage and potable water shortages at one extreme and floods on the other, synthetic rainfall models have been developed. These models simulate long- and short-term rainfall and steam flow records [see Tschannerl (1971) and Fiering and Jackson (1971)].

In conclusion, linear and other forms of programming do not play a direct role in estimating the supply of and the demand for services. Normally, the demand for and the supply of services are constraints in programming models.

2. TRANSPORTING AND SCHEDULING GOODS AND SERVICES

Supply and demand rarely coincide in place or time. Accordingly, the movement of supplies must be efficiently scheduled and routed. Linear programming is one of a number of mathematical models used for these purposes.

Solid-waste management has a relatively brief but interesting record of attempts to apply models to routing trucks through the streets. Briefly, routing single and multiple vehicles through the streets in the most efficient manner has been modeled with programming and heuristic models. The programming models and several heuristic models are reviewed in Chapter 6. The heuristic models are mathematical models that use a series of intuitive computations to shortcut the lengthy iterative optimization programming procedures. For example, truck routes may be randomly selected, compared, and the most efficient route selected [see Bodner *et al.* (1970)]. The heuristic models save computer time but do not necessarily lead to an optimum solution.

Frankly, however, unless your community is very unusual, the routing of your garbage collection vehicles is being made by intuitive judgement. It is fair to conclude that while mathematical models could play a role in complex routing problems in populous urban communities, this role is relatively underdeveloped.

In water resources the analogy to moving garbage compactors efficiently through city streets is moving water from reservoirs through pipes to homes, stores, factories, and schools. Programming and other simulation models have been used to help design configurations for aqueducts and distribution networks (Chapter 7). In health, education, and law enforcement, the goal usually is to place the user as close as possible to the supply of hospitals, ambulance dispatch centers, schools, fire stations, and other public services. Programming and heuristic simulations have been used to route school buses and to divide cities into precincts, among other uses (Chapter 8). Again, however, intuitive decisionmaking predominates.

Programming has played an important role in selecting routes for pipelines and tankers. In turn, these applications have greatly influenced the relative economic development and environmental quality of many regions. Programming, modifications of the law of gravity, econometric equations, and other mathematical models have been used to allocate economic goods between regions and residences and other land uses within regions [Brown *et al.* (1972); Ohls and Hutchinson [(1975)]. Chapters 9 and 10 will review many of these models.

Efficient scheduling is an important task which usually parallels routing. The solid-waste manager has to schedule crews and pickups along specific routes. Intuition based on experience is his usual decisionmaking aid. Programming and simulation models have recently been developed to assist with scheduling. For example, programming and heuristic models (Chapter 6) have been used to schedule optimum paths and to assign crews.

Analogous uses are found in the scheduling of bus routes and classes, in the choosing of monitoring frequencies, and in the assigning of projected traffic between zones. Without doubt, the most interesting applications are in law enforcement, where routing and scheduling decisions must be made quickly and simultaneously. The goal is to reach the emergency as soon as possible. Some of the models use linear programming (Chapter 8). Most use simulation models that are based on projections of the probability of criminal activity occurring. One of the earliest projects had the goal of speeding up the response of patrol cars to emergency calls [Bennett and Dubois (1970)]. A model was developed that assigned patrol units to areas on the basis of projected crime rates. However, in order to clock the precise times when the cars would appear, the quotas of visits were assigned randomly with a roulette wheel. Data indicate that the Edina, Minnesota project met its goals. A variety of more complex models have been proposed for New York City, St. Louis, and Washington, D.C., to aid in the development of average and emergency response patterns.[2]

Optimization and simulation models have been developed as aids in allocating judicial resources. Some of the proposed uses include jury selection and scheduling and assigning court personnel and judges.[3]

[2] See Larson (1972) and Gass's "Models in Law Enforcement and Criminal Justice," Chapter 8, pp. 231–275 in Gass and Sisson (1975).

[3] See Gass and Sisson (1975), Chapter 8, pp. 231–275.

Recapitulating, if judged from the perspective of decisionmaking with the aid of models, mathematical programming has played an important role in routing decisions and a minor role in scheduling decisions. Programming models have shared the literature with heuristic simulation models in seeking efficient routes. Most of the scheduling models are probability-based simulations. A few are mathematical programming models. From the larger perspective, in which countless thousands of routing and scheduling decisions are made by public institutions and private organizations every day, the role of models has been small. However, as the complexity and monetary importance of these decisions increases and as more people are trained in these models, one is likely to see efforts made to routinize routing and scheduling decisions with mathematical models.

3. CHOOSING SITES, TECHNOLOGIES, AND SCALES

The decision to build a new manufacturing or public service facility is a multi-million-dollar decision. Alternatives usually include different technologies, different sized facilities, and different locations. If the decision is a complex one, consisting of hundreds of combinations, then simulation models and linear and integer programming are appropriate. If a single consideration is paramount, then complex models are probably inappropriate.

During the first half of the twentieth century, Weber (1929) and Lösch (1967) developed physical and mathematical methods for selecting sites. The physical model consists of a board that represents the geographical space and a series of weights and pulleys that represent the forces pulling and pushing the decisionmaker toward alternative sites. The importance of the decision factors (e.g., raw materials, markets, labor) are represented by different weights that displace the system of pulleys and weights to the least-cost site. The author has used a version of this physical model in teaching the principles of location analysis. While useful in teaching, the physical model cannot replicate the many combinations of sites, technologies, and scales that influence a decision.

One mathematical model uses lines of equal net cost to represent the economic advantages and disadvantages of locating at alternative sites. Cost contours centered on each of the potential sites are drawn. The contours from each site intersect one another. At the points of intersection, one can total the costs from each of the sites. The optimum site has the lowest net cost. Suboptimum sites and their costs are also identified on the map. The final result is a map with economic values for potential sites.

A second version of the mathematical map model assigns factor attractiveness values to grid cells that have been drawn on a map. Each cell is given a value for every location factor. The values are added by cell. The final map has a total value for each grid cell. The location factors may be economic, environmental, political, and others.

The physical model and the two mathematical models represent a level of sophistication above a verbal decisionmaking process. However, these models cannot treat the thousands of decision combinations that programming and heuristic models can encompass.

A typical programming formulation has an objective function that minimizes the total cost (capital, operating, and transportation) of building and operating the facility. The objective function coefficients in a linear programming version aggregate the total capital, operating, and transportation costs of each activity. Different runs are required to determine which site/technology/scale combination is optimal. Integer programming separates the capital from the operating and the transportation costs in the objective function. As a result of this formulation, either a complete facility or nothing is built at a site. Heuristic models usually can arrive at a solution more rapidly than the mixed linear and integer programming models. However, the heuristic models may not arrive at an optimum solution. The applications chapters, especially Chapters 6 and 8, review programming and heuristic models for choosing among technologies, scales, and sites.

4. EVALUATING THE IMPACT OF ACTIVITIES

The socioeconomic and environmental sciences contain many important opportunities for impact analysis. Many impacts are made without the use of mathematical or physical models. However, as the importance of the decision increases, the use of mathematical models increases.

Water resources provide numerous interesting examples. A standard impact study uses a mathematical model to simulate the impact of effluent discharges on a river. Specifically, a model based on calculus and statistics can estimate the downstream impact of an oxygen-demanding waste discharge [see Thomann (1972)]. If the water body is tidal, is an estuary, a lake, or an underground aquifer, the impacts and the models become considerably more complex. Spatial and temporal dispersion of the pollutants is more difficult to estimate. In the case of a river, water can be assumed to flow downstream at a regular rate. In the tidal reaches of steams, water flows downstream part of the time and upstream part of the time. Impacts of discharges into streams and lakes are influenced by winds, tides, and such factors as temperature and salinity at different depths. The impact, for example, of leachate from a landfill moving into the ground is influenced by such characteristics as porosity, slope of the bedrock, and location of water-bearing strata. Models have been developed to simulate the impact of the characteristics on water quality.

Water quality models are prominent among water resource models. Other efforts in modeling water resource impacts include, for example, estimating the impact of land-use alternatives on the quality and quantity of runoff [see Crawford and Linsley (1966) and Lager *et al.* (1971)] and the impact of the intrusions of man into the ecosystems of specific species [see Chen (1972)].

Analogous diffusion models are used in air quality. A simple box model is used to distribute area-wide emissions over a region. More complex models based on a Gaussian distribution of pollutants are used to study air emissions from factories and other points of discharge [see Epstein (1974)]. The impact of noise emissions is studied with the aid of diffusion models that take into account the distance between the source of noise and the receptor of noise. The impact of barriers that reflect and absorb noise can be tested with the models, and weather and other factors may be studied [see Kessler (1975) and Gordon *et al.* (1971)]. Overall, the environmental sciences use simulation models to estimate the spatial and temporal dispersion of pollutants and, in turn, their impact on the ambient environment. These simulation models typically interact with programming models. The simulation model specifies the effectiveness of the alternative methods of meeting the standards, and the programming model specifies the economic efficiency of the alternatives.

Statistics, causal models, and econometric and input–output models are used to estimate the demographic and economic impacts of water resources projects and of water, air, and emission control regulations on regional economies. The Corps of Engineers, for example, is required to project the socioeconomic impacts of their multipurpose projects. If they are constructing dams and powerhouses on a river, they will have to estimate the economic impact of the construction employment, the power generated at the powerhouses, any recreation benefits, and other potential economic impacts [see Kim (1974)].

With respect to the impact of environmental regulations on the economy, numerous studies that have used models have recently been published. For example, Chase Econometric Associates (1972a and b; 1975) modeled the impact of federal environmental laws on the American economy as a whole and on selected economic sectors. Miernyk and Sears (1973) used an input–output model to project the impact of the Clean Air Act on the West Virginia economy.

Models have been designed to simulate the impact of location decisions on regional development. For example, the Community Land Use Game (CLUG) is a table game in which the players construct buildings, negotiate contracts, and vote on municipal issues. The model simulates the reassessment of existing building, the assessment of new structures, and the condemnation of other structures based on the age of the buildings and expenditures by the players. More complex games such as, River Basin, City Model, and Apex use the players as major decisionmakers and the computer as the simulator of relationships between development, the environment, housing, voting, and other impacts of development.

A second group of economic development models is less dependent on user–computer interaction and more dependent on fixed mathematical relationships. These models use calculus, econometrics, and other mathematical models to simulate one or more impacts of regional economic development. Chapters 9 and 10 explore several of these models.

Impact models are in their infancy but are becoming important in health, education, and law enforcement. The goal of these models is to test the impact of a program on the clients. Promulgating a population growth policy is among the most controversial policies that a government can propose. Researchers [Grossman (1972) and Kane *et al.* (1972)] have developed models to test the impact of different government health service strategies on morbidity, mortality, and fertility. Other models have evaluated the relative effectiveness of programs for tuberculosis, mental health, alcoholism, family planning, and education [see University of the State of New York (1972), Miller (1970), Levy (1971), and Chorba and Sanders (1971)]. All these models seek the elusive relationship between resource inputs and outputs of educated students, healthy people, and rehabilitated criminals. Unlike the natural science relationships that govern the diffusion of effluents through a water body, albeit not always in a completely predictable manner, the path of funds through the health, education, and law enforcement systems is not at all clear. If suitable impact models can be developed in the socioeconomic sciences, these models can be used in conjunction with programming models to efficiently allocate resources.

5. EVALUATING THE EFFICIENCY OF ALTERNATIVE POLICIES

An intuitive blending of economic, legal, political, psychological, social, and other factors go into decisionmaking. While not the only factor, cost effectiveness is usually one of the most important inputs into a decision. Models used to test the economic effectiveness of different sites, technologies, scales, routes, and schedules are examples of cost-effectiveness studies that have previously been reviewed in this chapter. This brief section will add to this initial list of resource allocation decisions.

In solid-waste management, a typical goal is to find the cheapest combination of truck and crew sizes that the unions will permit and manufacturers of trucks will sell. In water and air resources management, least-cost reasoning will often guide the choice among methods of financing a project (taxes, bonding, etc.), different procedures for managing a reservoir, and different methods for meeting ambient air quality goals. In land-use and economic development, the critical choice among different transportation projects is frequently based on the cost effectiveness of alternative models. In health, education, and law enforcement planning, choosing among different neighborhood programs, different standard and special programs at the community and regional scale, and a variety of staff sizes and mixes are typical problems in which economic cost is weighed along with political and social factors to arrive at the decision. Mathematical programming models may be used in complicated cases where many alternatives must be evaluated. Chapters 6–11 present many of these programming applications.

Recapitulating, this second section of the application overview chapter has developed a perspective on how mathematical models in general and mathematical programming models in particular fit into five common types of decisions. Intuition is used to make the vast majority of decisions. Statistics and simulation models are valuable aids for studying present and projected demands for services, for scheduling activities, and for evaluating the impact of activities. Mathematical programming models aid in allocating resources to activities where many choices are feasible.

C. GUIDELINES FOR USING MATHEMATICAL PROGRAMMING

Books and lectures are good mathematical programming teachers when you are a beginner and when you need to learn a new application or method. Veteran users of mathematical programming will probably suggest that the experience of interacting with users is the key to learning the capabilities of the technique. This section presents guidelines that may help you avoid some of the problems which will turn a theoretically beautiful application into a nightmare of unhappy users and missed deadlines. As you study the applications in Chapters 6–11 keep the guidelines in mind and ask yourself this question: If I am faced with this problem, would I or should I use mathematical programming? Budgetary, professional, and information constraints are likely to lead you to answer yes in the case of some applications and no in the case of others.

1. *Get to know the users before you develop a model and stay in contact with them throughout the project.* User–analyst interaction is the most important step to take. You should determine whether the users want to learn something quite specific or if they have only a vague notion of what they want. In the first case, you had better design with a particular goal in mind. If the second is

the case, you can formulate and discuss several different outcomes with the users.

Most users are much more concerned with outcomes than with inputs and intermediate steps. If the users expect more than you can deliver, it's better to be aware and be able to discuss the problem early rather than shortly before the project is to be completed.

During the course of user–analyst interactions try to learn how important the solution of the problem is to the users. If the solution is important enough, you can expect to obtain help with data, contacts, and numerous other problems that can be labeled the bureaucratic headaches of mathematical modeling. If the attention of the potential users is always being drawn away by crises, you may not have the political and financial support to develop, test, and present a complex model.

2. *Do not use mathematical programming if a more simple method will suffice.* If you acquire a reputation as a person who fits the problem to the method rather than the method to the problem, you are likely to find your services terminated. Perhaps a group of knowledgeable people can arrive at a solution by studying the data and by applying their judgement to the problem. A session or two with experts will tell you either that a verbal model is satisfactory, or that the problem is sufficiently complicated to warrant mathematical programming.

If a simple and inexpensive heuristic model can be written, use it. However, get a number of estimates of the cost and time required to develop and use the heuristic model.

3. *If you use mathematical programming, use linear programming if at all possible.* Nonlinear programming and many heuristic models are costly to develop, costly to run, and difficult to explain. If the problem has nonlinearities, try to straighten them out by piecing together linear equations. If it has integer constraints or other indivisibilities, rounding off a solution may be satisfactory. In short, the step from linear to nonlinear and complex heuristic models should be made with great care.

4. *Determine who is responsible for defining the variables, parameters, validity, and alternative uses of the model.* Mathematical programming models have the following components: objective function weights, activities, resources that are available in limited amounts, and coefficients that indicate how much of each resource is used for one unit of each activity. Determine in quite specific terms who is responsible for selecting these components. It is especially important that the goals of the study are clearly translated into objective function weights. Your efforts will be frustrated if, in the eyes of the client, the objective function activities and weights are unreasonable indices of the goals. With respect to the model components, it is important to learn

which of the constraints and objective function activities can be controlled by the decisionmakers. Alternatives that should be tested by parametric programming will spring from this information.

Third, make sure that there is a consensus on how the model is to be validated. Some options include a single point in time (usually the present), a single or series of historical points, and a future period for which another set of projections exist.

5. *Quickly determine the quantity and quality of data.* A feasible study becomes infeasible if necessary data are not available or are available at too great a cost. When data do not exist, synthesis is possible. The user should be informed about this problem as soon as possible. Too much synthesized data may negate the value of the study in the eyes of the users and/or may lead you to recommend that a more simple model be developed.

If data exist but are tightly locked away by a private organization or a reluctant government agency, the client will have to unlock the data bank. Earlier discussions with the user will bear fruit at this time.

When it is known that data exist, determine how much it will cost to acquire and prepare it for machine-ready codes. Before the data go to cards or tape, make sure that a consensus has been reached about the geographical units and time periods that are appropriate. Do not be afraid to suggest aggregation, especially by geographical units. Aggregations can save hours of interpretation and computer time, money and time that could be spent on making additional tests of alternatives.

6. *Learn how the results are to be used.* Looking ahead to the completion of a study is a wise initial step. The user may want to use the study as a special report, as part of an annual report, and as a special display. You must be able to translate the output into conclusions, graphs, maps, etc. that are comprehensible to the decisionmakers. If the results are going to be mapped or graphed, steps can be taken to prepare and put aside appropriate information as it is developed.

If the model is judged to be a success, be prepared to lecture about it. Accordingly, the results should contain a full set of instructions. In the author's experience, a good method of developing a sense for appropriate instructional materials is to figuratively put yourself in the shoes of the person who must interpret your results and carry them to implementation.

REFERENCES

Bennett, W., and Dubois, J. (1970). The use of probability theory in the assignment of police patrol areas. Paper PR70-2, Washington, D.C., Law Enforcement Assistance Administration, U.S. Department of Justice.

Bodner, R., Cassell, E., and Andros, P. (1970). Optimal routing of refuse collection vehicles. *J. Sanit. Eng. Div., Amer. Soc. Civil Eng.* **96** SA4, 893–904.
Brown, H. J., Ginn, J., James, F., Kain, J., Straszheim, M. (1972). "Empirical Models of Urban Land Use." National Bureau of Economic Research, Washington, D.C.
Chase Econometric Associates, Inc. (1972a). "The General Economy." National Technical Information Service, Springfield, Virginia.
Chase Econometric Associates, Inc. (1972b). "The Economic Impact of Pollution Control." National Technical Information Service, Springfield, Virginia.
Chase Econometric Associates, Inc. (1975). "The Macroeconomic Impacts of Federal Pollution Control Programs." National Technical Information Service, Springfield, Virginia.
Chen, C. (1972). "Ecologic Simulation for Aquatic Environments." National Technical Information Service, Springfield, Virginia.
Chorba, R., and Sanders, J. (1971). Planning models for tuberculosis control programs. *Health Services Research* **6**(2), 144–164.
Crawford, N., and Linsley, R. (1966). "Digital Simulation in Hydrology: Stanford Watershed Model IV," Report No. 39. Stanford Univ. Press, Stanford, California.
Epstein, A. (1974). "A Guide for Considering Air Quality in Urban Planning." National Technical Information Service, Springfield, Virginia.
Fiering, M., and Jackson, B. (1971). "Synthetic Streamflows," Vol. 1. American Geophysical Union Water Resources Series, Washington, D.C.
Gass, S., and Sisson, R. (1975). "A Guide to Models in Governmental Planning and Operations." Sauger, Potomac, Maryland.
Gordon, C. (1971). "Highway Noise—A Design for Highway Engineers," National Cooperative Highway Research Program Report 117. Bolt, Baranek, and Newman, Los Angeles, California.
Greenberg, M., Carey, G., Zobler, L., and Hordon, R. (1971). A geographical systems analysis of the water supply networks of the New York metropolitan region. *Geogr. Rev.* **LXI**(3), 339–354.
Grossman, M. (1972). "The Demand for Health: A Theoretical and Economic Investigation." Columbia Univ. Press, New York.
Hale, S. (1972). "Residential Solid Waste Generated in Low-Income Areas." National Technical Information Service, Springfield, Virginia.
Isard, W., Choguill, C., Kissin, J., Seyfarth, R., Tatlock, R. (1972). "Ecologic–Economic Analysis for Regional Development." Free Press, New York.
Jaffee, A. (1969). "Handbook of Statistical Procedures for Long Range Projections of Public School Enrollment." U.S. Dept. of Health, Education, and Welfare, Office of Education, Washington, D.C.
Kane, J., Thompson, W., and Vertinsky, I. (1972). Health care delivery: a policy simulation. *Socio-Econ. Planning Sci.* **6**(3), 283–293.
Kessler, F. (January–February 1975). Assessment of acoustic impact of a proposed steam generating plant in New York state. *Noise Contr. Eng.*, pp. 6–12.
Kim, U. (1974). "Evaluation of Interregional Input–Output Models for Potential Use in the McLellan–Kerr Arkansas River Multiple Purpose Project Impact Study." National Technical Information Service, Springfield, Virginia.
Lager, A., Pyatt, E., and Shubinski, R. (1971). "Storm Water Management Model," Vols. 1–4. Water Pollution Contr. Res. Ser. 11024, U.S. Environmental Protection Agency, Washington, D.C.
Larson, R. (1972). "Urban Police Patrol Analysis." MIT Press, Cambridge, Massachusetts.
Levy, L. (1971). An evaluation of a mental health program by use of selected operating statistics. *Amer. J. Pub. Health* **61**(10). 2038–2045.

Linaweaver, F., Jr., Geyer, J., and Wolff, J. (1967). "A Study of Residential Water Uses." U.S. Government Printing Office, Washington, D.C.

Lösch, A. (1967). "The Economics of Location." Wiley, New York.

Mathematica, Inc. (1971). "Enrollment and Financial Aid Models for Higher Education." U.S. Office of Education, Bethesda, Maryland.

Miernyk, W. (1965). "The Elements of Input–Output Analysis," Random House, New York.

Miernyk, W., and Sears, J. (1973). "Air Pollution Abatement and Regional Development." Economic Development Administration, Washington, D.C.

Miller, J. (1970). An indicator to aid management in assigning programming priorities. *Pub. Health Rep.* **85**(8), 725–730.

Ohls, J., and Hutchinson, P. (1975). *In* "A Guide to Models in Governmental Planning and Operations" (S. Gass and R. Sissons, eds.), Chapter 6, pp. 165–200. Sauger, Potomac, Maryland.

Theil, H. (1971). "Principles of Econometrics." Wiley, New York.

Thomann, R. (1972). "Systems Analysis and Water Quality Management." Environmental Sciences Services Division, New York.

Tschannerl, G. (1971). Designing reservoirs with short stream flow records. *Water Resour. Res.* **7**(4), 827–833.

University of the State of New York (1972). "New York State Performance Indicators in Education." Bureau of School Programs Evaluation, Albany, New York.

Weber, A. (1929). "Theory of the Location of Industries." Univ. of Chicago Press, Chicago, Illinois.

Chapter 6

Solid Waste

Mathematical models are newcomers to the field of solid-waste management. The vast majority of the modeling algorithms and applications have been since 1970.[1] Linear, integer, and heuristic programming comprise between one-third and one-half of these applications. The growing relevance of modeling in general, and programming specifically, in solid-waste management is due to a chain of events that have made erstwhile garbage a product of potential value and waste disposal an increasingly expensive operation, especially in urban regions. An overview of the solid-waste system should help place the models presented in this chapter in the context of their applications.

For the purposes of this chapter, solid waste may be divided into residential and nonresidential streams. Residential solid waste, which is the focus of nearly all of the applications, includes waste generated in households, small commercial and business establishments, and institutions such as schools and government offices. The residential waste stream contains paper, ferrous and nonferrous metals, food waste, yard waste, and other products. Wastes produced by industrial establishments, at mines and on farms, and through construction, sewage treatment, and street cleaning processes are excluded. These nonresidential wastes are more difficult to handle, often are hazardous, and they are usually collected by private carters with special contracts. There

[1] General treatments of solid-waste models include Marks and Liebman (1970) and Vasan (1974).

is no theoretical reason why the nonresidential stream could not be modeled. In reality, however, the composition and origin of the nonresidential solid-waste stream in most regions is not known, and the recovery technologies may not be amenable to the disposal of much of this stream without prior separation. Accordingly, this chapter will concentrate on the management of residential solid waste. As more information about the nonresidential stream becomes available, the models reviewed here should become applicable to analysis of nonresidential wastes.

A. THE PROBLEMS

The major components of residential solid-waste management can be divided into two parts: (1) collection, including storage, transfer, and initial transportation, and (2) ultimate disposal. The collection and initial transport of solid waste accounts for 70–90% of total system costs.[2] Optimization and other programming models have been proposed to aid in the scheduling of labor and the routing of trucks. While the initial collection and transport processes accounts for the vast majority of the cost, the most troublesome components of solid-waste management and the components most amenable to programming revolve around ultimate disposal. Four conditions have modified standard solid-waste disposal practices, creating this new opportunity for the application of programming models: (1) a large and increasing solid-waste load, (2) the unavailability of inexpensive landfilling sites, (3) the high cost of energy and other materials, and (4) the location of solid-waste sources near energy users.

1. A LARGE AND INCREASING SOLID-WASTE LOAD

Since 1920, when the United States became an urban nation, the volume of solid waste has substantially increased. In urban regions, some of this increase may be attributed to the nation's population increase and to the urbanization of the population. Between 1920 and 1970, the urban population of the United States increased from about 55 million to almost 150 million.

A second factor has been the affluence of a society that has encouraged the marketing of throwaway products. The amount of waste generated on a daily per capita basis has risen from 2.75 pounds in 1920 to about 5.3 pounds

[2] See Salvato (1972) for a basic discussion of cost. See also Young (1972) and ABT Associates, Inc. (1975).

in 1968. The metropolitan region's large amount of solid waste should enable it to realize economies of scale in transfer stations, regional landfills, and large recovery facilities which less populated regions cannot support.

2. UNAVAILABILITY OF INEXPENSIVE LANDFILL SITES

More than 90% of the waste material has been deposited on the land, where rain and natural leaching have moved contaminants into surface water and groundwater. Some of the trash has been burned, contributing to air pollution. Open dumps have been a breeding ground for rat populations and other disease vectors.

Landfills have become considerably more expensive to operate because of increasing land costs and environmental regulations designed to minimize the environmental degradation of the landfill. Moreover, marginal, vacant land, readily accessible to urban centers for low-cost disposal, has become a scarce commodity. When urban communities find potential sites, they are usually faced by a set of seemingly insurmountable problems beyond the normal processes of collection, processing, and disposal. The sites may be too close to residential developments, which engender high land costs and an irate citizen reaction, especially if the waste is to come from another political jurisdiction. Alternatively, the proposed site may be near water sources and therefore face opposition from environmental interests and water supply managers. It may be near one of the urban airports, in which case the site may be dismissed a priori because of a fear of plane crashes caused by bird populations. The rare site that raises little opposition may be unacceptable for the following reasons: the terrain is unsuitable, the site is unaccessible to the hundred or more heavy trucks that would haul to a 200- to 300-acre site, or it is too remote from a source of clean fill used to cover the waste. Landfilling will continue to be a key element in the solid-waste system. Acceptable landfill sites, however, will become more difficult to find.

3. HIGH COST OF ENERGY AND OTHER MATERIALS

The Department of Commerce (1974) discussed the chronic problems of industrial nations trying to obtain natural resources for meeting short-term and long-term needs. The Department listed 29 commodities that were in short supply, among them major sources of secondary materials and their virgin counterparts. The department saw shortages as increasingly more common, with the short-term "solution" of sharp price rises and the consequent frustration of business conditions, with possible unemployment and bankruptcy. Fossil fuels stood out as an acute problem.

The Environmental Protection Agency (1974) stated that, in the absence of technological, economic, and institutional constraints, municipal waste could generate 4–5 million barrels of oil a day, or 1% of U.S. annual needs, and meet the following percentages of our annual national demands: iron, 7%; aluminum, 8%; tin, 20%; paper, 14%. The National Commission on Materials Policy (1973) found that use of the aluminum, steel, and paper available from municipal waste, in lieu of primary sources by the respective industries, would save an additional 386 billion KWh, or 2% of the energy demand.

4. LOCATION OF SOLID-WASTE SOURCES NEAR ENERGY USERS

With a cautionary note that the energy recovery technologies are being improved, at the present time the steam, dry fuel, gas, or oil products cannot be shipped over long distances without reducing their competitive position in comparison to fossil fuels. Plans are to pipe the steam and gas within 1 or 2 miles of the energy recovery facility. The oil and dry fuel may be able to go longer distances by either pipe or by truck.

In short, the ideal model of a region that can realize the energy or materials recovery system would have a large, densely clustered population, high-value land, strict environmental codes, and a large manufacturing base. Such a region might suffer an economic penalty unless it tested alternative locations, sizes, and types of fixed facilities with an optimization or heuristic model. It is this type of application that will be the focus of this chapter.

B. FIXED-FACILITY ALGORITHMS

1. LINEAR MODEL

The transportation of the waste from a set of communities to a set of alternative disposal sites and the processing of the wastes at these sites may be written as a linear programming transportation problem. The user can test the economic efficiency of alternative locations, technologies, and size facilities.

The objective function in the linear model is

$$\text{Minimize:} \quad Z = \sum_i \sum_j C_{ij} X_{ij} + \sum_i \sum_k C_{ik} X_{ik} + \sum_j \sum_k C_{jk} X_{jk} \tag{1}$$

where C is the cost in dollars per ton associated with all possible travel links in the system, and C includes the cost of transporting and disposing of the waste. It is important to emphasize that in the linear model the link cost C includes the travel cost, which depends on travel time, and the charge at the facility (capital and operating), minus any revenue from sale of the waste products, which are constants for each link. The variable X refers to the amount of waste or recovered product transported from a source to an intermediate facility, from a source to a disposal facility, and from an intermediate facility to a disposal facility. The subscript i denotes a source of waste (a waste shed), j refers to an intermediate facility for waste processing (gas pyrolysis, incinerator, dry-fuel plant), and k denotes an ultimate disposal site (a landfill or electric generating station).

The second equation represents the movement of waste away from the source nodes. In each case, all waste must leave the origin and be conveyed over the transportation links to either intermediate facilities j or final sites k.

$$\sum_j X_{ij} + \sum_k X_{ik} = S_i \qquad \text{for} \quad i = 1 \ldots m \tag{2}$$

where S_i is the amount of waste generated at each source of solid waste input and m the number of sources in the system.

In Eq. (3), we require that each intermediate facility in the system cannot accept more waste than its physical capacity allows. The physical capacity may be equipment, space, entrance from highways, and others.

$$\sum_i X_{ij} \leq K_j \qquad \text{for} \quad j = 1 \ldots n \tag{3}$$

where K_j is the capacity in tons per week for each intermediate facility in the system and n the number of intermediate facilities.

The same constraint is also true for ultimate sites:

$$\sum_i X_{ik} + \sum_j X_{jk} \leq K_k \qquad \text{for} \quad k = 1 \ldots u \tag{4}$$

where K_k is the capacity in tons per week for each ultimate facility in the system and u the number of ultimate sites.

Here, waste can arrive directly from a source i or as residue from an intermediate process j. The total acceptable quantity cannot exceed the capacity K of the ultimate site.

Equation (5) states that, whatever quantity of waste exists at an intermediate facility after processing, it must be disposed of at an ultimate facility or be removed from the system.

$$\sum_i \alpha X_{ij} - \sum_k X_{jk} = 0 \qquad \text{for} \quad j = 1 \ldots n \tag{5}$$

where α is the percent of waste remaining at site j after intermediate processing has been completed, an amount that may be usable recovered material, residue, or both, and n the number of intermediate facilities.

For example, an α of .6 indicates that 60% of the original waste stream remains to be dealt with and that 40% of that initial amount has been removed from the system. The recovered materials may include such products as glass, paper, and ferrous metal. Consumers of these products can be included in the system as final sites. For example, if gas pyrolysis techniques can convert 80% of a waste stream into a usable gas product, the other 20% is extracted and sold to external industries. In such a case, α would equal .8 since it is the remaining gas product that must now be dealt with in the system. Some of the technologies that can be tested require two or more balance constraints.

$$X_{ij};\ X_{ik};\ X_{jk} \geq 0 \tag{6}$$

The linear formulation has one drawback—the calculation of the cost values in the objective function. Briefly, since hauling costs vary with each source-to-site link, one has to prepare and hand manipulate different decks for different-sized facility plans to test the economic cost of different scale facilities (e.g., 1000, 2000, 3000 tons per day). While the solutions are perfectly satisfactory, the amount of labor involved is substantial. A good programmer can reduce the effort by writing a computer program that will automatically change some of the data and call linear programming as a subroutine. The supposed drawback has the advantage of producing multiple solutions which may be of great benefit for decisionmakers. A sample problem is to be found at the end of the chapter, and Chapter 11 includes an application of this model.

2. MIXED-INTEGER MODEL

If, however, the user is faced with many choices among sites, technologies, and scale operations, a mixed-integer programming algorithm may be necessary as a screening device [Marks and Liebman (1970); Helms and Clark (1971)].

The objective function of the mixed integer problem is

$$\text{Minimize:} \quad Z = \sum_i \sum_j C_{ij} X_{ij} + \sum_i \sum_k C_{ik} X_{ik} + \sum_j \sum_k C_{jk} X_{jk} + \sum_j F_j Y_j + \sum_k F_k Y_k \tag{7}$$

where F_j and F_k are the fixed cost of constructing the intermediate and final facilities (these two may also be written as F_m, where m represents the fixed cost of all fixed facilities), and Y_j and Y_k are 0 if the facility is not constructed at site j or k and 1 if it is constructed there. The other variables are as previously defined. New constraints are added to specify that, if a facility is constructed, its capacity will not be exceeded.

$$K_j Y_j - \sum_i X_{ij} \geq 0 \tag{8}$$

and

$$K_k Y_k - \sum_i X_{ik} - \sum_j X_{jk} \geq 0 \tag{9}$$

Finally,

$$Y_j;\ Y_k = (0, 1) \tag{10}$$

This mixed-integer programming model has the capability of choosing among multiple technologies and operating scales at a single site in a single run. Each combination of technology and size is made into one technology–location–scale choice. For example, the user can model a 1000-, 2000-, and 3000-tons-per-day resource recovery facility at the same site. Since no more than a one facility is desired at a single site, the user inserts the following type of constraint:

$$Y_1 + Y_2 + Y_3 \leq 1 \tag{11}$$

The Y's represent $j = 1,2,3$, the three choices at the first site. A similar type of constraint may be used to make sure that a group of interdependent facilities are all built or are not all built. For example, many of the new resource recovery facilities require a shredder and a transfer station at the site in addition to the facility that processes the waste. Haddix (1977) has used a mixed linear and integer model in applications in southeast Virginia and in the Minneapolis–St. Paul region, and Kühner and Harrington (1976) have used a similar version accompanied by an interpreting program that indicates the value of every run for each interest group.

A heuristic model developed for the U.S. Environmental Protection Agency by Roy F. Weston, Inc. (1973), and subsequently modified by Argonne National Laboratory and the MITRE Corporation adds the potential of flexibility in determining plant size [see Berman (1973) and Yaffe (1974)]. The EPA model makes use of the Walker (1973) algorithm, a heuristic which is a modification of the simplex method. The user may specify a constant, a single line, a piecewise set of linear lines, or nonlinear lines for each technology–location combination. These equations can approximate the concave cost functions associated with economies of scale. They should replace the multiple technology–location–scale runs of the linear model.

All versions of the EPA model have promised a dynamic solution. The user would be able to simultaneously develop solutions for projected years under variable conditions of waste generation, operating and capital cost, and other important considerations. This dynamic capability would lessen the criticism that these models are limited to short-sighted and static perspectives.

If the EPA model is made widely available, it could supplant the linear and mixed linear and integer models as a facility-siting tool. The author has made use of the first two versions of the EPA model and must report that neither model worked satisfactorily. More recently at the JSGOMRAM conference, Berman (1977) demonstrated the successful use of the MITRE version of the model.

Criticism of all site selection models in this and other chapters should caution the reader against overoptimism on the basis of mathematical elegance. Site selection involves numerous considerations that cannot be modeled. The user can obtain a global optimum location–technology scale only to be rewarded with an emphatic no on the basis of political realities. Indeed, if the author's experience in the field of siting disposal facilities is typical, the reader should be prepared to develop sets of alternative solutions with the aim of presenting the decisionmaking body with economic opportunity costs for choosing a suboptimum solution.

In the case of resource recovery technologies, the success of the effort depends more on finding a buyer for the recovered products (i.e., fuel or ferrous metals) than it is on isolating the best or even the best set of locations from the cost perspective. Finally, the data sets in the field of solid-waste management are poorer than any of the comparable sets in other fields that the author has examined. Even the most elegant model will produce dubious results from dubious data. The case study in Chapter 11 emphasizes data problems.

These political realities, and the poor quality of solid-waste data, have led the author to conclude that the inelegant but inexpensive linear model is a satisfactory tool for site selection in most cases. It can be fine tuned by the EPA models, or, if the solution from the linear model is not obviously in favor of a particilar decision, a fixed-charge or the EPA heuristic models may be used.

Linear and mixed-integer programming models have also been proposed or applied to the siting of transfer stations [Marks and Liebman (1970)] and the location of garaging facilities [Clark and Helms (1970); Heaney (1971)], and they could be applied to expansion of existing fixed facilities.

3. THE MITRE MODELS: WRAP AND RAMP[3]

At the previously mentioned 1977 JSGOMRAM conference, analysts and potential users debated the relative strengths and weaknesses of linear, mixed-integer, and heuristic models. Chapter 11 presents an application of the linear model. While there has not been a comparative test of the relative utility of the different models, applications of a heuristic model by the MITRE Corporation will be presented at this point to provide the reader with some basis of comparison.

The MITRE models represent a most ambitious optimization modeling project. The Waste Resources Allocation Program (WRAP), sponsored by the EPA, helps select technologies, sites, and scales for between one and four time periods. Recovery and Market Planning (RAMP) incorporates market considerations into the model.

WRAP is a fixed-charge, linear programming heuristic. The model has a preprocessor, an optimizer, and a postprocessor. One important processor component is a CROW–FLY program. If the user does not have transportation costs, the CROW–FLY option has the capability of generating transportation cost data. The only inputs are the coordinates in longitude and latitude (in degrees, minutes and tenths of minutes) for all sources, and intermediate and ultimate sites. This tempting convenience should be avoided, if possible. In this author's opinion, CROW–FLY programs are not usually accurate reflections of regional transportation costs. For example, many regions have directional biases in transportation because of mountains and rivers. If the user desires unbiased transportation costs, they will have to be calculated from maps—or even better, collected from carters.

The MITRE optimization process has four phases. The first two generate an optimal linear programming solution. The second two use the Walker algorithm to impose fixed charges. The equation structure for a single time period is presented in the following.

The first equation requires that all waste leave the waste shed:

$$\sum_k T_{ik}^{15} + \sum_j T_{ij}^{25} = G_i \tag{12}$$

where T_{ik}^{15} is transportation in thousand tons per year from source i to ultimate site k, T_{ij}^{25} the transportation in thousand tons per year from source i to intermediate site j, and G_i the total tons per year generated at site i.

[3] The following reports document WRAP and RAMP and are available from the MITRE Corporation, Bedford, Massachusetts: Williams (1976), Berman (1976a,b,c), Berman *et al.* (1976), and Hensey (1976).

The second constraint limits the input to intermediate facilities:

$$\sum_{p}\sum_{l} a_{jp} P_{jpl} \leq K \tag{13}$$

where P_{jpl} is processing in tons per year at intermediate site j, process p, linear segment l; K is the capacity constraint; $a_{jp} = K/K_{jp}$ is the capacitation coefficient for intermediate site j, process p; K_{jp} is the capacity of process p at intermediate site j, assuming site j were totally devoted to process p.

The third constraint is the analogous restriction for ultimate disposal sites:

$$\sum_{p}\sum_{l} a_{kp} P_{kpl} \leq K \tag{14}$$

Every ultimate site has one constraint. If the site has no process constraints, a land requirement constraint is added.

$$\sum_{p}\sum_{l} d_k P_{kpl} \leq K \tag{15}$$

where $d_k = dK/L_k$ is the land requirement coefficient for ultimate site k, d the land requirement in acre-feet per thousand tons per year of sanitary landfill, and L_k the available land at ultimate site k in acre-feet divided by the number of years in the planning period. If constraints (14) and (15) are both relevant, the more imposing constraint is used.

The next three equations ensure that no waste is lost in the system. The fifth equation (16) balances the inputs to the intermediate facilities. All the incoming waste is processed. The model chooses which process and which process scale to use.

$$\sum_{i} T_{ij}^{25} + \sum_{i}\sum_{p} T_{ipj}^{35} - \sum_{p}\sum_{l} P_{jpl} = 0 \tag{16}$$

where T_{ipj}^{35} is transportation in thousand tons per year from intermediate site i, process p, to intermediate site j.

The sixth equation is the analogous input balance equation for the ultimate disposal sites.

$$\sum_{i} T_{ik}^{15} + \sum_{j}\sum_{p} c_p T_{jpk}^{45} - \sum_{p}\sum_{l} P_{kpl} = 0 \tag{17}$$

where T_{jpk}^{45} is transportation in tons per year from intermediate site j, process p, to ultimate site k, and c_p the density coefficient, specifically the ratio of the effective density of raw refuse in place in a landfill (assuming 750 lbs/yd^3) to the effective density in place of the output of process p, assuming that the output were to be landfilled without further processing.

The seventh constraint requires that all processed waste leave the intermediate facilities.

$$\sum_k T^{45}_{jpk} + \sum_i T^{35}_{jpi} - b_p \sum_l P_{jpl} = 0 \tag{18}$$

where b_p is the output coefficient for process p or the tons of nonsaleable output per ton of input and T^{35}_{jpi} the transportation in thousand tons per year from intermediate site j, processes p, to intermediate site i.

Finally, the model contains an equation for numbers and capacities of trucks.

$$e_p(\sum T_s^1 + \sum T_s^2) + e_t(\sum T_s^3 + \sum T_s^4) \leq M_s \tag{19}$$

where T_s^c is the sum of transportation activities of category c, in thousands of tons, that either arrive at or originate from any of the specified sites of set s; e_p is the reciprocal of tons per packer truck, as input; e_t is the reciprocal of tons per transfer vehicle, as input; and M_s is the maximum number of trucks, in thousands per year, which are permitted to service the set of sites s. Equation (19) is used for each site or group of up to three sites.

The *objective function* is divided into two major categories: transportation and processing. The transportation costs are defined as cost per ton minute. The Chapter 11 case study will elaborate on transportation costs. The processing costs are assumed to be a concave function. The model approximates these as three linear segments. Each linear segment represents a capital intercept and slope and operating intercept and slope per ton per year. The static model adds the operating and capital costs. Revenue per ton of input is entered for each site–process and is subtracted from the operating slope of each segment of the site–process. A site preparation cost (annual amortized cost) is entered for each site by adding it to the capital intercept of each linear segment of each process at that site. Other processing costs are entered by process.

The dynamic model adds a few interesting components, particularly the capability of abandoning capacity during the useful life of the facility if system costs, annual discount rates, and inflation rates can be lowered. Overall, WRAP has four outputs: (1) choice of sites, (2) selection of processing and disposal technologies at each site, (3) choice of scale, and (4) selection of transportation links between waste sources and intermediate and final disposal facilities.

RAMP includes market saturation effects in the model. It represents the impact of saturated markets by declining prices and limited market sizes. The model generates multiple commodities at multiple sites, with multiple market segments, and defines transportation links between processing sites and markets.

Table I WRAP Computer Time Summary[a]

Information	Minimum	Maximum	Median
Number of rows	60	115	60
Number of columns (without slacks)	148	512	149
Number of slacks	2	20	7
CPU time, sec	53.18	811.71	250.04

[a] From E. B. Berman (1976a).

WRAP and RAMP are available through the MITRE Corporation and WRAP through the U.S. Environmental Protection Agency. Program manuals and test run data are available. Computer time is a problem. Dr. Berman, the MITRE project leader, indicated an average cost of about $200.00 per run. One of the MITRE reports [Berman (1976a)] presents the dimensions and running times of 23 runs for the St. Louis and eastern Massachusetts studies. Some of these results are summarized in Table I.

The main message coveyed by the table is that relatively small problems from the linear programming perspective are expensive in computer calculation (CPU) time when this heuristic is used. Linear programming problems of similar size take between 4 and 9 seconds of CPU time. When translated into dollars, the median 250-second run costs between $55.00 and $95.00 in CPU dollars. The linear programming model of 6.5 seconds costs between $1.50 and $2.58.

MITRE has reported three applications of their models: St. Louis, eastern Massachusetts, and a region in Indiana, Ohio, and Kentucky (INOKY). The last two regions are different with respect to characteristics that are quite important to solid-waste management. Berman *et al.* (1976) state that the eastern Massachusetts and INOKY regions are representative of about 80% of the United States with respect to population and waste generation density. The two studies are seen as demonstrating the utility of the MITRE model and the feasibility of resource recovery. Nine single-period computer runs were made for the eastern Massachusetts study area and seven for the INOKY region. Table II summarizes key information from these studies.

The eastern Massachusetts region consists of eastern Massachusetts (from Worcester east, excluding southeastern Massachusetts and Cape Cod) and including a small part of New Hampshire. The INOKY region consists of sourthern Indiana, north and central Kentucky, and the Cincinnati, Ohio (southwest Ohio), area. The INOKY region has 6.6 million people and produces 15,000 tons of solid waste per day. The population of the eastern

Massachusetts area is over 5.0 million and produces about 12,000 tons of waste per day.

While the INOKY region produces more solid waste, this production occurs at a density only one-eighth of the density in eastern Massachusetts. The 22 source regions in eastern Massachusetts are much more homogeneous in waste production and compact than their 31 INOKY counterparts (Table II). The largest waste shed in eastern Massachusetts produces about 15 times as much waste as the smallest shed and less than 7 times as much as the median-sized shed. Solid-waste production in the INOKY region is relatively concentrated in a few areas. The largest waste shed generates almost 80 times as much waste as the smallest and 25 times as much as the median-sized source area. This difference tends to increase hauling costs relative to processing costs in the INOKY region. It follows that the INOKY region should require less capital-intensive technologies designed for smaller operating scales.

This expectation is reinforced by the potential markets for recovered materials. Oil is the major fuel for electric generators and industrial boilers in eastern Massachusetts. Incineration with steam-electric generation (ISEG) is the dominant energy recovery option in the area. ISEG is expensive on a per ton basis and is a large-scale technology. The MITRE group also tested ECOFUEL II, a high-quality dried, shredded fuel and gas pyrolysis which produces a gas fuel. These technologies could produce fuels for existing electric generating stations, but they are not proven. Coal is fired in boilers in the INOKY region. Therefore, relatively low-cost dry shredded waste (RDF) can be used as fuel. Overall, the resource recovery option in the eastern Massachusetts area depends more on large-scale technologies than does the INOKY region.

The combination of relatively dispersed waste sources and markets in INOKY and concentrated sources and large-scale markets in eastern Massachusetts are evident in the inputs to the computer tests. The INOKY facilities are predominantly transfer stations and RDF-fired electric utility stations. The eastern Massachusetts facilities include a wider variety of energy recovery facilities than INOKY, but they are oriented to large-scale incineration.

The third important difference between the eastern Massachusetts and INOKY regions is that inexpensive landfilling is a more viable option for INOKY because of the availability of land. Summarizing, the eastern Massachusetts study area has a greater waste production density, markets requiring greater capital investment, and a more expensive landfilling alternative than INOKY.

The three differences are manifested in the test run results. The processing facilities in the eastern Massachusetts area are, in general, larger. In one

Table II Summary Data from Eastern Massachusetts and INOKY Studies[a]

	Eastern Massachusetts	INOKY
Number of computer runs	9	7
Regional population, millions	>5.0	>6.6
Population density per square mile	1350[b]	171
Tons of solid waste per year	3.8 million	4.7
Tons of solid waste per day	12,000	15,000
Number of source zones	22	31
Waste generation per source (1000 tons per year)		
maximum	898.5	1237.3
minimum	56.9	15.7
median	133.4	51.3
Number of facility sites	21	31
transfer stations	8	23
processing/transfer	7	8
processing	6	0
Process tested:		
incineration/steam generation	yes	yes
incineration/electric power generation	yes	no
ECOFUEL II (a high-quality dried, shredded fuel)	yes	no
PUROX process (gas pyrolysis) with and without methanation	yes	no
steam to electric power conversion	yes	yes
transfer, packer to van	yes	yes
transfer, packer/van to rail	yes	yes
Refuse-derived fuel with residue recovery	no	yes
Processing facility size		
maximum	10,992	4315
minimum	379	294
median	3000	1476
System Cost, dollars per ton		
maximum	7.03	3.80
minimum	2.25	1.53
median	4.62	2.96
Percentage of waste to resource recovery		
maximum	95.1	97.4
minimum	36.1	69.3

[a] Data for this table were derived from tables in Berman (1976c) and supplement and from Berman *et al.* (1976).

[b] Estimated by the author.

run almost 11,000 tons per day were attracted to an incineration/electric generating site at Haverhill, Massachusetts. The typical, viable facility for eastern Massachusetts identified by the computer model was about 3000 tons per day. In the INOKY area, existing dispersed, coal-fired electric generating stations dominated the solution. The largest and average facilities in the INOKY results were less than half the size of the eastern Massachusetts counterparts.

The large-capital-intensive incineration facilities and the lack of an inexpensive landfilling option are reflected in much higher system costs in the eastern Massachusetts computer runs. The INOKY per ton system costs range from $1.53 to $3.80. The eastern Massachusetts runs identify costs between $2.25 and $7.03 per ton. The $7.03-per-ton system cost run restricts any single facility to 3000 tons per day. This limit does not allow major, assumed economies of scale to occur. Therefore, almost two-thirds of the waste goes to landfills. Since there are no facilities in the world as large as 3000 tons per day, competitive resource recovery disposal in eastern Massachusetts depends on the successful operation of huge facilities, which have no operating histories at desired scales. The coal-fired plants in the INOKY region offer far less expensive and complicated solutions. The worst case sent almost 70% of the waste to resource recovery even with relatively favorable landfill prices and a 2000-ton-per-day limit on the recovery facilities. Both the INOKY and eastern Massachusetts studies also rely on cooperative transfer station projects. Cooperative efforts in the movement of solid waste have a mixed record in the United States.

With respect to the model, these case studies demonstrate that this heuristic does work and that each run produces useful information. However, in the author's opinion, similar information can be secured from a linear programming formulation in which the wealth of information produced by a single expensive run of the heuristic model is more than equalled by a number of inexpensive linear programming runs.

C. COLLECTION ALGORITHMS

Minimizing the cost of the collection service depends on such considerations as type of container, type of truck, crew size, crew motivation, crew wages, equipment maintenance, and frequency of pickup. Some of these considerations are a matter of delicate labor negotiations, others are unnegotiable demands imposed upon the manager by the community. For example, many communities retain backyard rather than curbside collection despite an additional 5–25% collection cost. Overall, while much of the collection cost is set by factors that are not or cannot be modeled, scheduling and routing (two important aspects of collection cost) have been modeled with optimization techniques. Scheduling includes choosing manpower and mobile equipment placement in place and time. Tanaka and Quon (1971) have proposed a linear programming model that minimizes short-run labor costs and the costs of not meeting a schedule subject to available personnel and equipment constraints. As yet, however, this model and other possible equipment and personnel assignment formulations have not been adequately tested by solid-waste managers. A general model structure for this type of problem will be reviewed in the law enforcement chapter.

Clark and Helms (1972) have proposed and tested an equipment replacement model. They assume that a proportion of the existing fleet (given as 15%) is to be replaced. The objective is to determine if larger truck–crew combinations are efficient replacements for the existing combination of smaller truck–crew teams. The objective function minimizes the average daily operating cost of the existing fleet of trucks and the average daily costs of replacements:

Minimize:
$$Z = \sum_k D_k T_k + \sum_i \sum_k C_k X_{ik} \tag{20}$$

where D_k is the average daily crew and amortization cost of the replacement trucks, T_k the number of each type of replacement truck, C_k the average daily operating cost of the alternative truck types, and X_{ik} the number of trucks of type k assigned to a collection district i.

Four constraints are defined:

$$\sum_k A_{ik} X_{ik} \geq R_i \tag{21}$$

where A_{ik} is the collection coefficient representing the average number of residences per day that truck type k can service in district i, and R_i the average daily number of residences that must be served in each collection district i. This constraint requires that enough trucks–crews be assigned to a collection district to pick up the waste.

A second constraint represents the number of smaller (in this case 16-yd) trucks.

$$\sum_i X_{i_1} = T_1 + W_1 \tag{22}$$

where X_{i_1} is the number of 16-yd trucks assigned to district i, T_1 the number of new 16-yd^3 trucks that will be added to the fleet, and W_1 the number of 16-yd^3 trucks that will not be replaced.

A third constraint represents the number of larger trucks (20 and 25 yd in this case).

$$\sum_i X_{ik} = T_k \qquad (k = 2,3) \tag{23}$$

where $\sum_i X_{ik}$ is the number of 20- and 25-yd trucks that will be assigned to the collection districts, and T_k the number of replacement vehicles of type 2 and 3 (20 and 25 yd) that will be purchased.

Finally,

$$W_1 = C_1 \tag{24}$$

is the number of trucks of the existing fleet that will not be replaced. The value C_1 is given by the solid-waste manager and is based on experience and budgetary considerations. The linear programming solution is reported in fractions of trucks and can be rounded off to the nearest integers or tested by iteration as explained in Chapter 3.

The authors tested the model in a metroplitan area and report a considerable potential savings by replacing 16-yd trucks with 25-yd trucks. This linear programming model, in essence, verifies the advantages of a large truck–small crew combination. Suffice it to say that equipment replacement problems are common to every field considered in this volume.

A research problem in the field of solid-waste management has been the development of models for routing solid-waste vehicles from the garages to their pickup points, from the pickup points to the disposal points, and from the disposal sites back to the garages. Two approaches to this routing problem are the "traveling salesman problem" and the "Chinese Postman's problem." In the traveling salesman problem, individual residences are aggregated into collection points, and a path is traced based on distance minimization. Whereas the traveling salesman models treat the path as a set of discrete stops, the Chinese Postman's problem treats the sources of waste as a continuum along the length of a network.

The simplest versions of these problems can be formulated in a linear programming format, although they contain integer route choices. For example, Balinski and Quandt (1964) have presented a model that can be solved in a linear programming format.

Minimize:

$$Z = \sum_j C_j X_j \tag{25}$$

where C_j is the cost of the jth feasible route and X_j signifies the existence of a route,

subject to

$$\sum_j A_j X_j = E \tag{26}$$

where A_j is a column vector of size m (the number of places on the collection route) by 1, and E is an m by 1 column vector of 1's.

$$X_j = (0,1) \tag{27}$$

Constraint (27) imposes an integer requirement. Each possible route is represented as a column vector with a 1. The problem finds the set of X_j's that minimizes cost while meeting the constraint that each city be visited only once. Constraint set (27) makes the problem into an integer programming problem. The problem can, however, be run as a linear programming problem with $0 \leq X_j \leq 1$ replacing (27).

Balinski and Quandt present the model as a means of solving small, multiroute truck dispatching problems. The problem has a limit on the number of paths. To be used in solid-waste planning, this formulation would require preselection of feasible routes. Indeed, most routing algorithms, including the IBM vehicle scheduling program, which is the best available heuristic model, have constraints on the number of possible collection sites that can be considered.[4] In short, a weakness of the routing algorithms is that all the possible vehicle paths cannot be tested. Furthermore, the weight of the literature seems to be on the side of simulation models and other special routing algorithms as more efficient means of solving routing problems than linear and related programming models. Finally, Shuster and Schur (1972) have argued that efforts to find an optimum route are not rewarding for four major reasons:

(1) Data requirements of the models are unrealistic.

(2) Nonquantifiable factors are more important than quantifiable factors.

(3) Routes are subject to rapid change because of changes in such factors as truck size.

(4) A set of flexible guidlines is an adequate tool for route assignment.

In response to these needs, and in particular the need for reliable information, the U.S. Environmental Protection Agency has developed the Collection Management Information System (COLMIS). COLMIS is a computerized information system that guides the community to the collection, storage, processing, and reporting of productivity and cost data. The system requires two types of inputs: background and daily. The background information includes the following: number of routes; normal workday, overtime, and lunch time hours; collection frequency; wages of drivers and collectors;

[4] Clarke and Wright (1964) has been modified as an IBM Vehicle Scheduling Program available for 360 computers. Marks and Liebman (1970) present an interesting integer programming algorithm.

vehicle type and size; and fuel, oil, maintenance and depreciation rates. Daily information is recorded for every operating day by the crew driver.

The COLMIS system produces six reports: (1) an editing report that shows the completeness of the data, (2) a route information report that shows how the crew works the route, (3) a collection information report that summarizes crew productivity on the basis of time and weight, (4) a collection cost information report that provides average daily cost figures, (5) a collection system operation summary that summarizes information from the other reports, and (6) a management analysis report that rates crew performance on ten performance indicators such as homes served and weight collected.

The COLMIS system has been applied in about forty communities. Some of these communities report that savings of more than 20% have been achieved by converting the most inefficient routes [see Shuster (1974)]. It is too early to know if COLMIS will establish the data base for the application of programming models. Whether it does or not, COLMIS is certainly a badly needed step.

Overall to date, the large metropolitan regions that should benefit the most from scheduling and routing problems have not benefited because of their great complexity. Most communities are struggling with four problems: inefficient collection systems that are based on the historical development of the community; failure to have sound vehicle maintenance and replacement programs; failure to have enough data to prepare similar work routines for all crews; and serious fiscal constraints.

D. A FIXED-FACILITY EXAMPLE

The following fixed-facility example, which can be solved with the aid of a computer, illustrates the first algorithm presented in this chapter. It contains three sources of waste, one dry-fuel facility, two landfills, and one electric generating station. The total system of nine possible links is illustrated by Fig. 1. Cost data for the problem are listed in Tables III and IV.

The example can be used as a prototype to test the hypotheses that a dry-fuel system is cost effective in comparison to landfills. We are assuming that a moderate-sized dry-fuel facility is going to convert 60% of the residential waste into a dry-fuel product and send 40% to a landfill. These are conservative numbers.

The calculations for this model are simple. You should, however, experiment with the parameters and constraints of the model through parametric programming to acquire an understanding of the economic tradeoffs involved. For example, one could assume that a reduction in solid waste

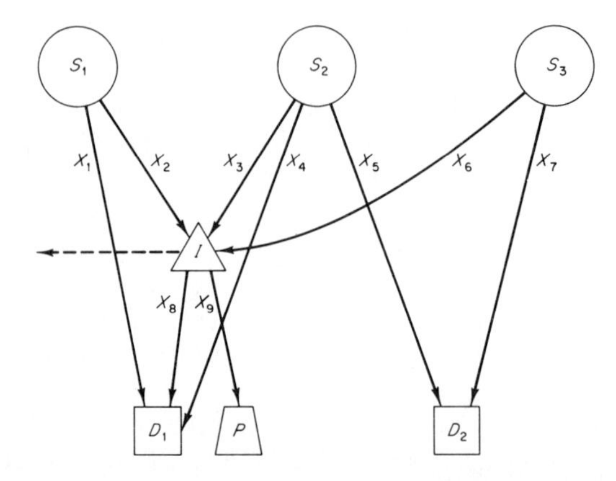

Fig. 1 Sample solid-waste problem. *S*, source; *I*, intermediate facility; *D*, landfill; *P*, generating facility.

occurs because of bottling legislation or due to economic incentives to collect newspaper. Such a policy would reduce the number of tons and also affect the .6 and .4 α values. You would seek to determine the sensitivity of the dry-fuel facility to changes in the available waste load. Second, one could systematically increase the disposal cost at the dry-fuel facility to determine the threshold disposal charge for successful operation at 5000 tons per week. Third, you might expand the problem by increasing dry-fuel capacity at the current dry-fuel site. A second 4000-ton-per-week facility with the same costs as the first is recommended. In short, even this simple model can be used to test important solid-waste policy alternatives. See Chapter 11 and Greenberg *et al.* (1976) for a case study of the application of this model.

Table III Data for Objective Function

	Dollars per ton		
Link	Transportation	Disposal	Total
X_1	3.00	4.00	7.00
X_2	2.50	1.50	4.00
X_3	2.70	1.50	4.20
X_4	3.50	4.00	7.50
X_5	3.70	3.10	6.80
X_6	4.00	1.50	5.50
X_7	3.10	3.10	6.20
X_8	1.00	—	1.00
X_9	0.00	—	0.00

Table IV Data for Constraints

Tons per week
$S_1 = 4000$
$S_2 = 9000$
$S_3 = 10{,}000$
$I \leq 5000$
$D_1 \leq 10{,}000$
$D_2 \leq 10{,}000$
$\alpha_1 = .6$ For energy facility
$\alpha_2 = .4$ For disposal facility
$P \leq 4000$

The following problem is an illustration of equation set (1-6) presented earlier in the chapter. You should solve it and try some variations as suggested in the preceding. The system is described by the linear model:

$$\text{Minimize:} \quad Z = 7.00X_1 + 4.00X_2 + 4.20X_3 + 7.50X_4 + 6.80X_5 + 5.50X_6 + 6.20X_7 + 1.00X_8 + 0X_9 \qquad \text{(P1.1)}$$

subject to

$$X_1 + X_2 = 4000 \qquad \text{(P1.2)}$$

$$X_3 + X_4 + X_5 = 9000 \qquad \text{(P1.3)}$$

$$X_6 + X_7 = 10{,}000 \qquad \text{(P1.4)}$$

$$X_2 + X_3 + X_6 \leq 5000 \qquad \text{(P1.5)}$$

$$X_1 + X_4 + X_8 \leq 10{,}000 \qquad \text{(P1.6)}$$

$$X_5 + X_7 \leq 10{,}000 \qquad \text{(P1.7)}$$

$$X_9 \leq 4000 \qquad \text{(P1.8)}$$

$$.6(X_2 + X_3 + X_6) - X_9 = 0 \qquad \text{(P1.9)}$$

$$.4(X_2 + X_3 + X_6) - X_8 = 0 \qquad \text{(P1.10)}$$

$$X_1, \quad X_2, \ldots, \quad X_9 \geq 0 \qquad \text{(P1.11)}$$

Solution: $Z = 143{,}000$; $X_1 = 4000$; $X_3 = 5000$; $X_4 = 4000$; $X_7 = 10{,}000$; $X_8 = 2000$; $X_9 = 3000$.

E. SUMMARY

Prior to 1970, modeling efforts in solid-waste management were extremely rare. Management decisions were made on an *ad hoc* basis by each community based on experience and intuition. The increasing complexity of management alternatives and the availability of computer capacity has begun to change the decisionmaking process. Transfer station and intermediate and ultimate disposal facility siting is a promising arena for the application of mathematical programming models. Collection practices would seem to be eminently amenable to modeling with optimization and heuristics. However, such applications await suitable data and a willingness on the part of interested parties to agree upon the goals and limitations of the research.

REFERENCES

ABT Associates, Inc. (1975). "Evaluation of the Feasibility and the Economic Implications of Pricing Mechanisms in Solid Waste." National Technical Information Service, Springfield, Virginia.

Balinski, M. L., and Quandt, R. E. (1964). On an integer program for a delivery problem. *Oper. Res.* **12**, 300–304.

Berman, E. B. (1973). "A Model for Selecting, Sizing and Locating Regional Solid Waste Processing and Disposal Facilities." MITRE Corp., Bedford, Massachusetts.

Berman, E. B. (1976a). "WRAP—A Model for Regional Solid Waste Management Planning: User's Guides," MTR-3222. MITRE Corp., Bedford, Massachusetts.

Berman, E. B. (1976b). "WRAP—A Model for Regional Solid Waste Management Planning: Documentation of Operational and Exercise Runs," MTR-3219. MITRE Corp., Bedford, Massachusetts.

Berman, E. B. (1976c). "Resource Recovery Regional Design Analysis for the Eastern Massachusetts Region," MTR-3221. MITRE Corp., Bedford, Massachusetts.

Berman, E. B., Rofé, R., Williams, E. C., and Yaffe, H. J. (1976). "Evaluation of Policy Issues in Resource Recovery: An Application of RAMP to the INOKY Region," MTR-3320. MITRE Corp., Bedford, Massachusetts.

Berman, E. B. (1977). *33rd National Meeting of JSGOMRAM, April 12–14.* MITRE Corp., Bedford, Massachusetts.

Clark, R., and Helms, B. (1970). Decentralized solid waste collection facilities, *J. Sanit. Eng. Div., Amer. Soc. Civil Eng.* **96** (SA 5, Proc. Paper 7594), 1035–1043.

Clark, R., and Helms, B. (1972). Fleet selection for solid waste collection systems. *J. Sanit. Eng. Div., Amer. Soc. Civil Eng.* **98** (SA 1), 71–78.

Clarke, G., and Wright, J. W. (1964). Scheduling of vehicles from a central depot to a number of delivery points. *Oper. Res. Soc. Amer.* **12**, 568–581.

Greenberg, M. R., Bottge, M., Caruana, J., Horowitz, D., Krugman, B., Masucci, N., Milewski, A., Nebenzahl, L., O'Neill, T., Skypeck, J., and Valente, N. (1976). "Solid Waste Planning in Metropolitan Regions." Center for Urban Policy Research, New Brunswick, New Jersey.

Haddix, G. (1977). "Application of a Solid Waste Planning Model with Resource Recovery." Paper presented at 33rd National Meeting of JSGOMRAM, April 12–14, Bedford, Massachusetts.

Heaney, J. P. (1971). Discussion of decentralized solid waste collection facilities. *J. Sanit. Eng. Div., Amer. Soc. Civil Eng.*, **97** (SA 2), 240–241.
Helms, B., and Clark, R. M. (1971). Selecting solid waste disposal facilities. *J. Sanit. Eng. Div., Amer. Soc. Civil Eng.* **97** (SA 4, Proc. Paper 8280), 443–451.
Hensey, V. (1976). "WRAP—A Model for Regional Solid Waste Management Planning: Programmer's Manual," MTR-3224. MITRE Corp., Bedford, Massachusetts.
Kühner, J., and Harrington, J. J. (1976). Mathematical modeling for regionalization of resource recovery. *In* "Mathematical Models for Environmental Problems" (C. A. Brebbia, ed.). Pentech Press, London, 27–41.
Marks, D., and Liebman, J. (1970). "Mathematical Analysis of Solid Waste Collection." U.S. Dept. of Health, Education, and Welfare, U.S. Public Health Service, Washington, D.C.
National Commission on Materials Policy (1973). "Materials Needs and the Environment Today and Tomorrow," Final Report. U.S. Government Printing Office, Washington, D.C.
Salvato. J. (1972). "Environmental Engineering and Sanitation," p. 408. Wiley, New York.
Shuster, K. A. (1974). "A Five Stage Improvement Process for Solid Waste Collection Systems." U.S. Environmental Protection Agency, Washington, D.C.
Shuster, K. A., and Schur, D. A. (1972). "A Heuristic Approach to Routing Solid Waste Collection Vehicles." U.S. Environmental Protection Agency, Washington, D. C.
Tanaka, M., and Quon, J. E. (1971). A Linear Programming Model for the Selection of Refuse Collection Schedules." Paper presented at 40th annual meeting of the Operations Research Society of America.
U.S. Department of Commerce (1974). "U.S. Industrial Outlook 1975 with Projections to 1980." Domestic and International Business Administration, U.S. Government Printing Office, Washington, D.C.
Vasan, K. S. (1974). "Optimization Models for Regional Public Systems." University of California, Berkeley.
Walker, W. (1973). "A Heuristic Adjacent Extreme Point Algorithm for the Fixed Charge Problem." Rand Institute, New York.
Weston, Roy F., Inc. (1973). "Development of a Solid Waste Allocation Model." Roy F. Weston, Environmental Scientists and Engineers, West Chester, Pennsylvania.
Williams, E. C. (1976). "WRAPping up the Solid Waste Management Problem: A Model for Regional Solid Waste Management Planning," M 76-207. MITRE Corp., Bedford, Massachusetts.
Yaffe, H. J. (1974). "Note on Equation Structures and Sizings of the EPA and MITRE Solid Waste Allocation Models." MITRE Corp., Bedford, Massachusetts.
Young, D. (1972). "How Shall We Collect the Garbage?" Urban Institute, Washington, D.C.

Chapter 7

Water Resources Management

During the past 25 years, more than 50 articles, reports, and books have reported the use of linear programming in water resources research. This rich literature ranges in scale from macrogeographic studies of regional water supplies and demands to studies aimed at determining the most efficient size pipe diameters for a small network. To simplify the presentation of this literature, it has been dichotomized into two main issues: water quantity and water quality.

A. WATER QUANTITY

The goal of water quantity analysis is to efficiently allocate a scarce resource. The important decisions include but are not limited to building new or enlarging existing fixed facilities and pipes, and to weighing the importance of conflicting demands. The linear programming applications can be conveniently divided into four groups: (1) evaluation of multipurpose uses, (2) transfer of water from areas of surplus to areas of deficit, (3) siting of fixed facilities, and (4) analysis of the availability of water resources.

1. EVALUATING MULTIPURPOSE USES

Determining the most efficient uses of water is perhaps the most controversial application of linear programming to problems of water quantity allocation. The contending parties include fishermen, boaters, swimmers, electric utility managers, water supply managers, flood control engineers and water quality managers. At one extreme, the white-water enthusiast is opposed to any man-made control of a river. At the other extreme, the local utility cannot secure a permit for a nuclear generating station without a series of on-stream reservoirs that will be drawn down as the demand rises and falls.

Optimization models can be applied to broadly isolate the tradeoffs among alternative uses and to narrowly define rules for fixed-facility operation. The goal is to maximize the benefits to as many users as possible. Two of the earliest multipurpose water resource studies are by Mannos (1955) and Manne (1962). Mannos formulated a linear programming model to maximize the energy output from a series of dams on the Missouri River. The decision variable was reservior storage. Manne developed a model with the goal of maximizing the benefit to electric power, irrigation, and flood protection uses. The constraints were the demands, often conflicting, of the users, and the variability of rainfall measured as the probability of receiving a given amount of inflow. Manne's article is especially useful because he provides a sample problem and constraints.

Other interesting papers on multipurpose uses and the derivation of operating rules related to water quantity are by Thomas and Revelle (1966), Males (1968), Loucks (1968), Revelle *et al.* (1969), and Revelle and Kirby (1970). Thomas and Revelle (1966) sought to optimize the use of the Aswan Dam for hydroelectric power and irrigation. This application demonstrated the great difficulty of achieving acceptable solutions in the case of conflicting uses. On the one hand, the demand for water power is relatively constant throughout the year. On the other hand, the need for irrigation is highest during the low-flow season. It is during this season of low flow that the two uses seriously conflict.

Males (1968) demonstrated the complementary use of linear programming and simulation models. He obtained a set of reservoir operating rules for multipurpose reservoirs from linear programming runs, and he used these rules as inputs to simulate the operation in detail.

The papers by Loucks (1968), Revelle *et al.* (1969), and Revelle and Kirby (1970) demonstrated the complementary use of optimization and probability. Loucks (1968) sought a set of operating rules based on a set of desired flows and on the current and expected volume of inflows into the reservoir system. The inflows are represented as stochastic elements. The papers by Revelle *et al.* (1969) and Revelle and Kirby (1970) will be reviewed in detail because they combine most of the elements of the other multipurpose water resources papers and add the dimension of comparing deterministic and stochastic formulations.

The authors argue that the complex problem of storing and releasing waters from reservoirs is amenable to a treatment that seeks to satisfy user demands while minimizing the size of the dam. Revelle *et al.* (1969) sought an optimal reservoir size and a set of parameters which would instruct the manager on monthly release requirements. The monthly parameters are defined as

$$b = s - p \tag{1}$$

where b is the decision parameter that is sought, s the storage at the conclusion of the previous month, and p the release during reservoir operation.

The authors argue that the parameter b should be linear rather than non-linear (power or fraction), because linear models are easier to solve. The linear decision parameters are formulated in two models: the deterministic and the stochastic. The deterministic model treats as given the amount of inflow for 240 months. The 20 years of monthly records may be an actual record or a synthesized record. A second formulation treats the inflow as random variables. The important difference between the two approaches is the constraints that pivot around the postulated expected inflow. The deterministic model uses the most limiting inflow among the 20 years of record as right-hand-side values. This formulation has the disadvantage of not providing the manager with the probability of the reservoir failing to meet the needed storage and outflow requirements. The stochastic model uses an inflow value of a given reliability. The stochastic formulation enables the manager to test the tradeoffs between dam size (cost) and reliability of meeting needed water releases (damage function).

Both the deterministic and the stochastic linear programming approaches may be written in a similar format:

Minimize:

$$Z = X \tag{2}$$

where X is the reservoir capacity to be determined and is the same for both formulations (the b activities become zero-weighted X values),

subject to

$$X - b_i \geq \max_n (r_i + 12n) + V_i \tag{3}$$

where b_i is the linear decision parameter for the ith month to be determined in the model, r_i the given input in the ith month, and V_i the flood storage capacity required at the end of the ith month. In essence, this equation says that a freeboard of at least V_i must be available for purposes of flood control. Theoretically, 20 constraint equations, one for each of the years of record, are required for each month. The authors suggest, however, that the user choose the most restrictive of the 20 equations, which are identical except for their right-hand-side values. Equation set (3), therefore, has 12 equations, which represent a monthly division of the year.

In the stochastic model, Eq. (3) becomes

$$X - b_i \geq r_i^{90} + V_i \tag{4}$$

where r_i^{90} is the flow that should be exceeded in period i only 10% of the time. If the manager is willing to take a greater risk, perhaps because the potentially flooded area is relatively undeveloped, an r_i value of .80 or .85 might be selected. In an urban region, where flood damage might be extremely costly, an r_i of .95 or higher might be selected. In short, the stochastic model is a superior tool for the user because safety and dam costs can be tested for cost effectiveness. Accordingly, the remaining constraints will be presented only in their stochastic form.

$$a_m X - b_i \leq r_i^{10} \tag{5}$$

This constraint requires that the reservoir maintain a minimum storage. In Eq. (5), a_m is the minimum requirement, expressed as a fraction of the total capacity, and r_i^{10} the flow it can be expected to fall below only 10% of the time. A region that requires low flow augmentation to avoid high costs for pumping out low-lying water, replete with assorted troublesome insects, may choose a level of r_i^{05} or lower.

$$\begin{aligned} b_{i-1} - b_i &\geq q_i - r_{i-1}^{10} \\ &\vdots \\ b_{12} - b_i &\geq q_i - r_{12}^{10} \\ a_0 X - b_i &\geq q_1 \end{aligned} \tag{6}$$

These 13 equations require that the release exceed a monthly minimum. In Eq. (6), q_i is the minimum release required in the ith month of the year, and a_0 the initial storage expressed as a fraction of the reservoir capacity.

Finally,

$$\begin{aligned} b_{i-1} - b_i &\leq f_i - r_{i-1}^{90} \\ &\vdots \\ b_{12} - b_1 &\leq f_1 - r_{12}^{90} \\ a_0 X - b_i &\leq f_i \end{aligned} \tag{7}$$

These 13 equations require that the release *not* exceed a monthly maximum. In Eq. (7), f_i is the maximum allowable release in the ith month of the year. The f_i and r_i values reflect the cost of channel erosion, flooding, and other costs of too much water.

The solution of the stochastic model is a reservoir size (X) and 12 linear parameters (b's). The inputs to the model are 50 equations. A sample problem appears at the end of the chapter.

Revelle and Kirby (1970) later improved the model in two respects. First, alternative formulations for the objective function are presented. Second, the stochastic constraints are reformulated, and the effects of evaporation losses and possible failures to meet release requirements are considered.

While the chance-constrained linear programming model has been considered in the context of water resources, the reader should note that its applicability extends to any system in which one of the constraints is stochastic. In the case of solid-waste management, the problem might be to find an optimum-size resource recovery facility. The facility would produce the fuel product on which the electric utility is depending and over which it can demand a penalty if the product does not arrive. In the cases of law enforcement, health, and transportation, the goal might be to minimize the number of workers and mobile equipment, subject to the constraint that enough personnel and equipment be present to avoid politically costly public service crises on a daily basis.

2. TRANSFER OF WATER FROM AREAS OF SURPLUS TO AREAS OF DEFICIT

Regional management of water supplies is an issue that until recently engendered little enthusiasm. Local governments and private interests have been able to meet average daily demands and have coped with droughts and equipment malfunctions by employing demand-reducing restrictions. The local approach to managing water supply is rapidly drawing to an end, especially in metropolitan regions. Regions have begun to see the advantages of regional decisionmaking, or at least cooperative efforts in the face of increasing base and peak demands, increasing capital and operating costs, and decreasing availability of nearby high-quality, inexpensive supplies of water.

Optimization models can help determine the cost effectiveness of using alternative supplies and transfer facilities. Three of the earliest studies were reported by Dracup (1966), Clyde *et al.* (1967), and Heaney (1968). These three studies focus, not surprisingly, on arid regions. Dracup (1966) described a linear programming model given the formidable task of testing alternative means of supplying the demands for public potable, irrigation, and industrial water demands and the artificial recharge of groundwater supplies in the San Gabriel Valley of Southern California. The author sought to juggle five sources of water: locally available surface water, groundwater, imported Colorado River water, imported Feather River water, and reused wastewater. Repeated iterations of the constraints were used to measure the cost effectiveness of the alternatives.

Clyde *et al.* (1967) used linear programming to find an optimum conjunctive use of groundwater and surface water. The objective of the model was to maximize the benefits for irrigation, subject to constraints on dry season supply and demand, wet season supply and demand, fixed-facility capacity, and continuity constraints. Finally, Heaney's linear programming model was aimed at meeting multipurpose uses in the Colorado River Basin.

The unprecedented drought during the 1960s has brought forth a number of applications in the eastern United States. DeLucia (1971) developed a linear programming model to investigate alternatives for meeting water supply requirements for the North Atlantic region. On a smaller geographical scale, Greenberg *et al.* (1971) and Greenberg and Hordon (1975) have used linear programming to study the hypothesis that a regional approach to public potable water resources would lessen intraregional shortages. The linear model was constructed to minimize the retail cost of meeting projected water demands.

Minimize: $$Z = \sum_i \sum_j C_{ij} X_{ij} \qquad (8)$$

where C_{ij} is the retail cost of water for the j demand agencies and X_{ij} the amount of water transferred from a source i to an area of demand j,

subject to

$$\sum_i X_{ij} = D_j \qquad (9)$$

The first constraint requires that each agency's projected demands D_j are satisfied.

$$\sum_j X_{ij} \leq S_i \qquad (10)$$

The second constraint limits the amount of water that can be taken from a source.

$$X_{ij} \leq T_{ij} \qquad (11)$$

The third constraint limits the transfer of water to the capacity of the pipe network and pumps.

$$X_{ij} \geq 0 \qquad (12)$$

These five equations were applied to determining the economic cost of meeting projected potable water demands in northeastern New Jersey, a major metropolitan region of over 4.5 million people, for the years 1975–2000. Parametric programming was performed with the three constraint equations to determine the tradeoffs among increasing the supply [Eq. (10)], controlling the demand [Eq. (9)], and increasing the capacity for interagency water transfers [Eq. (11)]. A sample problem appears at the end of this chapter.

A more sophisticated version of this five-equation model can be written for regions that have a better data set. First, from the management perspective, C_{ij} should represent the actual engineering cost of diverting and transferring water from a source to a sink. When C_{ij} represents the retail cost, the model tends to produce multiple optima because some of the key agencies can receive water from many agencies. A more useful result would inform the manager precisely which links cost more to use, either due to physical costs (e.g., pumping or energy) or administrative considerations (contracts among agencies).

If a superior data set were available, source-to-sink links could be split into source to sink, source-to-transfer node (a transfer point in the network), and node to sink. Equations (8)–(11) treat all links as if the water goes directly from a source of water to a sink. Differences in the cost of moving the water are assumed to be reflected in the price of the water C_{ij}. A more detailed representation of choices should be beneficial to the user, to the extent that the operator can choose among specific supplies and links. If source-to-node and node-to-sink links are included in the model, one additional constraint is required. The amount of water entering the transfer node N_p must leave the node.

$$\sum_i X_{ip} - \sum_j X_{pj} = 0 \tag{13}$$

This more complex formation requires the redefinition of Eqs. (8)–(11) into i sources, p transfer nodes, and j sinks.

$$\text{Minimize:} \quad Z = \sum_i \sum_j C_{ij} X_{ij} + \sum_i \sum_p C_{ip} X_{ip} + \sum_p \sum_j C_{pj} X_{pj} \tag{14}$$

$$\sum_i X_{ij} + \sum_p X_{pj} \geq D_j \tag{15}$$

$$\sum_j X_{ij} + \sum_p X_{ip} \leq S_i \tag{16}$$

$$X_{ij} \leq T_{ij} \tag{17}$$

$$X_{pj} \leq T_{pj} \tag{18}$$

$$X_{ip} \leq T_{ip} \tag{19}$$

Constraint (13) is identical to a constraint in a solid-waste system in which the waste is moved from a waste shed to a transfer station and then by a larger vehicle to a recovery or disposal site. If recovery occurs in the process of transfer, the equations require adjustment coefficients as in Eq. (5) in the solid-waste chapter.

If the problem involves the construction of new facilities, it may be formulated as a fixed-charge model. The capital costs of the proposed new facilities are represented as fixed-charge elements. In the linear formulation the fixed charges can be evaluated by testing the alternative proposed facility sizes in successive runs.

Finally, another improvement in the model could be achieved by treating the demand and supply constraint equations in a chance-constrained format. The user prepares demand and supply estimates based on alternative demographic, economic, pricing, and water conservation scenarios and on the hydrologic record. Combinations of these constraints are inserted as right-hand-side values and are used to define an optimal transfer pattern under a wide range of conditions. The failure to meet a demand by a given amount can be estimated by a damage function. The result is a comparison of damage costs and costs of additional supply and network capacity.

3. THE SITING OF FIXED FACILITIES

The siting and sizing of new and the expansion of existing fixed facilities is an activity common to the application chapters in this volume. The general approach may consist of a linear model, with repeated runs to test alternative locations, scales, and technologies, or a fixed-charged model that can perform the same analysis with fewer enumerations. The linear and the fixed-charge approaches were reviewed in the solid-waste chapter and will not be repeated. Instead, models that were developed for water resources applications will be considered, and their applicability to other systems will be outlined. Specifically, two problems will be explored: the selection of optimum pipe sizes and heads for a water network, and the scheduling of improvements in fixed facilities.

Papers dealing with the selection of pipe sizes were prepared by Ramaseshan (1962) and Karmeli *et al.* (1968). Ramaseshan's model sought to maximize the benefits from a water network, subject to constraints on water demand, pipe capacity, inflows and outflows from the reservoir, reservoir storage capacity, and minimum storage requirements. The model assumed reservoir capacity and the net benefits and costs of different aqueduct capacities as given, and it tested different aqueduct capacities for different sections of the aqueduct. The tradeoffs between the costs of aqueduct capacity and the benefit of water are tested by enumeration. Clearly, the user might also choose to test the tradeoffs among benefits and costs of reservoir capacity, the size of pumping facilities, and the option of controlling demand.

Karmeli *et al.* (1968) developed a model that chooses the minimum pipe diameters and least-expensive energy requirements for a water distribution system. The model assumes that the configuration of the pipes and lengths of the pipes are fixed, that required flows are known, and that the system draws on a single source of water. The authors considered the total friction loss in pipes and the relationship between pressure head and energy costs. Finally, they provided data for an actual problem.

At first the reader might be inclined to relegate the Ramaseshan and Karmeli articles to a category of highly specialized water system applications. A broader interpretation of their articles would include any system that channels energy between nodes. From this perspective, pipelines carrying oil, gas, chemicals, and wastewater may be seen as directly analogous systems. Road and rail networks and canal, dredged-river, and energy grid systems are also analogous energy transmitting systems with the potential of requiring a cost effectiveness analysis of different size transmission systems.

Two studies with far more ambitious goals than the network, capacity size models are the facility scheduling models offered by Facet and Marks (1972) and by Young *et al.* (1970). Both studies propose linear and integer programming to examine problems of monumental size. One Facet and Marks algorithm chose the facility sizes and construction times. Unfortunately, to solve the equations, the number of time periods must be severely restricted. Their second model assumes that the facility size is known. Integer programming is then used to decide on an optimal sequence.

The Young *et al.* (1970) project for sequencing the construction of reservoirs and canals in Texas may be used to illustrate why optimization models have been most effective in testing the cost effectiveness of narrowly limited alternatives. Young *et al.* (1970) sought a simulation model to test the economic efficiency for reservoir and canal starting times. The study considers a prespecified network that is not subject to alteration. Inflows and demands on the system are also prespecified. Even with these two assumptions, the problem size is staggering. The time horizon is 50 years of monthly data. The system potentially contains 30 fixed facilities: 10 reservoirs and 20 canals. Overall, this system potentially contains 120,000 variables ($600 \times 10 \times 20$)! By means of an additional set of simplifying assumptions, the use of a simulation model and integer programming, and the use of random sampling, the authors report that staging decisions can be modeled. Nevertheless, the purpose of briefly reporting the Facet and Marks (1972) and the Young (1970) studies in this volume is to indicate that the all-encompassing optimization model is so large as to frustrate even the most Herculian computer. In the case of water quantity analysis, one could prepare a set of equations which would select technologies, sites, operating scales, pipe diameters, pump sizes, a myriad of other fixed and mobile equipment, and a sequencing of all of these costs. The constraints might include monthly chance constrained inflows, chance constrained demand estimates, chance constrained system breakdown estimates, budget constraints, and other limitations. While such a model might win a prize for mathematical elegance, it could not be easily solved. In short, while the literature is filled with models of increasing elegance, the biggest payoffs for the models reported in this book will continue to be in making decisions that were made, and in most cases are still made, by informed intuition.

4. ANALYSIS OF THE AVAILABILITY OF WATER RESOURCES

Linear programming has also been used to estimate hydrologic parameters: Kleinecke (1971) and Deininger (1969). In both cases the authors were troubled by the inability of regression models and other routine statistical techniques to easily solve models that required inequalities and nonnegativity constraints. Deininger's model represents runoff as a function of rainfall in previous time periods.

$$R_j = X_0 P_j + X_1 P_{j-1} + X_2 P_{j-2} + \cdots + X_m P_{j-m} \tag{20}$$

where R_j is the runoff during the jth time interval and P_j the effective precipitation during the $(j-i)$th period. The X_m parameters are moving averages between the time periods and are to be estimated by the model. Equation (20) is turned into a linear programming format and is solved for the X_m values that minimize the absolute deviations between time periods.

$$\text{Minimize:} \quad Z = \sum_j |R_j - X_0 P_j - X_1 P_{j-1} - \cdots - X_m P_{j-m}| \tag{20$'$}$$

Three constraints are utilized. One constraint requires that the sum of the moving average values equal unity.

$$\sum_i X_i = 1 \tag{21}$$

The second and third constraints are the justification for the linear programming formulation. Equation (22) requires that the X_m parameters decrease monotonically with increasing i.

$$X_i - X_{i+1} > 0 \tag{22}$$

The basis for this constraint is the assumption that the contribution of rainfall to runoff decreases with time. Finally, Deininger recognizes that effective precipitation is nonnegative with a third constraint.

$$X_i \geq 0 \tag{23}$$

The author also develops a model for minimizing the maximum absolute deviation and tests these models and a least-square model. The results demonstrate the advantageous properties of the linear programming model.

B. WATER QUALITY

Water quality can be enhanced directly by control of the water resource and indirectly through land-use controls. The direct activities include the following: dilution of polluted waters by low-flow augmentation, redirection of wastes from sections of a water body without excess aerobic capacity to sections with surplus capacity, artificial reaeration of the water bodies, locating new and upgrading existing treatment facilities, management of storm water, removal of accumulated bottom deposits, and allocation of the assimilative capacity of the waste. The indirect controls aim at preventing water quality degradation by controlling the land uses in the drainage basin. These land-use controls include but are not limited to zoning, transfer of development rights, building bans, floodplain management, and fragile terrain restrictions. The direct activities are the focus of this chapter. Models relating to indirect controls are considered in the land-use chapter because they relate to the preservation of the air, noise, land, and water environments.

1. ALLOCATION OF ASSIMILATIVE CAPACITY

The most widespread applications of linear programming to water quality have been efforts to allocate the assimilative capacity for oxygen-demanding wastes. The pioneering work of Thomann and Sobel (1964) and Thomann (1965) on the Delaware Estuary has been followed by different formulations and refinements of Thomann's approach [Sobel (1965), Revelle *et al.* (1968), and Anderson (1968)].[1]

The general problem has been to minimize the waste treatment costs for point dischargers subject to water quality constraints. The waste treatment parameter is BOD_5, the in-stream parameter is DO. The model seeks the

[1] An excellent and terse treatment is found in Smith and Morris (1969).

cheapest combinations of BOD_5 removal that will not violate water quality standards. Three basic versions of the model have been put forward. The simplest minimizes the economic cost of BOD removal subject to water quality constraints.

Minimize:
$$Z = \sum_j C_j X_j \tag{24}$$

where C_j is the cost of removing the waste, and X_j the amount of waste to be removed in pounds of BOD. Three constraints are required:

$$\sum_j R_{ij} X_j \geq S_i \tag{25}$$

where $R_{ij} X_j$ is the change in water quality at point i in the water body resulting from a waste discharge of X_j at source j, and S_i the minimum required improvement in quality at point i in the water body of DO in ppm. This constraint requires a water quality model with the ability to make a steady-state transfer in ppm per pound.

$$X_j \geq 0 \tag{26}$$

Equation (26) is a nonnegativitly constraint requiring that no industry reduce its treatment level.

$$X_j \leq B_j \tag{27}$$

where B_j is the upper limit of removal of the waste at source j in pounds of BOD.

This cost minimization model has been criticized as potentially unfair because it does not take into account the current efforts by the j waste sources. Briefly, industries already making major efforts may be asked to increase their expenditures. Other industries immediately adjacent to the first group and presently making little effort to reduce discharges may not be asked to make any improvements if the cost to the recalcitrant industries of starting a new system exceeds the cost to the other dischargers of improving their already existing facilities. A method of reallocating the cost among the dischargers is theoretically possible but may not be politically feasible.

The converse of the economic efficiency model discussed in the preceding is a model which ensures equity by requiring every discharger to remove the same percentage of waste from the effluent, irrespective of location and existing level of treatment. In addition, the uniform removal policy is much easier to administer than the efficiency policy.

Minimize:
$$Z = \sum_j C_j X_j \tag{28}$$

where X_j is the amount of currently discharged waste to be removed at the sources in pounds of BOD. In essence, the model minimizes the uniform percentage removal of the waste. Five constraints are required:

$$\sum_j R_{ij} X_j \geq S_i \tag{29}$$

where R_{ij} is the unit change of water quality at point i in the water body resulting from a unit change in the waste discharge at source j.

$$X_j = B_j(U - E_j) \qquad \text{if} \quad U \geq E_j \tag{30}$$

$$X_j = 0 \qquad \text{if} \quad U < E_j \tag{31}$$

where E_j is the proportion of waste currently removed, B_j the total loading of waste produced at j, and U the uniform treatment percentage.

$$X_j \geq 0 \tag{32}$$

$$X_j \leq B_j \tag{33}$$

If no treatment facilities exist, E_j becomes U. If treatment facilities exist, Eqs. (30) and (31) are used. Equation (30) is used when an existing source is not meeting the required uniform level of treatment U. If an existing source is exceeding the desired proportional removal, then it does not have to remove an additional percentage of its waste. The solution is obtained by systematically increasing the value of U until the constraints are satisfied. As the user increases the value of U, constraints formerly written as Eq. (31) may be changed to Eq. (30).

The major criticism of Eqs. (28)–(33) is that the assimilative capacity of the stream is treated as irrelevant. Industries, therefore, do not have an incentive to locate at sites that have excess assimilative capacity. Indeed, the penultimate illogical outcome of the uniform treatment approach is that it could lead to increasingly higher cost solutions, even in streams with the highest water quality.

If industries can choose a location within a basin irrespective of water effluent, they may—and historically have—selected locations that result in agglomeration. Agglomeration tends to produce an increasingly geographically concentrated waste load. In turn, the value of uniform treatment U would have to increase for the entire basin. Dischargers removed from this agglomeration would justifiably charge that the increased level of treatment of their waste is illogical, because their areas have continued to meet the standard. A sample problem appears at the conclusion of the chapter.

The advantages of Eqs. (24)–(27) and (28)–(33) have been combined into a hybrid. The body of water is divided into water quality zones that have uniform water quality and uniform removal levels. The objective function of this model is

Minimize: $$Z = \sum_j C_j X_j \tag{34}$$

subject to

$$\sum_j R_{ij} X_j \geq S_i \tag{35}$$

$$X_j \geq 0 \tag{36}$$

$$X_j \leq B_j \tag{37}$$

$$X_j = B_j(U_k - E_j) \qquad \text{if} \quad U_k \geq E_j \tag{38}$$

where U_k is the uniform treatment level in zone k, and j includes all the sources in zone k.

$$X_j = 0 \qquad \text{if} \quad U_k < E_j \tag{39}$$

Johnson (1967) tested these approaches for the Delaware Estuary and reports that model (34)–(39), modified to be an effluent charge model, is the best choice. Ackerman and Sawyer (1972) have argued that the Delaware study, and by inference all three models, are seriously questionable. They question the accuracy of the input data, the accuracy of the water quality model that was used to transform pounds of BOD_5 into DO in ppm, and the absence of nonpoint sources of waste and accumulated bottom wastes in the model. Furthermore, they contend that BOD_5 may not be the most important indicator in a slow-moving estuary like the Delaware. Overall, however, the load-allocation model approach has been applied to many basins and extended to other indicators such as nutrients and temperature. Chapter 9 notes applications to air quality.

Once the treatment levels have been chosen, the technology, scheduling, and financing of the treatment facilities must be chosen. Lynn *et al.* (1962) developed a linear programming model for designing a treatment plant. Processes within the facility were treated as the nodes and the links of a transportation system. Process costs are assigned to each potential link and minimized subject to continuity constraints and requirements for BOD removal. The continuity constraints resemble the constraints used in the solid-waste model [Eq. (5)] to redistribute waste which has been reduced in volume at the site.

A few attempts to design specific technologies with linear programming have been developed. Galler and Gotaas (1966) used linear programming to aid in the design of an optimum trickling filter system. Dostal *et al.* (1966) prepared a model to optimize the efficiency of a carbon absorption system.

Finally, Lynn (1964) developed a general linear programming model to estimate the timing and financing of treatment plants.

2. ALTERNATIVES TO AND COMPLEMENTS OF LOAD ALLOCATION

The fortuitous circumstance of the Delaware Estuary study combining water quality and optimization modeling capabilities, when a federal law requiring water quality standards had been promulgated, gave impetus to point source load allocation models. Few articles have appeared detailing alternatives to allocating point source waste discharges. However, a sufficient number of these alternatives have been modeled to illustrate the potential use of optimization models to test their economic efficiency.

An alternative to diluting the quality of waste in the effluent stream by treatment is to dilute the impact of effluent by increasing the quantity of water through augmentation. Rogers and Gemmell (1966) used a linear programming approach to test the economic efficiency of waste treatment and flow augmentation. They concluded that flow augmentation was a useful complement to waste treatment in their area of study. However, they were not prepared to argue for the relative efficiency of flow augmentation in every case. Grantham *et al.* (1970) developed a model that can be used to compare flow augmentation, waste entering a reach from upstream, and waste treatment of waste entering in a reach. The authors define flow augmentation benefits as the avoidance of treatment costs. The overall modeling process is described as a combination of simulation and optimization.

A second alternative to effluent treatment based on load allocations is to transfer the waste to reaches of a water body that can tolerate a higher load. Graves, *et al.* (1969) present a linear programming model and later a nonlinear version [Graves (1972)] to accomplish this task and to apply it to the Delaware Estuary. They recognize the political and legal problems of trying to bypass waste from one stretch of a water body to another. They argue, however, that ultimately the most efficient combination may be bypass piping to regional treatment plants. The Graves *et al.* (1969) article has the additional merit of providing a substantial data set which of and by itself is worth examination.

A complement to point source load allocations is a model for capturing and treating runoff and other nonpoint waste sources. Kirshen *et al.* (1972) prepared and tested a complex linear programming model to select storm water control alternatives. The authors sought to minimize the cost of the following system alternatives: the capacity and length of new pipes, short-term storage facilities, special treatment facilities, and regular treatment facilities. System constraints include continuity limits for the temporary storage tanks and the in-system storage pipes, capacity constraints on all of the fixed facilities, and water quality limits. Kirshen *et al.* tested the model in the Bloody Run drainage basin in Cincinnati, Ohio. While the application produced useful results, the substantial data input requirements imply that the model will find its greatest use in urban regions with serious water quality problems.

C. SAMPLE PROBLEMS

1. A WATER NETWORK EXAMPLE

The following water network problem illustrates the simplest regional transfer model [Eqs. (8)–(12)]. It contains four water agencies and six interagency links. The total system of nine links is illustrated by Fig. 1. Data for the example are listed in Tables I and II.

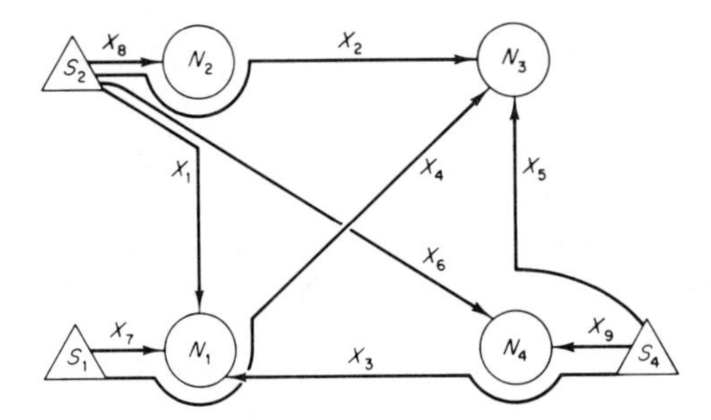

Fig. 1 Sample water network problem.

Table I Data for Water Network Examples: Links

Link	Cost ($/mg)	Capacity (mgd)
X_1	250	5
X_2	300	3
X_3	270	10
X_4	270	15
X_5	340	10
X_6	325	5
X_7	200	20
X_8	175	10
X_9	225	10

Table II Data for Water Network Example: Nodes

Node	Supply controlled	Demand
N_1	20	10
N_2	10	8
N_3	0	10
N_4	10	5
Total	40	33

Agency 1 is in the comfortable position of having a water surplus and a low price. Agency 3 does not have a supply and is completely dependent on transfers.

The system is described by the linear model:

$$\text{Minimize:} \quad Z = 250X_1 + 300X_2 + 270X_3 + 270X_4 + 340X_5 + 325X_6 + 200X_7 + 175X_8 + 225X_9 \qquad \text{(P1.1)}$$

subject to

$$\begin{array}{lllllllllrlll}
 & & & & & & & & & & & \text{For} & \\
X_1 + & & + X_3 + & & & & + X_7 & & & = & 10 & ① & \text{(P1.2)} \\
 & & & & & & & + X_8 & & = & 8 & ② & \text{(P1.3)} \\
 & + X_2 + & & X_4 + X_5 & & & & & & = & 10 & ③ & \text{(P1.4)} \\
 & & & & & X_6 + & & & X_9 & = & 5 & ④ & \text{(P1.5)} \\
 & & & + X_4 + & & & + X_7 & & & \leq & 20 & \triangle_1 & \text{(P1.6)} \\
X_1 + X_2 + & & & & & + X_6 & & + X_8 & & \leq & 10 & \triangle_2 & \text{(P1.7)} \\
 & & + X_3 & & + X_5 & & & & + X_9 & \leq & 10 & \triangle_4 & \text{(P1.8)} \\
X_1 & & & & & & & & & \leq & 5 & \overset{1}{\rightarrow} & \text{(P1.9)} \\
 & X_2 & & & & & & & & \leq & 3 & \overset{2}{\rightarrow} & \text{(P1.10)} \\
 & & X_3 & & & & & & & \leq & 10 & \overset{3}{\rightarrow} & \text{(P1.11)} \\
 & & & X_4 & & & & & & \leq & 15 & \overset{4}{\rightarrow} & \text{(P1.12)} \\
 & & & & X_5 & & & & & \leq & 10 & \overset{5}{\rightarrow} & \text{(P1.13)} \\
 & & & & & X_6 & & & & \leq & 5 & \overset{6}{\rightarrow} & \text{(P1.14)} \\
 & & & & & & X_7 & & & \leq & 20 & \overset{7}{\rightarrow} & \text{(P1.15)} \\
 & & & & & & & X_8 & & \leq & 10 & \overset{8}{\rightarrow} & \text{(P1.16)} \\
 & & & & & & & & X_9 & \leq & 10 & \overset{9}{\rightarrow} & \text{(P1.17)} \\
X_1, & X_2, \ldots, & & & & & & & X_9 & \geq & 0 & & \text{(P1.18)}
\end{array}$$

Solution: $Z = \$7225$; $X_4 = 10$; $X_7 = 10$; $X_8 = 8$; $X_9 = 5$.

Once an initial solution has been obtained, you can test the impact of price changes on system flows. New links can be added. Safe yields can be increased or decreased. Finally, demands can be increased or decreased. In short, the more you test the sensitivity of a solution to changes in right-hand sides and cost values, the greater will be your grasp of how to test the impact of policy alternatives on system performance.

2. A LOAD ALLOCATION EXAMPLE

A load allocation model with one reach and three discharges has been written in the uniform removal format.

Figure 2 and Table III provide the necessary data for Eqs. (28)–(33). The numbers are purely for convenience and do not reflect actual relationships.

The stream has three dischargers. Each currently discharges 1000 pounds of BOD_5 waste B_j. Discharger 3 is already removing 95% of the BOD, while discharger 2 is only removing 60% of its waste E_j. These differences are reflected by the per unit removal costs C_j. The desired improvement in DO is exactly 1 ppm.

The system assumes the following form and the uniform level for the first run U is set at 0.8.

Minimize:

$$Z = 60X_1 + 50X_2 + 100X_3 \qquad \text{(P2.1)}$$

subject to

$$.0040X_1 + .0055X_2 + .0020X_3 \geq 1 \qquad \text{(P2.2)}$$

$$X_1 = 1000(.8 - .7) \qquad \text{(P2.3)}$$

$$X_2 = 1000(.8 - .6) \qquad \text{(P2.4)}$$

$$X_3 = 0 \qquad \text{(P2.5)}$$

$$X_1, X_2 \geq 0 \qquad \text{(P2.6)}$$

$$X_1, X_2, X_3 \leq 1{,}000 \qquad \text{(P2.7)}$$

Solution: $Z = 16{,}000$; $X_1 = 100$; $X_2 = 200$; $X_3 = 0$.

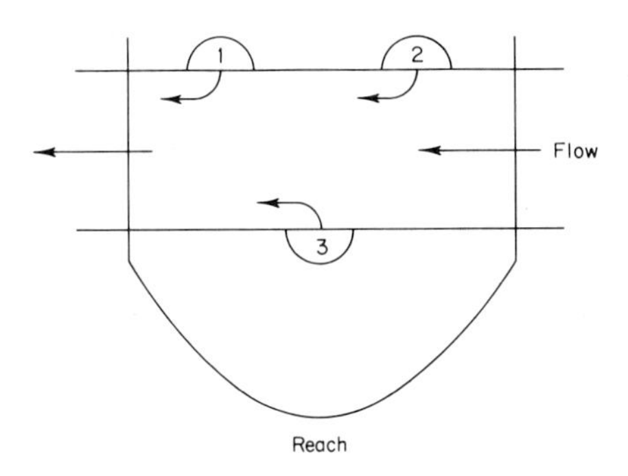

Fig. 2 Representation of load allocation model.

Table III Data for Load Allocation Example

Discharger	C_j	R_{ij}	S_i	B_j	E_j	U
1	60	.0040	1	1000	.70	.8
2	50	.0055	1	1000	.60	.8
3	100	.0020	1	1000	.95	.8

If the uniform treatment level of 0.8 does not satisfy Eqs. (P2.2), (P2.6), and (P2.7), systematically raise the value of U until the smallest U and, in turn, C_jX_j results. This can be done with parametric programming.

The user should then solve the problem using the least-cost model [Eqs. (24)–(27)]. Other exercises would involve adding a second reach and using the multizoned model [Eqs. (34)–(39)], and testing the sensitivity of the model to changes in constants (e.g., R_{ij}, E_j, and S_i). If you have access to a water quality model that can calculate the R_{ij} values on the basis of changes in the quality of the effluent or quantity of water, then work with both models. Next, you can bypass waste between facilities to create a "regional" wastewater system. A multireach model is, however, recommended in the regional system case. Suffice it to say that each of these exercises involves policy decisions in the real world that can mean millions of dollars in expenditures.

3. A RESERVOIR MANAGEMENT EXAMPLE

The following sample problem is a wet and dry season version of the first equation set [Eqs. (2)–(7)] in the chapter. The input data are in Table IV.

Table IV Data for Reservoir Management Example

Parameter	Wet season	Dry season
Activity	X_2	X_3
r_i^{90}	50	20
v_i	10	10
r_i^{10}	5	3
q_i	2	2
a_m	.2	.2
a_0	.4	.4
f_i	50	50

In this example, X_1 is the reservoir size to be chosen, X_2 the decision rule for the wet season, and X_3 the decision rule for the dry season.

Minimize: $$Z = X_1 + 0X_2 + 0X_3 \quad \text{(P3.1)}$$

subject to

$$X_1 - X_2 \geq 60 \quad \text{(P3.2)}$$

$$X_1 - X_3 \geq 30 \quad \text{(P3.3)}$$

$$.2X_1 - X_2 \leq 5 \quad \text{(P3.4)}$$

$$.2X_1 - X_3 \leq 3 \quad \text{(P3.5)}$$

$$X_2 - X_3 \geq -3 \quad \text{(P3.6)}$$

$$-X_2 + X_3 \geq -1 \quad \text{(P3.7)}$$

$$.4X_1 - X_3 \geq 2 \quad \text{(P3.8)}$$

$$X_2 - X_3 \leq 0 \quad \text{(P3.9)}$$

$$-X_2 + X_3 \leq 30 \quad \text{(P3.10)}$$

$$.4X_1 - X_3 \leq 50 \quad \text{(P3.11)}$$

Solution: $Z = 68.75$; $X_1 = 68.75$; $X_2 = 8.75$; $X_3 = 10.75$.

D. SUMMARY

This chapter has focused on the use of linear programming models in the management of water quantity and quality. The common characteristic of the water quantity and quality applications is the goal of meeting an economic management objective while allocating a scarce resource. The scarce resources are water and assimilative capacity. The spatial scale and temporal scope of the problems range from the pipe to the metropolis and from the month to the half century.

As a tool, linear programming is one of the more pervasive in a field where mathematical models are, in general, pervasive. Nevertheless, the use of optimization models has been hampered by the sparsity of reliable data and by the host of debatable models that have been used to measure the benefits of water resources, the impact of a discharge on water quality, the water requirements for small regions, and numerous other inputs to optimization models. In all candor, the widespread application of linear programming, in contrast to its current academic acceptance, awaits harder data sets and more reliable simulation models.

REFERENCES

Ackerman, S., and Sawyer, R. (1972). The uncertain search for environmental policy: scientific factfinding and rational decision making along the Delaware River. *U. Penn. Law Rev.* **120**(3), 419.

Anderson, M. W. (1968). Regional mangagement of water quality—a systems approach. *J. Water Pollut. Contr. Fed.* **40**(10), 1679–1687.

Clyde, C. G., Jensen, B. C. and Milligan, J. H. (1967). Optimizing conjunctive use of surface water and ground water. *Proc. Symp. Groundwater Development in Arid Basins*, Utah State University, Logan, Utah.

Deininger, R. A. (1969). Linear programming for hydrologic analyses. *Water Resour. Res.* **5**(5), 1105–1109.

deLucia, R. J. (1971). "The North Atlantic Regional Supply Model." Amer. Geophysical Union 52nd Meeting, Washington, D.C.

Dostal, K. A., Harrington, J. J., Clark, R. M., and Roebeck, G. G. (1966). Development of optimization models for carbon bed design. *J. Amer. Water Works Ass.* **58**(9), 1171–1186.

Dracup, J. A. (1966). "The Optimum Use of a Ground-Water and Surface-Water System: A Parametric Linear Programming Approach." Technical Report HEL 6-24, Water Resources Center, Univ. of California, Berkeley, California.

Facet, T., and Marks, D. (1972). "Scheduling and Sequencing in Water Resource Investment Models." Paper presented at International Symposium on Water Resources Planning, Districto Federale, Mexico, December 1–8.

Galler, W. S., and Gotaas, J. B. (1966). Optimization analyses for biological filter design. *J. Sanit. Eng. Div., Amer. Soc. Civil Eng.* **92**(SA 1), 163–182.

Grantham, G. R., Pyatt, E. E., Heaney, J. P., and Carter, B. J., Jr. (1970). Model for flow augmentation—an overview. *J. Sanit. Eng. Div., Proc. Amer. Soc. Civil Eng.* **96**(SA 5), 1045–1055.

Graves, G. W. (1972). "Extensions of mathematical programming for regional water quality management." UCLA, Los Angeles.

Graves, G. W., Hatfield, G. B., and Whinston, A. (1969). Water pollution control using by-pass piping. *Water Resour. Res.* **5**(1), 13–47.

Greenberg, M. R., Carey, G. W., Zobler, L., and Hordon, R. M. (1971). "A geographical systems analysis of the water supply networks of the New York metropolitan region. *Geograph. Rev.* **LXI**(3), 339–354.

Greenberg, M. R., and Hordon, R. M. (1975). "Water Supply Planning: A Case Study and Systems Analysis." Center for Urban Policy Research, Rutgers University, New Brunswick, New Jersey.

Heaney, J. P. (1968). Mathematical programming analysis of regional water resource systems. *Proc. Nat. Symp. Analysis Water Resour. Systems.*, American Water Resources Association, pp. 231–240.

Johnson, E. L. (1967). A study in the economics of water quality management. *Water Resour. Res.* **3**(2), 291–305.

Karmeli, D., Gadish, Y., and Meyers, S. (1968). Design of optimal water distribution networks. *Proc. Soc. Amer. Civil Eng.* **94**(PL 1), 1–10.

Kirshen, P. H., Marks, D., and Schaake, J. C., Jr. (1972). "Mathematical Model for Screening Storm Water Control Alternatives." MIT, Dept. of Civil Engineering, Cambridge, Massachusetts (see especially pp. 46–74).

Kleinecke, D. (1971). Use of linear programming for estimating geohydrologic parameters of groundwater basins. *Water Resour. Res.* **7**(2), 367–374.

Loucks, D. P. (1968). Computer models for reservoir regulation. *J. Sanit. Eng. Div., Amer. Soc. Civil Eng.* **94**(SA 4), 657–669.

Lynn, W. R. (1964). Stage development of wastewater treatment works. *J. Water Pollution Countr. Fed.* **36**(6), 722–751.

Lynn, W. R., Logan, J. A., and Charnes, A. (1962). Systems analysis for planning wastewater treatment plants. *J. Water Pollution Contr. Fed.* **34**(6), 565–579.

Males, R. M. (1968). "Optimal Operation Rules for Multipurpose Reservoir Systems." Unpublished Ph.D. thesis, Massachusetts Institute of Technology, Cambridge, Massachusetts.

Manne, A. S. (1962). Product-mix alternatives. flood control, electric power and irrigation. *Int. Econ. Rev.* **3**(1), 30–59.

Mannos, M. (1955). An application of linear programming to efficiency in operation for a system of dams. *Econometrica* **33**(3). 335–336.

Ramaseshan, S. (1962). Aqueduct capacity under an optimum benefit policy. *Proc. Amer. Soc. Civil Eng.* **88**(IR 2), 97–100.

Revelle, C., and Kirby, W. (1970). Linear decision rule in reservoir management and design, 2, performance optimization. *Water Resour. Res.* **6**(4), 1033–1044.

Revelle, C., Loucks, D.P., and Lynn, W. (1968). Linear programming applied to water quality management. *Water Resour. Res.* **4**, 1–9.

Revelle, C., Joeres, E., Kirby, W. (1969). The linear decision rule in reservoir management and design, 1, development of the stochastic model. *Water Resour. Res.* **5**(4), 767–777.

Rogers, J. R., and Gemmell, R. (1966). "Economic Evaluation of Flow Augmentation: A Systems Analysis Case Study." Proc. 21st Industrial Waste Conference, Purdue University, Lafayette, Indiana.

Smith, E. T., and Morris, A. R. (1969). Systems analysis for optimal water quality management. *J. Water Pollution Contr. Fed.* **41**(9), 1635–1646.

Sobel, M. J. (1965). Water quality improvement programming problems. *Water Resour. Res.* **1**(4), 477–487.

Thomann, R. (1965). Recent results from a mathematical model of water pollution control in the Delaware Estuary. *Water Resour. Res.* **1**(3), 349–359.

Thomann, R., and Sobel, M. J. (1964). Estuarine water quality management and forecasting. *J. Sanit. Eng. Div., Proc. Amer. Soc. Civil Eng.* **90**(JA 5), 9–36.

Thomas, H., and Revelle, R. (1966). On the efficient use of the High Aswan Dam for hydropower and irrigation. *Management Sci.* **12**(8), B296–B311.

Young, G. K., Moseley, J. C., and Evenson, D. E. (1970). Time sequencing of element construction in a multi-reservoir system. *Water Resour. Bull.* **6**(4), 528–541.

Chapter 8

Health, Education, and Law Enforcement

The purpose of this chapter is to review the application of optimization models to the broad range of concerns encompassed by three public services. The literatures of these three fields is largely devoid of cross references. Nevertheless, my review of these independently conceived literatures suggests that they are so similar as to demand coordinated treatment. Nearly all of the literature can be broadly trichotomized: (1) allocation of people, including personnel and clients, (2) allocation of funds, and (3) allocation of capital.

A. ALLOCATION OF PEOPLE

Optimization is used in education, health, and law enforcement to allocate the faculty, research staff, and students of schools, the staff and patients of health facilities, and the judges, officers, inmates, and victims of the criminal system to appropriate locations. In the field of education, Plessner *et al.* (1968) use linear programming to allocate faculty to alternative teaching, research and administrative duties. The most complicated model has two levels of objectives: departmental and administrative. The dean of the college uses the objectives of the departments as constraints for allocating faculty resources. McNamara (1973) reports on other applications to personnel allocation in higher education. Childs and Wolfe (1972) use linear programming to allocate professional personnel to work on a series of research projects. The problem maximizes evaluations of professional effectiveness made by management subject to constraints on personnel availability and project scheduling.

In comparison to education, law enforcement applications have been limited in number. Lyons (1971) seeks to maximize the number of legal cases to be heard by judges subject to constraints on the availability of judges and the need to adjudicate a backlog. Gass (1975) reports the use of linear programming and heuristic programming models to allocate patrolmen to beats. Clearly, the allocation of law enforcement as well as fire protection personnel by optimization could increase in the future.

Optimization has been used to allocate health care personnel. Warner and Prawda (1972) minimize the cost of a shortage of nurses in a hospital. Constraints include the availability of nursing personnel of different skill classes and the expected demand for nursing skills by wards, shifts, and skills. The solution is obtained by a mixed-integer, quadratic programming algorithm for a 6-ward, 600-bed general hospital. Warner and Prawda suggest that a much more simple linear programming formulation would work almost as well.

Liebman *et al.* (1972) assign nurses to patients who require special care.

Maximize:
$$Z = \left(\frac{1}{e}\right) \sum_{i} \sum_{j} \sum_{k} C_{jk} t_{ik} X_{ijk} \quad (1)$$

where C_{jk} is a measure of the effectiveness of assigning nursing personnel type j to task k. The C_{jk} values were obtained by statistical analysis of interviews. The t_{ik} values are estimated times required to perform task k on patient i. The X_{ijk}'s are the assignments of personnel type j to task k on patient i. The X_{ijk} value is 1 if nurse type j does task k on patient i, or 0 otherwise. The constant e is the highest estimated possible effectiveness. It normalizes the effectiveness values C with respect to time t. The value of e is obtained from $e = \sum_{i} \sum_{k} (\max_j C_{jk})(t_{ik}/100)$. Two constraints are defined:

$$\sum_{j} X_{ijk} = 1 \quad (2)$$

This constraint requires that only one staff member perform a procedure on a patient at a single time.

$$\sum_{i} \sum_{k} t_{ik} X_{ijk} \leq S_j \quad (3)$$

where S_j is the available amount of time of personnel type j.

Shuman *et al.* (1971) treat the delivery of health care services as a problem of optimizing the use of manpower and equipment. They assume the need for j different types of health services to be provided by i different personnel types (doctors, nurses, technicians) at k locations using m different technologies.

The model seeks to determine the mix of personnel, services, levels of technology, and amount of community investment in construction necessary to minimize the total cost to society of meeting health service requirements. The notation is the following:

q_{ijkm}	the number of personnel i providing service j at location k using technology m at the end of the planning period
q_{ijko}	the number of personnel providing the service at the beginning of the planning period
h_{ijk}	the number of personnel hired
f_{ijk}	the number of personnel fired
b_{ijkm}	productivity coefficients denoting the average number of type j services produced by type i personnel at k using technology m
X_{jk}	the maximum number of personnel the average physician can supervise for service j at location k
P_i	the number of type i personnel available during a planning period
p_i	the additional number of i type personnel allocated by the region in a particular planning period
g_k	the cost of constructing facilities
y_k	the additional amount of facility area to be added at the kth facility
Y_k	total available space
B_k	the additional financial support given to the kth facility to meet operating expenses
S_{ik}	the personnel support cost
C'_{ijk}	the hiring and on-the-job training cost
C''_{ijk}	the cost of releasing personnel
r_{jk}	the patient's fee
u_i	the cost of developing personnel
e_j	the incentive cost of having people avail themselves of needed services
v_j	the cost of not providing a needed service
d_j	the demand for services
N_j	the need for type j services ($N_j \geq d_j$)
w_{jk}	the space requirement for each unit of service j at location k
ϕ_{ijkm}	the cost of a given level of technology
a_k	the proportion of the budget allocated for meeting and operating expenses
C_S	the total amount of money that the community is willing to spend

Minimize:

$$Z = \sum_i \sum_j \sum_k \sum_m (r_{jk} - v_j) b_{ijkm} q_{ijkm} + \sum_k (B_k + g_k y_k) + \sum_i u_i p_i \quad (4)$$

The first part of the equation measures the cost of providing services. The costs include the patient's fee minus the cost of not providing health service weighted by the productivity of the personnel. The second part minimizes the cost of financial support for operating (B_k) and constructing new facilities ($g_k y_k$). The final portion of objective function seeks to minimize the cost of bringing in new personnel.

$$\sum_{i=2}^{I} q_{ijkm} - X_{jkm} q_{1jkm} \leq 0 \quad (5)$$

where $i = 1$ designates physicians and $i = 2$ designates auxiliary personnel. This constraint indicates the maximum number of auxiliary personnel that the average physician can supervise. In essence, Eq. (5) is a quality of service constraint.

$$\sum_i \sum_j \sum_m b_{ijkm} q_{ijkm} \geq d_j \quad \text{or} \quad \leq N_j \quad (6)$$

The second constraint requires that service demands are met. The third constraint recognizes personnel limitations.

$$\sum_j \sum_k \sum_m q_{ijkm} - p_i \leq P_i \quad (7)$$

The fourth constraint recognizes facility capacities.

$$\sum_i \sum_j \sum_m w_{jk} b_{ijk} q_{ijkm} - y_k \leq Y_k \quad (8)$$

Two equations define financial limitations. One is for operating expenses:

$$\sum_i \sum_j (C'_{ijk} h_{ijk} + C''_{ijk} f_{ijk}) + \sum_m [(S_{ik} + \phi_{ijkm} - a_k r_{jk} b_{ijk}) q_{ijkm}] - B_k \leq 0 \quad (9)$$

The first part of the financial constraint takes the hiring, training, and releasing of workers into account. The second part is the operating cost. The amount B_k is that which the community must pay if operating deficits are incurred.

The seventh equation limits the amount of money that the community is willing to spend.

$$\sum_k (B_k + g_k y_k) + \sum_i u_i p_i + \sum_i \sum_j \sum_k \sum_m e_j b_{ijkm} q_{ijkm} \leq C_s + \sum_j d_j e_j \quad (10)$$

Finally, a continuity constraint for personnel is required.

$$q_{ijkm} - h_{ijk} + f_{ijk} = q_{ijko} \quad (11)$$

Shuman *et al.* use this complex linear programming model for planning a health care center. The center is to serve about 80,000 people. It will utilize two personnel types (i = physicians and assistants), two services (j = diagnostic and theraputic), and two technology levels (m = high and low). A maximum of $1 million is available for the project. The optimal solution requires 10.2 physicians and 40.7 assistants. The total cost is about $2.7 million.

The authors conducted sensitivity analyses of policy alternatives, including budgetary, satisfaction of all demands for health care service, provision first of services to the indigent, and increases and decreases in the level of technology. Finally, suggestions are offered for incorporating personnel attrition into the model, for introducing stochastic elements in the health care delivery system into the model, and for developing dynamic solutions.

The clientele served by health, education, and law enforcement personnel include specialized groups of patients, students, and criminals and the general public. The last is to be protected from criminal activity and fires. The bulk of the optimization applications in education allocate students to schools with a transportation model. An early and useful example by Yeates (1963) allocates high school students in Grant Country, Wisconsin. The problem is to minimize the distance that the students must travel.

Minimize:
$$Z = \sum_i \sum_j d_{ij} X_{ij} \tag{12}$$

where d_{ij} is the distance from residential section i to high school j. The X_{ij} are the number of students in the ith residential zone assigned to the jth school. The d_{ij} values used by Yeates alternatively could be travel times or costs.

Yeates develops four constraints:

$$\sum_j X_{ij} = a_i \tag{13}$$

where a_i is the number of students to be assigned, the student demand.

$$\sum_i X_{ij} = b_j \tag{14}$$

where b_j is the number of available places in the school system. Alternatively, Eq. (14) could be a less than or equal to constraint.

$$\sum_i a_i = \sum_j b_j \tag{15}$$

The third constraint requires the supply of students and demands for students to be equal. This constraint should be redundant with the first two constraints.

Finally, negative pupil assignments are not permitted.

$$X_{ij} \geq 0 \tag{16}$$

After making the obvious allocations by inspection, a total of 726 students were assigned using the model. Yeates estimates that a savings of $3000 to $4000 per year could be realized if students were assigned on the basis of the revised district boundaries.

More complex versions of the original Yeates transportation model have been prepared by Newton and Thomas (1969) and Angel *et al.* (1972), who used modified traveling salesman heuristic algorithms to design schoolbus routes [see also Noonan and Whinston (1969)]. Both these models minimize mileage subject to constraints similar to those developed by Yeates. The authors caution the potential user that the algorithm was uneconomical for problems with several hundred bus stops, a caution that was also noted in the solid-waste chapter.

Potentially, the most politically sensitive applications of school assignment models have allocated students while constrained by desegregation requirements. Koenigsberg (1968), Clarke and Surkis (1968), and Heckman and Taylor (1969) have tackled this problem. Usually, these models minimize student travel time subject to specifications about racial composition levels in each school, school capacities, student sizes, student travel times, and the capacity of the transportation system.

The similarity of the Yeates and the latter three models may be illustrated by reviewing the Koenigsberg paper.

Minimize:
$$Z = \sum_i \sum_j \sum_k \sum_h t_{ij} X_{ijkh} \tag{17}$$

The objective function minimizes the travel time t_{ij} of the trip to school i from enumeration district j. The Yeates model did not distinguish among students by age or race. Koenigsberg allocates the students not only on the basis of school i and origin j, but also by race k and age group h.

Equation (18) is the supply of students that must be allocated. Constraint (18) is identical to Eq. (13).

$$\sum_i X_{ijkh} = a_{jkh} \tag{18}$$

Similarly, the schools' capacities for particular age groups must be respected. This constraint is analogous to Eq. (14).

$$\sum_k X_{ijkh} \leq b_{ih} \tag{19}$$

Next, negative student allocations are not permitted as in Eq. (16).

$$X_{ijkh} \geq 0 \tag{20}$$

Additional equations are required to define integration restrictions. Let $k = 1$ denote nonwhites, a_{1i} the minimum proportion of nonwhites desired in school i, and b_{1i} the maximum proportion of nonwhites desired in school i.

Then, minimum [Eq. (21)] and maximum [Eq. (22)] nonwhite restrictions for each school are set.

$$a_{i1} \sum_{j} \sum_{k} X_{ijkh} - \sum_{j} X_{ij1h} \leq 0 \tag{21}$$

and

$$\sum_{j} X_{ijkh} - b_{i1} \sum_{j} \sum_{k} X_{ij1h} \leq 0 \tag{22}$$

Additional constraints about travel budgets, travel time restrictions, and the capability of the transportation system may also be added.

Health planning journals have featured applications that do not resemble the previously described education examples. For example, Earickson (1970) discusses allocating patients to hospitals on the basis of distance minimization. The distance-based approach is rejected, however, because patients are thought to choose hospitals on the basis of their economic status, age, ethnic orientation, and their special medical needs rather than on distance minimization. Palmer (1975) reports papers by Scheffler and Chen and by Bush. Scheffler uses a nonlinear programming model to formulate optimal group practice sizes. Chen and Bush allocate people for treatment with the goal of minimizing the time spent in the unhealthiest categories.

Baligh and Laughhunn (1969) maximize the treatment of different classes of patients in a hospital subject to restrictions based on resources, technology, budgets, and policies. The objective function maximizes the value assigned by the hospital administration to different groups of regular and indigent patients.

Maximize:
$$Z = \sum_{i} w_i(X_i + Y_i) \tag{23}$$

where w_i are weights given to the ith class of patients, X_i the number of regular (paying) patients, and Y_i the number of indigent patients. A fixed-resource and technology constraint is defined:

$$\sum_{i} \sum_{j} (X_i + Y_i) a_{kj} R_{ij} \leq T_k \tag{24}$$

where T_k is the available fixed resources, R_{ij} the average requirement of a patient in the ith class for the jth service, and a_{kj} the number of units of a given fixed resource needed to produce one unit of the jth service. Equation (24) is required for such fixed resources as beds and various pieces of surgical apparatus. These cannot be altered during the time period.

The second constraint requires that the number of patients is limited to the number of available paying and indigent patients.

$$\sum_{i} (X_i + Y_i) \leq H_i \tag{25}$$

where H_i is the total number of available patients. A budgetary requirement is defined:

$$\sum_i \sum_j X_i R_{ij} p_j - \sum_i \sum_j \sum_1 c_1 b_{ij}(X_i + Y_i) R_{ij} \geq G + F - S \qquad (26)$$

The first portion of the equation is the revenue: number of paying patients X_i times the average requirement of a patient in the ith class for the jth service R_{ij} times the price p_j charged for one unit of service. The second portion of Eq. (26) is the total cost incurred by the hospital for variable resources. Variable resources include such items as food and supplies that can be changed during the study period. The scalar c_1 is the per unit cost of variable input 1. The coefficient b_{ij} is the number of units of a variable resource required in the production of a unit of the ith service. The value G is the target profit, F the cost of fixed resources, and S the external subsidy.

Next, policy constraints may be added. For example, the hospital may be required to treat a minimum proportion of indigent patients. Let V_i represent the minimum proportion of indigent patients. Then,

$$Y_i(1 - V_i) - X_i V_i \geq 0 \qquad (27)$$

Finally, the nonnegativity restrictions are required.

$$X_i, Y_i \geq 0 \qquad (28)$$

Summarizing, the models reviewed in this section allocate workers to their jobs and people to the schools, hospitals, courts, and other facilities that provide public services. To date, a variety of applications have been made in health and education and few applications in law enforcement and fire protection.

B. ALLOCATION OF FINANCIAL RESOURCES

In addition to the allocation of their employees and their clients, public service agencies allocate financial and capital resources. This section reviews optimization models used to allocate liquid capital resources to educational and health institutions. The education literature is the most extensive, featuring examples at international and local scales. McNamara (1973) reviews national level applications of optimization by Bowles and Golladay. Bowles uses dynamic linear programming to maximize the value of education expenditures to national income. The study area is northern Nigeria, and the planning horizon is eight years. Constraints include types of teachers, student enrollments, facility construction and utilization levels, teacher recruitment, and political or legal restrictions. Golladay develops a similar model for Morocco and Adelman (1966) another for Argentina.

With respect to these national level studies, McNamara notes that these models have been applied almost exclusively in developing countries that face severe fund shortages. An exception is the work of Benard (1967), who developed a model for optimum investment in the centralized French education system. The model treats education as that portion of economic system which produces skilled graduates subject to constraints on production capacity, maximum consumption, a minimum growth rate for education, and a budgetary ceiling.

In the United States, budget powers are concentrated at the state and school district levels. Bruno (1969a, 1971) and McNamara (1971) have studied state funding questions. Bruno (1969a) focuses on maximizing the utilization of financial resources distributed by the states to school districts. Budgetary and political constraints on tax rates, level of foundation support, state aid to districts, tuition, and state funds available from the legislature are defined. McNamara (1971) develops a linear programming model to allocate funds for vocational education. The model maximizes the number of students graduating from each county in the state. Each county has an upper-limit capacity and lower-limit demand and a budget constraint.

At the school district level, three studies stand out as unusually interesting. Bruno (1969b) develops a linear programming solution to determine a flexible and a minimum cost salary schedule. Nine characteristics are used in devising a consistent salary schedule: learning environment, subject matter or special skills, supervisory responsibility, highest academic degree obtained, work experience, special awards and distinctions, college credits completed in addition to degree, in-service units completed, and relative additional workload in the hierarchy. Each of these nine objective function characteristics is weighted on an ordinal scale. For example, a difficult learning environment receives a weight of 3 and an easy environment a weight of 1. A Ph.D. or Ed.D. is worth 5 units, while a B.A. is worth 2. The 42 weights prepared for the nine activities are multiplied by the number of employees possessing the characteristic. For example, 220 employees were considered to work in difficult learning environments. The objective function value for a difficult learning environment is 660 (220 employees times 3 points for difficult).

The model includes six types of constraints. One set specifies the number of employees required in each category (superintendent, administrator, department head, teacher, teacher's aide), a second type specifies minimum percentage salary spreads within each job classification, and a third requirement indicates minimum percentage overlaps between job classifications.

Next, minimum dollar spreads between the highest salaries of each job classification are specified. Fifth, school district funds to support the salary structure are indicated. Finally, upper and lower bounds are placed upon the value of each of the nine activities. This procedure ensures that a few activities will not account for the bulk of the salary.

The author provides an example that includes parametric programming and sensitivity analysis. Bruno argues that the model should not be used as a deterministic tool, but rather as a means of conducting negotiations on specific issues of concern to the school system, the teachers, and the public.

Holtzman (1968) views the school system as a business that maximizes profits. The profits are measured as a function of the number of students educated. The benefits of graduating students are assumed to be the present net value of future income associated with the attainment of a particular level of education. Constraints include the availability of teachers, classrooms, and students. Finally, Holtzman discusses the problem of measuring the income benefits of persons, chiefly women, who do not enter the labor force and the establishment of valid student–teacher ratios.

He applies the model to the 57,000-student Detroit school system with 1963–1964 data. The high school system is found able to produce benefits of more than $20 million. Social science and foreign language teachers are in short supply; vocational and science teachers are found over abundant in the system. The author notes the policy implications of these findings and of the implications of relaxing some of the constraints.

Perhaps the most comprehensive paper on funding education is by Kuch (1975). It draws on the models developed by Bowles, Adelman, and others. Kuch discusses the problems of deciding on a single measure of effectiveness for the educational system and of measuring the benefits of graduates who migrate from the region in which they were educated.

The model portrays the education process as a series of training activities that, through inputs of financial, human, and physical resources, transform the student into a more valuable entity. Constraints include budgets, the supply of students, and the availability of equipment and faculty.

The objective function is

$$\text{Maximize:} \quad Z = \sum_{j=1}^{m} \sum_{t=1}^{T} [q_j(Y_{j,t+sj-1} - Y^*_{j,t-1} - C_{jt}) + (1 - q_j)P_j(Y_{j,t+sj-1} - Y^*_{j,t-1} - C_{jt})]X_{jt} - \sum_{k=1}^{K} \sum_{t=1}^{T} V_{kt} K_{kt'} \tag{29}$$

where the following notation is used:

Z the total of net social benefits derived from the educational system
m the number of possible training activities
T the length of the planning period, measured in years
q_j the fraction of those admitted to activity j who can be expected successfully to complete it
Y_{jt} the present value (discounted to year one) of the earnings stream associated with having completed activity j in year t
Y^*_{jt} the present value (discounted to year one) of the earnings stream foregone by enrolling in training activity j in year t
s_j the duration of training activity j
C_{jt} the recurrent cost of operating activity j for s_j years starting in year t, discounted to year one
P_j the fraction of the net benefits associated with activity j received by dropouts from j
X_{jt} the number of students admitted to activity j in year t
K the number of different types of school buildings
K_{kt} the number of student places of type k constructed in year t
V_{kt} the present value of the capital cost of providing an additional student place of type k in year t

The value of V_{kt} is estimated by using the relationship

$$V_{kt} = \sum_{\tau=t}^{t+20} r_t L_{kt}(1 + p)^{-\tau+1}$$

where L_{kt} is the cost of constructing a student place of type k in year t, r_t the interest rate on funds borrowed in year t, and p the discount rate used throughout the objective function.

The first part of the objective function is the value associated with those who complete the activity. The second part defines the value of those who begin but do not complete the activity. The third part of the equation specifies the cost of facilities.

Seven types of constraints are defined. One specifies the availability of teachers.

$$\sum_{j=1}^{m} \sum_{\tau=t+1-sj}^{t} a_{ijt} X_{j\tau} = S_{i0}(1 - d_i)^{t-1} + \sum_{\tau=1}^{t} R_{i\tau} \qquad (30)$$

where

a_{ijt} the minimum input of teachers of type i in the year t required to accommodate one student in activity j

S_{i0} — the stock of teacher input i available to the system in the base year
d_i — the annual natural attrition rate of teachers of type i
$R_{i\tau} > 0$ — the number of teachers of type i recruited in year t
$R_{i\tau} < 0$ — the number of surplus teachers of type i dismissed in year t

An equation of this type [Eq. (30)] is required for each type of instructor. A similar constraint is required for buildings.

$$\sum_{j=1}^{m} \sum_{\tau=t+1-sj}^{t} b_{kj} X_{j\tau} \leq SK_{k0} + \sum_{\tau=1}^{t} K_{k\tau} \tag{31}$$

where $b_{kj} = 1$ if activity j uses building type k, 0 otherwise, and SK_{k0} is the available number of student places during the base year. One equation of this type [Eq. (31)] is required for each of the planning years.

$$\sum_{j=1}^{m} X_{jt} \leq \sum_{i=1}^{m} h_i q_i X_{i,t-s_i} \tag{32}$$

This equation is a pupil continuity constraint. The values of $X_{i,t-s_i}$ are admissions, s_i years before, to an activity that feeds into j. The value h_i equals $(1 - e_i)$, and e_i is the net proportion of graduates who continue their educations elsewhere. The scalar h_i may be positive or negative.

$$R_{it} \leq \sum_{i=1}^{m} h_i q_i X_{i,t-s_i} + I_{it} \tag{33}$$

Equation (33) is a teacher supply constraint. The value R_{it} is the number of teachers of type i recruited during time period t, and I_{it} is the number of teachers who must be imported.

Two constraints on funds are developed.

$$\sum_{k=1}^{K} L_{kt} K_{kt} \leq B_{ct} + D_t \tag{34}$$

The first [Eq. (34)] restricts the amount of funds for construction. The value L_{kt} is the cost of constructing a student place of type K in year t, B_{ct} the capital budget funds available in year t from sources other than borrowing, and D_t the amount borrowed in year t.

A second constraint restricts the amount of operating capital.

$$\sum_{j=1}^{m} \sum_{\tau=t+1-s_j}^{t} a_{fjt} X_{jt} + \sum_{i=1}^{m+n} \left[\sum_{\tau=1}^{t} W_{i\tau}(1 - \delta_{i\tau}) R_{i\tau} \right] + \sum_{i=1}^{m+n} W_i \times [S_{i0}(1 - d_i)^{t-1}] - \sum_{i=1}^{m+n} W_i \left[\sum_{\tau=1}^{t} \delta_{i\tau} R_{i\tau} \right] + \sum_{\tau=-18}^{t} r_\tau D_\tau \leq B_{0t} \tag{35}$$

where

a_{fjt} the current annual operating cost of activity j, less teachers' salaries, per student in year t
$W_{i\tau}$ the annual salary of teachers of type i who are initially hired in year t
W_i the average annual salary paid to all teachers of type i,
r_τ the interest rate on bonds initially issued in year t, assumed to be at most of 20-year duration
$\delta_{i\tau}$ a "zero–one" variable such that $\delta_{i\tau} = 0$ if $R_{i\tau} \geq 0$, and $\delta_{i\tau} = 1$ if $R_{i\tau} < 0$
B_{0t} the total amount of operational funds available to the system in year t from local, state, and federal sources, less mandatory debt retirement

One constraint is required for each year. The first part of the equation specifies current annual operating cost per student, excluding teachers' salaries. The second section is the annual wages of teachers recruited during the planning period, where $W_{i\tau}$ is the starting salary. The third part of the expression is the salary of the initial set of teachers. The fourth piece represents savings from teachers who have left the school system. The last indicates the cost of borrowing to pay operating expenditures.

Finally, an equation is defined to represent the fraction of the student population f_{ij} that has the ability to go from activity i to activity j.

$$\sum_{j=1}^{m} X_{jt} \leq \left(\sum_{j=1}^{m} f_{ij}\right) q_i X_{i,t-s_i} \tag{36}$$

Kuch notes that many students do not go to school solely to maximize their earnings. The model will, however, eliminate programs that do not produce high earnings, and it will allocate funds to programs with high net salaries for graduates. He recommends upper- and lower-bound constraints on each curriculum.

The Kuch model was applied to a region in the state of Minnesota. The model identifies the third and fourth years of entertainment training and the third and fourth years of research and design training as subjects that would be eliminated without lower boundary constraints. Courses relating to administration and organization, the first and second year of research and design, and the first and second years of health services are identified as curricula that would receive high funding levels.

Fund allocation applications in the field of health have concentrated on evaluating key alternatives such as cure or prevention and contraception or other methods of family planning. Correa (1967) optimizes the distribution of national funds among preventive care or curative treatment. He compares different indices that could be used to measure the output of the health system. The number of deaths is selected as the objective to be minimized. The optimum allocation between cure and prevention is modeled with an example, including several diseases under both static and dynamic conditions. The results imply that prevention should receive priority.

Correa and Beasley (1971) study the choice of technologies for population and family planning. The linear model includes six methods of contraception and the cost of contraceptive devices and of doctor's visits. With respect to the results, the authors caution the reader to the fact that cost minimization is only one criteria to be weighed in population planning programs.

Lawrence *et al.* (1972) study the allocation of resources to alternative birth control programs in Honduras. Their model seeks to minimize births through contraceptive acceptances subject to a reduction of the crude birth rate to a particular level, annual budget constraints, the availability of doctor's time, available capacity of the laboratories for conducting tests, and the maximum proportion of persons accepting contraceptives.

The model is used to address four important questions. Is it economically feasible to reduce the birth rate to desired levels? If the rate cannot be reduced to a desired level, which resources are constraining? If the desired level can be reached, the addition of which resources will yield the greatest benefits in further birth reductions? Fourth, which age groups will achieve the greatest reductions in birth rates?

The results are quite revealing. For example, they indicate that the ages 25–27 will yield the greatest return on investment in contraceptive devices. About 0.48 births are averted per acceptance in this age group. The values of the dual variables, likewise, are revealing. The number of additional births that can be averted for each cytology lab hour is 3.2. Overall, the study has three major conclusions. Birth rates will not fall to the desired 44 per thousand level by 1980 without a major family planning effort. They will remain at about 49 per thousand. An expenditure of $900,000 for the years 1975–1980 would provide the necessary capital base to reduce the birth rate to 44 per thousand by 1980. Large increases in budgets, doctors' time, and laboratory facilities would be necessary to reduce the birth rate to 42 per thousand by 1980.

Bodin *et al.* (1972) develop a quadratic programming model for mental health services that resembles the education fund allocation models. The most unusual aspect of the paper was that the authors worked closely with the New York State Legislature to develop a politically feasible solution.

Finally, the literature contains a few papers which boldly seek to allocate funds across the spectrum of public services. Moder and Nickl (1969) model the optimal utilization of financial resources with mixed linear and integer programming. They illustrate their simple model with the hypothetical case of an urban renewal project with a $5-million budget. The model is used to test preliminary allocations made to housing, health, and employment. While the model is quite general, the paper is useful because all of the equations are presented.

Bernstein *et al.* (1970) propose one of the most unusual applications of linear programming. They construct equations that interrelate neighborhood changes in housing, educational facilities, health facilities, and social–recreational facilities. The equations specify the impact of changes in any one of the four neighborhood indicators on the other three. Given these fixed equations, a linear programming model is formulated to maximize the net return on investment, subject to constraints on minimum levels for each of the four.

Tabb (1972) presents an interesting perspective on black–white income differentials. The author argues that black–white income differentials in the major metropolitan regions may be lessened, not by residential integration, but rather by concentration on economic development in the black regions of Standard Metropolitan Statistical Areas (SMSAs). Tabb constructs a linear programming model that seeks to minimize costs to consumers of goods and services. Twenty goods and services are defined in the objective function. Nine policy alternatives form a key portion of the objective function. Six transform labor through education, counseling, daycare, and wage subsidies. Three aim at increasing minority group ownership through investment credits, property tax abatement, and low-cost insured loans.

Cost is minimized subject to constraints on labor and capital supplies, government expenditures, demand for goods and services, intermediate inputs to the production processes, and minimum production levels in the black regions. While the model is not tested, Tabb argues that it enables a comparative test of competing programs.

C. CAPITAL ALLOCATION

This section reviews the siting of large, capital structures, the allocation of existing capital facilities among alternative uses, and scheduling. The solid-waste chapter contained linear and mixed linear and integer models that could help select sites for landfills and resource recovery facilities. This portion of the capital allocation section addresses site selection from a broader perspective that includes the location of any public or private facility.

We begin with a region that contains population concentrations linked by a transportation system. The goal is to locate public and private facilities to serve the population. In the literature, these facilities are factories, stores, warehouses, landfills, and often firehouses, schools and medical facilities. These models assume that, since the public must come to the facility or be served by the facility, the facility ought to be located as centrally as possible with respect to the mass of the population. Populations that must rely on mass transit systems are especially sensitive to facility locations. A solution is to locate a sufficient number of facilities so as to make the facilities ubiquitous. However, the ubiquity solution is not economically feasible. Accordingly, the usual goal of the central facility location problem is to site an unknown number of facilities that minimize total population travel time in such a way as to satisfy demands and not violate budgetary restrictions.

One of the first efforts at developing a general approach to central facility location was a linear programming model by ReVelle and Swain (1970). Their objective function minimizes the cumulative distance traveled by the population.

Minimize:
$$Z = \sum_j \sum_i a_i d_{ij} X_{ij} \tag{37}$$

where a_i are the populations of the i communities, d_{ij} is the shortest distance from community i to community j, and X_{ij} are yes or no assignments of entire communities to central facilities.

$$X_{ij} = \begin{bmatrix} 0 \text{ if community } i \text{ does not assign to community } j \\ 1 \text{ if community } i \text{ does assign to community } j \end{bmatrix}$$

A variation in the approach is to maximize the utilization of the facility [Dee and Liebman (1972); Tietz (1967)].

Four types of constraints are defined. One requires that each community is fully assigned.

$$\sum_j X_{ij} = 1 \tag{38}$$

Next, a constraint fixes the number of central facilities.

$$\sum_i X_{ii} = m \tag{39}$$

where m is the number of central facilities.

ReVelle and Swain tie the number of central facilities constraints to a budget constraint. In a later paper, Rojeski and ReVelle (1970) substitute a budget constraint for the fixed facility [Eq. (39)] constraint to deal with the more usual case where costs of operation are not equal at every site.

$$\sum_j f_j X_{jj} + \sum_j b_j \sum_i a_i X_{ij} \leq C \tag{40}$$

where f_j is the fixed cost of opening facility j, b_j a variable cost coefficient for expanding facility j, and C the budget limit.

Third, ReVelle and Swain develop a constraint that eliminates the possibility of cross haulage by restricting assignments to only communities which assign to themselves.

$$X_{ij} + \sum_{\substack{k=1 \\ k \neq j}}^{n} X_{jk} \leq 1 \qquad \begin{array}{l} i = 1, 2, \ldots, n \\ j = 1, 2, \ldots, n \\ i \neq j \end{array} \tag{41}$$

If community i is assigned to a central facility in community j ($X_{ij} = 1$), then community j cannot reassign the people to community k ($X_{jk} \leq 0$ for all k except $k = j$).

Equation (41) is simplified to

$$X_{jj} \geq X_{ij} \qquad \text{and} \qquad X_{jj} - X_{ij} \geq 0 \tag{42}$$

which is a simpler version to use. Next,

$$X_{ij} \geq 0 \tag{43}$$

These four constraints ensure that each community is assigned once to a community which assigns to itself and that the number of self-assigned communities is m or is limited by a total budget.

ReVelle and Swain raise two computational problems. First, fractional assignments are possible. The authors suggest that an integer programming algorithm be used. Second, an advantage of the linear programming formulation is that linear programming solutions are inexpensive. While linear programming is inexpensive to run, the number of constraints is substantial. For example, a region of 20 communities and 6 facilities could require about 800 iterations to reach a solution. One way to reduce the computation is to reduce the number of constraints. The authors suggest that the $X_{jj} \geq X_{ij}$ constraints [Eq. (42)] can be added as needed. This and other papers present evidence to support the conclusion that constraints should be added as needed.

Finally, Rojeski and ReVelle add a constraint which specifies that the "closest center" be used.

$$X_{ij} \geq X_{jj} - X_{ii} \tag{44}$$

for adjacent $i - j$ pairs.

An alternative to the ReVelle and Swain and the Rojeski and ReVelle linear programming models is a mixed-integer programming model. A typical objective function is

$$\text{Minimize:} \qquad Z = \sum_i \sum_j (b_j + d_{ij}) a_i X_{ij} + \sum_j f_j Y_j \tag{45}$$

The first portion of the objective function is the variable cost of production b_j plus transportation d_{ij} multiplied by the demand for the service a_i and the proportion of demand at i that is supplied at facility j. The second segment is the fixed cost of the facility, which is composed of the fixed cost of production at site j multiplied by Y_j, a zero–one location variable.

Four general constraints are required in the simplest form of the mixed-integer model. One is a demand constraint, which is analogous to Eq. (38).

$$\sum_j X_{ij} = 1 \tag{46}$$

Next, a constraint is defined to ensure that the fixed costs of production j are counted.

$$Y_j \geq X_{ij} \tag{47}$$

A third constraint is the zero–one integer constraint.

$$Y_j = [0,1] \tag{48}$$

Finally, a nonnegativity requirement is added.

$$X_{ij} \geq 0 \tag{49}$$

Recent efforts in central facility location have aimed at refining the objective function and introducing applications. Toregas *et al.* (1971), Toregas and ReVelle (1972), and Walker (1974) have reviewed applications to the siting of emergency facilities. Their models seek to minimize the number of facilities required to cover every emergency region within a specified time. The emergency regions would include, for example, firehouses, police dispatch stations, and emergency ambulance depots. The goal of the models is to isolate the tradeoff between assigning new equipment, constructing additional facilities, and budget limitations.

Holmes *et al.* (1972) have used a linear programming model to help locate public daycare centers in a 120-census-block area in Columbus, Ohio. Runs were made specifying from 5 to 45 central facilities and the threshold distance from $\frac{1}{4}$ to 1 mile. Some runs included the locations of existing centers and some disregarded existing facilities.

Trifon and Livnat (1973) site schools in four medium sized towns in Israel (population 8200–16,400). Wagner and Falkson (1975) argue that, with the exception of Holmes *et al.* (1972), the central facility models do not recognize the effect of distance on demand. They develop an objective function that explicitly takes into account the value of the service minus the travel cost. Finally, Austin (1974) sets forth a dynamic programming approach to facility location that concentrates on the social and environmental equity issues implied by location instead of cost–benefit criteria.

Another group of capital allocation models is directed at optimizing the use of existing or projected capital resources. The most common examples are allocating space among different users. Graves and Thomas (1971) develop a linear programming model that maximizes the benefits of classroom size and location choices subject to cost constraints. The authors are given construction costs and budgeted quantities of classroom and lab spaces as constraints. From faculty surveys they derive objective function preference coefficients for different sizes, locations, and arrangements of space.

Maximize:
$$Z = \sum_i \sum_j (p_{ij} X_{ij} + q_{ij} Y_{ij}) \tag{50}$$

where X_{ij} is the number of classrooms of capacity type j for department i that are to be mixed with labs and faculty offices, Y_{ij} the number of classrooms of capacity type j for department i that are to be located in an aggregation of classrooms, p_{ij} a preference coefficient for classroom type X_{ij}, and q_{ij} a preference coefficient for classroom type Y_{ij}.

Two types of constraints are defined. One is for budgets.

$$\sum_i \sum_j (c_{ij} X_{ij} + d_{ij} Y_{ij}) \leq C \tag{51}$$

where c_{ij} is the cost of constructing facility type X_{ij}, d_{ij} the cost of constructing facility type Y_{ij}, and C the total budgeted funds for construction.

A second constraint requires that allocations authorized for each department are made.

$$X_{ij} + Y_{ij} = T_{ij} \tag{52}$$

where T_{ij} is the number of classrooms of category j that are authorized for department i and budgeted for construction.

The authors applied the model to the construction of facilities for the Faculty of Engineering and Applied Science of the State University of New York at Buffalo. Graves and Thomas present tables that demonstrate the major differences involved in different sizes, mixes, and locations of facilities.

The Graves and Thomas model is not unique. McNamara (1973) reports studies by Shapely and others that used linear programming to allocate idle classroom space and by Lawrie to schedule the construction of educational facilities. Dickey and Hopkins (1972) demonstrate how TOPAZ, a heuristic algorithm that uses linear programming, can be used to locate campus buildings.

In health administration, Delon *et al.* (1970) have reviewed the use of optimization and other models to determine the impact of hospital layout and design on efficiency and the care for patients. Gurfield and Clayton (1969) use optimization to study the impact on the operation of a hospital installing a cardiac unit. Menu preparation is a favorite optimization problem. Gue and Liggett (1966) review alternative algorithms for minimizing the cost of feeding patients subject to constraints on nutritional requirements and on menu variety. McNamara (1973) reports that Lutz has used a dynamic programming model to construct a monthly cafeteria menu that meets legal requirements and contains provisions for ethnic foods.

Finally, while optimization applications usually evoke little interest among the general public, I have found instant attention greets my review of a paper by Gould and Magazine (1971). Their highly complex linear and dynamic programming models seek to schedule the insertion of contraceptive devices (IUDs). Specifically, the model indicates the minimal number of IUDs that should be inserted during each period to meet specified birth rates.

D. SAMPLE PROBLEMS

1. A STUDENT ALLOCATION PROBLEM

This section presents an illustration of one of the few simple models in this chapter—a student allocation problem. The region has six pickup districts, and 1020 students. The problem seeks to minimize the distance that the students must travel subject to three types of constraints. The first constraint requires that all the students must be sent to a school [Eqs. (P1.2)–(P1.7)]. The second limits the capacities of the schools [Eqs. (P1.8)–(P1.9)], and the third requires that at least a specified proportion of students from two of the pickup districts must go to one of the schools.

Table I lists the input data, and Fig. 1 is a simple diagram.

Table I Data for Sample Student Allocation Problem

Pickup region	Number of students	Distance d_{ij} and activity number			
		School A	Activity		School B
1	100	$d_{1A} = 1.5$	1	2	$d_{1B} = 3.0$
2	200	$d_{2A} = 0.5$	3	4	$d_{2B} = 1.8$
3	150	$d_{3A} = 1.5$	5	6	$d_{3B} = 1.2$
4	120	$d_{4A} = 1.8$	7	8	$d_{4B} = 2.5$
5	200	$d_{5A} = 1.0$	9	10	$d_{5B} = 1.2$
6	250	$d_{6A} = 1.7$	11	12	$d_{6B} = 0.5$
Total	1020				

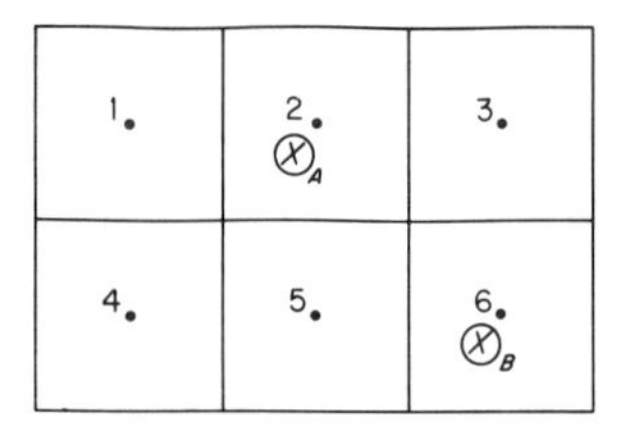

Fig. 1 Diagram of school district (⊗, school; ·, pickup region centroid).

$$\text{Minimize:} \quad Z = 1.5X_1 + 3.0X_2 + 0.5X_3 + 1.8X_4 + 1.5X_5 + 1.2X_6 + 1.8X_7 + 2.5X_8 + 1.0X_9 + 1.2X_{10} + 1.7X_{11} + 0.5X_{12} \quad \text{(P1.1)}$$

subject to six demand constraints

$$X_1 + X_2 = 100 \quad \text{(P1.2)}$$

$$X_3 + X_4 = 200 \quad \text{(P1.3)}$$

$$X_5 + X_6 = 150 \quad \text{(P1.4)}$$

$$X_7 + X_8 = 120 \quad \text{(P1.5)}$$

$$X_9 + X_{10} = 200 \quad \text{(P1.6)}$$

$$X_{11} + X_{12} = 250 \quad \text{(P1.7)}$$

and two capacity constraints

$$X_1 + X_3 + X_5 + X_7 + X_9 + X_{11} \leq 700 \quad \text{(P1.8)}$$

$$X_2 + X_4 + X_6 + X_8 + X_{10} + X_{12} \leq 500 \quad \text{(P1.9)}$$

An integration constraint that requires that a minimum number of the student body from both regions 5 and 6 go to school A. Once your have solved the problem, try to express Eqs. (P1.10) and (P1.11) as percentages.

$$X_9 \geq 40 \quad \text{(P1.10)}$$

$$X_{11} \geq 50 \quad \text{(P1.11)}$$

Solution: $Z = 1031; X_1 = 100; X_3 = 200; X_6 = 150; X_7 = 120; X_9 = 200; X_{11} = 50; X_{12} = 200.$

2. CENTRAL FACILITY SITING EXAMPLE

The following example reflects ReVelle and Swain's (1970) set of equations. The goal is to locate a central facility on the basis of distance minimization. Table II presents the distance matrix.

Table II Distances Between Four Regions and Student Population Size Data

Region	Distances (d_{ij}) 1	2	3	4	Student population (a_i)
1	0	1.5	1.7	2.4	100
2	1.5	0	2.6	1.3	200
3	1.7	2.6	0	1.4	150
4	2.4	1.3	1.4	0	120

Table III Objective Function Weights: $a_i d_{ij}$

Region	1	2	3	4
1	0	150	170	240
2	300	0	520	260
3	255	390	0	210
4	288	156	168	0

Given the weights from Table III, the problem assumes the following activity matrix (Table IV):

Table IV Activity Matrix

Region	1	2	3	4
1	$0X_1$	$150X_2$	$170X_3$	$240X_4$
2	$300X_5$	$0X_6$	$520X_7$	$260X_8$
3	$255X_9$	$390X_{10}$	$0X_{11}$	$210X_{12}$
4	$288X_{13}$	$156X_{14}$	$168X_{15}$	$0X_{16}$

The problem should be run three times. The first time set the number of central facilities m at 1. Then solve with $m = 2$ and $m = 3$.

Minimize:

$$Z = 0X_1 + 150X_2 + 170X_3 + 240X_4 + 300X_5 + 0X_6 + 520X_7 + 260X_8 + 255X_9 + 390X_{10} + 0X_{11} + 210X_{12} + 288X_{13} + 156X_{14} + 168X_{15} + 0X_{16} \quad \text{(P2.1)}$$

subject to

$$X_1 + X_2 + X_3 + X_4 = 1 \tag{P2.2}$$

$$X_5 + X_6 + X_7 + X_8 = 1 \tag{P2.3}$$

$$X_9 + X_{10} + X_{11} + X_{12} = 1 \tag{P2.4}$$

$$X_{13} + X_{14} + X_{15} + X_{16} = 1 \tag{P2.5}$$

$$X_1 + X_6 + X_{11} + X_{16} = 1 \tag{P2.6}$$

In subsequent runs, set the Eq. (P2.6) row constraint to 2 and 3.

$$\begin{aligned} X_1 - X_5 &\geq 0; & X_{11} - X_3 &\geq 0 \\ X_1 - X_9 &\geq 0; & X_{11} - X_7 &\geq 0 \\ X_1 - X_{13} &\geq 0; & X_{11} - X_{15} &\geq 0 \\ X_6 - X_2 &\geq 0; & X_{16} - X_4 &\geq 0 \\ X_6 - X_{10} &\geq 0; & X_{16} - X_8 &\geq 0 \\ X_6 - X_{14} &\geq 0; & X_{16} - X_{12} &\geq 0 \end{aligned} \tag{P2.7}$$

Solutions: if $m = 1$, then $Z = 696$; $X_2 = 1$; $X_6 = 1$; $X_{10} = 1$; $X_{14} = 1$.
if $m = 2$, then $Z = 306$; $X_2 = 1$; $X_6 = 1$; $X_{11} = 1$; $X_{14} = 1$.
if $m = 3$, then $Z = 150$; $X_2 = 1$; $X_6 = 1$; $X_{11} = 1$; $X_{16} = 1$.

3. HEALTH CARE DELIVERY EXAMPLE

The following example is drawn from the Shuman *et al.* (1971) health delivery problem. Even a simplified version of this problem is substantial. Accordingly, to reduce the size of the problem technology, the quantity (m, ϕ_{ijkm}) has been assumed to be constant. Second, changing space requirements (g_k, y_k) have been set at 0.

Minimize:
$$Z = \sum_i \sum_j \sum_k (r_{jk} - v_j) b_{ij} q_{ijk} + \sum_k B_k + \sum_i u_i p_i \tag{3.1}$$

Fees: r_{jk}

	diagnostic	therapeutic
town a	\$20	\$35
town b	\$25	\$40

Not providing service: v_j is \$10.

Annual productivity: b_{ij}

	diagnostic	therapeutic
physicians	1000	600
assistants	500	200

Cost of developing personnel: u_i, physicians \$8000, assistants \$3000.

Net cost of providing service for eight possible activities:

q_i	personnel	function	community	$(r_{jk} - v_j)b_{ij}$
q_1	physician	diagnostic	town a	(20 − 10)1000 = 10,000
q_2	physician	diagnostic	town b	(25 − 10)1000 = 15,000
q_3	physician	therapeutic	town a	(35 − 10) 600 = 15,000
q_4	physician	therapeutic	town b	(40 − 10) 600 = 18,000
q_5	assistants	diagnostic	town a	(20 − 10) 500 = 5000
q_6	assistants	diagnostic	town b	(25 − 10) 500 = 7500
q_7	assistants	therapeutic	town a	(35 − 10) 200 = 5000
q_8	assistants	therapeutic	town b	(40 − 10) 200 = 6000

Final objective function:

Minimize:

$$Z = 10{,}000q_1 + 15{,}000q_2 + 15{,}000q_3 + 18{,}000q_4 + 5000q_5 + 7500q_6 + 7500q_6 + 5000q_7 + 6000q_8 + B_1 + B_2 + 8000p_1 + 3000p_2 \tag{P3.1}$$

$$\sum_{i=2}^{I} q_{ijk} - X_j q_{1jk} \leq 0 \tag{P3.2}$$

Quality of service constraint: Four assistants per physician for diagnostic work; two assistants per physician for therapeutic work:

$$4q_1 - q_5 \geq 0; \qquad 2q_3 - q_7 \geq 0$$

$$4q_2 - q_6 \geq 0; \qquad 2q_4 - q_8 \geq 0$$

$$\sum_i b_{ijk} q_{ijk} \geq d_{jk} \tag{P3.3}$$

Demand for health services:

	diagnostic	therapeutic
town a	25,000	18,000
town b	30,000	30,000

$$1000q_1 + 500q_5 \geq 25{,}000; \text{ diagnostic for town a}$$
$$1000q_2 + 500q_6 \geq 30{,}000; \text{ diagnostic for town b}$$
$$600q_3 + 200q_7 \geq 18{,}000; \text{ therapeutic for town a}$$
$$600q_4 + 200q_8 \geq 20{,}000; \text{ therapeutic for town b}$$

$$\sum_j \sum_k q_{ijk} - p_i \leq P_i \qquad \text{(P3.4)}$$

Supply of personnel: physicians = 100, assistants = 300.

$$q_1 + q_2 + q_3 + q_4 - p_1 \leq 100; \text{ physicians constraint}$$
$$q_5 + q_6 + q_7 + q_8 - p_2 \leq 300; \text{ assistants constraint}$$

$$\sum_i \sum_j w_j b_{ijk} q_{ijk} \leq Y_k \qquad \text{(P3.5)}$$

Supply of space: w_j: space per diagnostic visit: 36.5 ft^2, space per therapeutic visit: 73 ft^2, productivity per day $(b_{ij}/365)$.

	diagnostic	therapeutic
physicians	2.73	1.64
assistants	1.37	.55

$$(b_{ij}/365)(w_j)$$

$$q_1 = (36.5)(2.73) \simeq 100 \qquad q_2 = (36.5)(2.73) \simeq 100$$
$$q_3 = (73)(1.64) \simeq 120 \qquad q_4 = (73)(1.64) \simeq 120$$
$$q_5 = (36.5)(1.37) \simeq 50 \qquad q_6 = (36.5)(1.37) \simeq 50$$
$$q_7 = (73)(.55) \simeq 40 \qquad q_8 = (73)(.55) \simeq 40$$

Available capacity: town a 8000 ft^2, town b 10,000 ft^2.

$$100q_1 + 120q_3 + 50q_5 + 40q_7 \leq 8000$$
$$100q_2 + 120q_4 + 50q_6 + 40q_8 \leq 10{,}000$$

$$\sum_i \sum_j [(C'_{ijk} h_{ijk} - C''_{ijk} f_{ijk}) + (S_{ik} - a_k r_{jk} b_{ijk}) q_{ijk}] - B_k \leq 0 \qquad \text{(P3.6)}$$

Operating Expenses: a_k is set at .4. Of the budget, 40 % is set aside for operating expenses. S_{ik} is personnel support cost. Set at $12,000 for physicians and $5000 for assistants. C'_{ijk}; C''_{ijk}: cost of hiring and firing.

	physicians		assistants	
	diagnostic	therapeutic	diagnostic	therapeutic
hiring	$5000	10,000	3000	6000
firing	1000	1000	600	500

For town a:

$$5000H_1 + 10{,}000H_3 + 3000H_5 + 6000H_7 + 1000F_1 + 1000F_3 + 600F_5 + 500F_7 + 4000q_1 + 3600q_3 + 3000q_5 + 4200q_7 - B_1 \leq 0 \qquad \text{(P3.6)}$$

For town b:

$$5000H_2 + 10{,}000H_4 + 3000H_6 + 6000H_8 + 1000F_2 + 1000F_4 + 600F_6 + 500F_8 + 2000q_2 + 2400q_4 + 0q_6 + 1800q_8 - B_2 \leq 0 \qquad \text{(P3.6)}$$

$$\sum_k B_k + \sum_i u_i p_i \leq C \qquad \text{(P3.7)}$$

Community budget constraint: Set at $900,000.00.

$$B_1 + B_2 + 8000p_1 + 3000p_2 \leq 900{,}000.00$$

$$q_{ijk} - h_{ijk} + f_{ijk} = q_{ijk0} \qquad \text{(P3.8)}$$

Personnel continuity constraint:

	town a		town b	
q_{ijk0}	diagnostic	therapeutic	diagnostic	therapeutic
physicians	10	10	10	10
assistants	20	20	20	20

$$q_1 - H_1 + F_1 = 10 \qquad q_5 - H_5 + F_5 = 20$$

$$q_2 - H_2 + F_2 = 10 \qquad q_6 - H_6 + F_6 = 20$$

$$q_3 - H_3 + F_3 = 10 \qquad q_7 - H_7 + F_7 = 20$$

$$q_4 - H_4 + F_4 = 10 \qquad q_8 - H_8 + F_8 = 20$$

Solution: $Z = \$2{,}543{,}000.00$; $q_1 = 15$; $q_2 = 10$; $q_3 = 23.333$ (note: this should be an integer); $q_4 = 26.667$ (q_4 should also be an integer); $q_5 = 20$; $q_6 = 40$; $q_7 = 20$; $q_8 = 20$; $H_1 = 5$; $H_3 = 13.333$; $H_4 = 16.667$; $H_6 = 20$; $B_1 = 446{,}333.33$; $B_2 = 346{,}666.7$.

E. SUMMARY

This chapter has reviewed the application of optimization models to the allocation of people, funds, and capital equipment resources in health, education, and law enforcement. To date, the principal applications have been in allocating students to schools, personnel to different jobs, funds to educational and health systems, facilities to sites, and space to different users. Other than normal data problems, the most important operational problem facing analysts in the fields of health, education, and law enforcement is defining a single measure of success. Misallocation of limited resources may not only result from errors or from the lack of optimizing behavior, but also from pursuing objectives that cannot be adequately measured by a single indicator of value.

REFERENCES

Adelman, I. (1966). "A Linear Programming Model of Education Planning." *In* "The Theory and Design of Economic Development" (I. Adelman and E. Thorbeck, eds.), pp. 385–417. Johns Hopkins Press, Baltimore, Maryland.

Angel, R., Caudle, W., Noonan, R., and Whinston, A. (1972). Computer-assisted school bus scheduling. *Management Sci.* **18B**(6), 279–288.

Austin, C. M. (1974). "The evaluation of urban public facility location: an alternative to benefit–cost analysis. *Geogr. Anal.* **6**(2), 135–145.

Baligh, H., and Laughhunn, D. (1969). An economic and linear model of the hospital. *Health Serv. Res.* **4**(4), 293–303.

Benard, J. (1967). "General Optimization Model for the Economy and Education." *In* "Mathematical Models in Education Planning." Organization for Economic Cooperation and Development, Paris, France.

Bernstein, S. J., Ahmadi, H., and Rashba, S. (1970). The rehabilitation of urban residential neighborhoods: a policy approach employing linear programming. *Proc. Amer. Stat. Assoc. Bus. Econ. Stat. Sect.*, pp. 254–261.

Bodin, L., Carroll, T., Lee, A., and Stout, S. (1972). Financing mental-health services in the state of New York. *Oper. Res.* **20**(5), 942–954.

Bruno, J. (1969a). A mathematical programming approach to school finance. *Socio-Econ. Planning Sci.* **3**, 1–12.

Bruno, J. (1969b). Using linear programming salary evaluation models in collective bargaining negotiations with teacher unions. *Socio.-Econ. Planning Sci.* **3**, 103–117.

Bruno, J. (1971). An alternative to uniform expenditure deductions in multiple resource state financing programs. *Management Sci.* **17B**, 569–587.

Childs, M., and Wolfe, H. (1972). A decision and value approach to research personnel allocation. *Management Sci.* **18B**(6), 269–278.

Clarke, S., and Surkis, J. (1968). An operations research approach to racial desegregation of school systems." *Socio-Econ. Planning Sci.* **1**(2), 259–272.

Correa, H. (1967). Health planning. KYKLOS **20**(4), 909–923.

Correa, H., and Beasley, J. (1971). Mathematical models for decision-making in population and family planning. *Amer. J. Pub. Health* **61**(1), 138–151.

Dee, N., and Liebman, J. (1972). Optimal location of public facilities. *Naval Res. Logistics Quarterly* **19**(4), 753–759.

Delon, G., and Smalley, H. (1970). "Quantitative Methods for Evaluating Hospital Designs," Final Research Report, National Center for Health Services. National Technical Information Service, Springfield, Virginia.

Dickey, J., and Hopkins, J. (1972). Campus building arrangement using TOPAZ. *Transportation Res.* **6**, 59–68.

Earickson, R. (1970). "The Spatial Behavior of Hospital Patients." Research Paper No. 124, University of Chicago, Dept. of Geography, Chicago, Illinois.

Gass, S. (1975). "Models in Law Enforcement and Criminal Justice." *In* "A Guide to Models in Government Planning and Operations" (S. Gass and R. Sisson, eds.), pp. 231–275. Sauger Books, Potomac, Maryland.

Gould, F., and Magazine, M. (1971). "A mathematical programming model for planning contraceptive deliveries. *Socio-Econ. Planning Sci.* **5**, 255–261.

Graves, R., and Thomas, W. (1971). A classroom location–allocation model for campus planning. *Socio-Econ. Planning Sci.* **5**, 191–204.

Gue, R., and Liggett, J. (1966). Mathematical programming models for hospital menu planning. *J. Ind. Eng.* **17**, 395–400.

Gurfield, R., and Clayton, J. (1969). "Analytic Hospital Planning: A Pilot Study of Resource Allocation Using Mathematical Programming in a Cardiac Unit," RAND Memorandum RM-5893. Rand Corporation, Santa Monica, California.

Heckman, L., and Taylor, H. (1969). School rezoning to achieve racial balance: a linear programming approach. *Socio-Econ. Planning Sci.* **3**(2), 127–133.

Holmes, J., Williams, F., and Brown, L. (1972). Facility location under a maximum travel restriction: an example using day care facilities. *Geogr. Anal.* **4**(3), 258–266.

Holtzman, A. (1968). Linear programming and the value of an input to a local public school system. *Publ. Finance* **23**(3), 429–440.

Koenigsberg, E. (1968). Mathematical analysis applied to school attendance areas. *Socio-Econ. Planning Sci.* **1**(4), 465–475.

Kuch, P. (1975). A Regional model for planning public investment in education. *Regional Sci. Urban Econ.* **5**, 375–397.

Lawrence, C., Mundigo, A., and ReVelle, C. (1972). A mathematical model for resources allocation in population programs. *Demography* **9**(3), 465–483.

Liebman, J., Young, J., and Bellmore, M. (1972). Allocation of nursing personnel in an extended care facility. *Health Serv. Res.* **7**(2), 209–220.

Lyons, N. (1971). "Analytic Models of Criminal Court Operations." Cornell Univ., School of Business and Public Administration, Ithaca, New York.

McNamara, J. (1971). A regional planning model for occupational education. *Socio-Econ. Planning Sci.* **5**, 317–399.

McNamara, J. (1973). Mathematical programming applications in educational planning. *Socio-Econ. Planning Sci.* **7**(1), 19–35.

Moder, J., and Nickl, J. (1969). An optimal allocation of resources for the capital budgeting problem. *Socio-Econ. Planning Sci.* **3**, 135–150.

Newton, R., and Thomas, W. (1969). Design of school bus routes by computer. *Socio-Econ. Planning Sci.* **3**, 75–85.

Noonan, R., and Whinston, A. (1969). An information system for vehicle scheduling. *Software Age*, December, p. 8.

Palmer, B. (1975). "Models in Planning and Operating Health Services." *In* "A Guide to Models in Government Planning and Operations" (S. Gass and R. Sisson, eds.), pp. 347–374. Sauger Books, Potomac, Maryland.

Plessner, Y., Fox, K., and Sanyal, S. (1968). On the allocation of resources in a university department. *Metro-economica* **20**, 258–271.

ReVelle, C. and Swain, R. (1970). Central facilities location. *Geogr. Anal.* **2**(1), 30–42.

Rojeski, P., and ReVelle, C. (1970). Central facilities location under an investment constraint. *Geogr. Anal.* **2**(4), 344–360.

Shuman, L., Young, J., and Naddor, E. (1971). Manpower mix for health services: a prescriptive regional planning model. *Health Serv. Res.* **6**(2), 103–119.

Tabb, W. (1972). Decreasing black–white income differentials: evaluating the evidence and a linear programming framework for urban policy choice. *J. Regional Sci.* **12**(3), 443–456.

Tietz, M. (1967). Toward a theory of urban public facility location. *Papers of the Regional Sci. Assoc.* **21**, 35–51.

Toregas, C., and ReVelle, C. (1972). Optimal location under time or distance constraints, *Papers Regional Sci. Assoc.* **28**, 133–143.

Toregas, C., Swain, R., ReVelle, C., and Bergman, L. (1971). The location of emergency facilities. *Oper. Res.* **19**, 1363–1373.

Trifon, R., and Livnat, A. (1973). The spatial allocation of schools over time in cities. *Regional and Urban Econ.* **2**(4), 387–400.

Wagner, J., and Falkson, L. (1975). The optimal nodal location of public facilities with price-sensitive demand. *Geogr. Anal.* **7**(1), 69–83.

Walker, W. (1974). Using the set-covering problem to assign fire companies to fire houses. *Oper. Res.* **22**, 275–277.

Warner, D., and Prawda, S. (1972). A mathematical programming model for scheduling nursing personnel in a hospital. *Management Sci.* **19**(4), 411–422.

Yeates, M. (1963). Hinterland delimitation: a distance minimizing approach. *Professional Geogr.* **15**(6), 7–10.

Chapter 9

Intraregional Land Use and Transportation

The optimization models reviewed in this chapter are applicable to a broad spectrum of public policy decisions: housing, zoning, intraregional transportation, environmental management, and many others. The comprehensive intent of the large-scale land-use models pushes the optimization concept close to, if not beyond, its limits. The theoretical and informational problems presented in the solid-waste, water resources, and health, education, and law enforcement chapters pale before the problems faced in modeling land use.

With respect to optimization theory, the three previous chapters concentrated on realistic, albeit complicated, efforts to explain and direct the behavior of a limited number of specialized decisionmakers. Water resources and solid-waste managers, law officers, and other public officials should have enough information and the professional responsibility to act in a manner that approaches an optimal solution. The land-use analyst boldly faces the task of explaining the complex human behavior patterns that influence decisions—from siting landfills to high-rise, luxury apartments. Moreover, the public is under no professional obligation to follow closely the modeler's logical solutions. Suffice it to say that many professionals seriously question the use of optimization theory to explain the human behavior mechanisms that have accounted for the entire urban–industrial complex.

With respect to data requirements, the large-scale land-use models demand enormous quantities of accurate data and introduce complicated interrelationships among the variables that are difficult, if not impossible, to eliminate from the tests. Indeed, the complexity of applying optimization models to large-scale land-use planning suggests that this volume should dichotomize between intra- and interregional applications.

On mathematical terms and on the broadest conceptual grounds, a split between intraregional land-use applications and interregional development studies is not warranted. Both the intra- and interregional models, in essence, ask the same policy questions. What types of flows should take place given the present distribution of labor, land, and capital resources? How should or would the flows in the system change if resource constraints were relaxed or were further constricted? Finally, at what rate will the system adjust to geographical shifts in the distribution of resources?

Four major characteristics of intra- and interregional models argue for distinguishing between intra- and interregional location models. Interregional flows typically are measured in long-duration time scales of months and years. The interregional migration of people, liquid capital, and productive capacity is measured by census at infrequent intervals. In contrast, the intraregional model time frame is normally the peak hour, the day, and the week. The critical cycle is the daily workplace commutation. Expensive, special surveys are almost always a requirement.

The role of land at the intra- and interregional scales is markedly different. At the interregional scale, the analyst focuses on important competitive advantages in interregional and international markets, costs of production, and transportation connectivity. At the intraregional scale, land capacity, local financial and infrastructure inducements, labor supply, and environmental considerations are the most important overall variables; while the role of land at both the intra- and interregional levels may be formally equivalent, the wide variety of alternative land-use choices that are at the heart of intraregional analyses are usually irrelevant to the interregional study. In the interregional study, the myriad of land uses within the region become aggregated to a centroid.

At the interregional level, capital is represented by the effect of government expenditures and of industrial location and relocation on development. Data are aggregated to a few general locations. At the intraregional scale, the focus is on capital goods—land, buildings and transportation systems—tied to specific sites. The fiscal and employment impacts are disaggregated to the county or local government level with important local political implications.

A final, important difference between the inter- and intraregional models is their treatment of population. Interregional-scale studies treat people as aggregate units of demand and labor supply. Intraregional analyses usually pivot around the place utility of commuters and the intraregional migration decisions of local residents.

Summarizing, while inter- and intraregional models have a surficial similarity because of similar formal mathematical structures, there are numerous important differences that imply that they should be treated separately. Accordingly, this chapter will focus on intraregional land-use models, and the next chapter will present interregional economic development and transportation applications.

The remainder of the intraregional land-use and transportation chapter is divided into four parts: (1) place utility—location theory studies, (2) criticisms of large-scale land-use optimization models, (3) problem-oriented land-use applications, and (4) examples of intraregional land-use models.

A. PLACE UTILITY—LOCATION THEORY

The goals of the place utility models were to operationalize economic theory and to apply the models to testing the implications of alternative public policies. The first goal was derived from the economic theories developed by Alonso (1960), Von Thunen [in Hall (1966)], Hurd (1903), and Haig (1927)[1]. The availability of the linear programming computer algorithm and the theory of optimization permitted analysts to operationalize theories, some of which had been developed in the early nineteenth and twentieth centuries. In addition to the welcomed possibility of validating theory, the models seemed to offer the hope of replacing "intuitive and artistic concepts" [Schlager (1965)] of land-use design with analytically based designs.

A sample of the real-world applications and aspirations of the researchers may be gleaned from the following remarks of Herbert and Stevens (1960) in their pioneering paper:

> The land use model is applicable to public policy decisions concerning zoning, transportation, redevelopment, public housing, segregation, and other areas of public interest. For example, the model could be run without zoning restrictions. Then restrictions could be applied to determine their effects, if any, on rents and residential patterns. Similarly, the effects of altering transportation costs could be tested by multiple runs of the model.
>
> Where the model gives high rents in areas with little available land and close to existing blighted areas, it may indicate that these areas are ripe for redevelopment. Moreover, the existence of high rent levels would indicate a low cost of writing down the land. Some areas may exhibit

[1] The full development of the theory was presented in Alonso (1964).

extremely low rent levels in comparison with surrounding areas. These may be ripe for redevelopment to raise amenity levels (and reduce amenity costs). Because the model can identify the "savings" that households might enjoy, not only on vacant land but also in partially improved and built-up areas, it should be of interest in problems of conservation and rehabilitation as well as redevelopment. In all of these applications it could be used as one type of cost and benefit analysis.

We may wish to use the model to consider real subsidies to households which have negative rent-paying ability in all or most areas. Tests could be made to determine whether it is more effective to raise these households' location budgets through direct subsidies (thereby shifting them into another household group) or to reduce their location cost through indirect subsidies (e.g., through public housing). The dual variables, together with knowledge of the total and locational budgets of such households, can provide a measure of how large such subsidies should be.

Notice that the model can recognize racial segregation and similar policies if they have measurable dollar consequences for rent-paying abilities. It can do this in much the same way as it recognizes the amenity value of particular areas. For example, we can ascribe to a predominantly white area high "amenity" costs for particular types of Negro household seeking particular locational bundles.

While Herbert and Stevens and others pointed out some of the limitations of their models, their goal was to rationally allocate land on the basis of economic principles of cost minimization or value maximization, subject to constraints on man-made and natural resources. As originally formulated by Herbert and Stevens (1960), modified by Harris (1962), and then modified by Senior and Wilson (1974), the economic land-use models assume the following general form:

Maximize:
$$Z = \sum_{k=1}^{u} \sum_{i=1}^{n} \sum_{j=1}^{m} X_{ijk}(b_{ij} - C_{ijk}) \qquad (1)$$

where X_{ijk} is the number of households of group i using housing j located in area k, b_{ij} the value of housing j to group i, and C_{ijk} the cost to household group i of housing j in area k. The model maximizes the net value of the housing minus location to the buyer. The net difference $(b_{ij} - C_{ijk})$ is the bid rent of the housing site.

Equation (2) is a housing–land capacity constraint:

$$\sum_{i=1}^{n} \sum_{j=1}^{m} S_{ij} X_{ijk} \leq L_k \text{ or } L_j \qquad (2)$$

where S_{ij} is the area occupied by a type j house in zone k, and L_k the area of available residential land or number of units in zone k.

The third constraint requires the model to fulfill the demands for households:

$$\sum_{k=1}^{u} \sum_{j=1}^{m} X_{ijk} = H_i \tag{3}$$

where H_i is the number of households to be located:

$$X_{ijk} \geq 0 \tag{4}$$

Finally, no negative assignments are permitted.

The dual problem is to

Minimize: $$Z = \sum_{k=1}^{u} r_k L_k + \sum_{i=1}^{n} V_i N_i \tag{5}$$

where r_k is the annual rent per housing site, and V_i the annual subsidy per household for all households of group i. The value V_i can be either a surplus value to the consumer or a deficit that must be paid by a subsidy. Normally, V_i is assumed to be zero. The V_i weight is necessary because each constraint in the primal problem requires a dual variable. The u land constraints [Eq. (2)] in the primal are associated with the u variables r_k. The n demand constraints [Eq. (3)] correspond to the n variables V_i. The requirement that all households be located requires the subsidy variables V_i, which are mathematical devices that may or may not have economic meaning.

Constraint (6) keeps the unit rent above the rent-paying ability of the potential applicant.

$$S_{ij} r_k - V_i > b_{ij} - C_{ijk} \tag{6}$$

Finally, one constraint prevents the rent value from being negative

$$r_k \geq 0 \tag{7}$$

and another allows the subsidy to be either positive or negative.

$$V_i \gtrless 0 \tag{8}$$

The original Herbert and Stevens (1960) model defined L_k in Eq. (2) as available land. Harris included structures on the land as well as the land (L_j). A second important difference between the three models is nonresidential land use, which was not included in the Herbert and Stevens model. Harris (1962) and Senior and Wilson (1974) included nonresidential land uses that compete with residential land for locations.

Finally, Harris (1962) and Senior and Wilson (1974) explicitly include transport costs in the objective function. The addition helps widen the interpretation of the dual to include the value of a unit–site combination.

In the solid-waste, health, education, law enforcement, and water resources chapters, the value of the dual variable is a unit–site–technology combination.

The land-use allocation models should produce a rent surface for a region. Given a map of a region, one could sketch general contours, use shading, or develop three-dimensional surfaces that reflect the unit rent. A single-centered region with uniform terrain and transportation access appears with a single-peaked center. The high-rent peak slopes downward with distance from the center. The more usual multiactivity-centered region with an irregular transportation system and terrain presents clear discontinuities. Multiple peaks and ridges representing, respectively, employment nodes and high-value lands along transportation arteries should be apparent.

The pioneering works of Herbert and Stevens and of Harris have been modified and extended. For example, Henderson (1968) and Stevens (1968) developed linear programming formulations of Von Thunen's and other agricultural land use theories. Stevens (1961) and Von Boventer (1961) tied together location rent and the linear programming transportation problem. More recent studies pivoting around the transportation problem and land use are by Morrill (1967), Ochs (1969), Wheeler (1966, 1970), Brown and Longbrake (1969), and Brown *et al.* (1970). The Ochs, Wheeler, and Brown studies will be reviewed in some detail.

Ochs (1969) implicitly proposes parametric programming to test the impact of alternative transportation network designs on place utility. The model minimizes the combination of housing and travel costs subject to land availability restrictions, employment requirements, and the capacity of the transportation system. Different values of place utility are obtained by relaxing the network capacity. In essence, tradeoffs are obtained between land value and network capacity.

The model formulation is restricted to a predetermined configuration of the transport network, and test data are not presented in the paper. Real-world applications would seek a broader emphasis, including basically different configurations for the transportation network (rail, high-speed, limited access) and alternative distributions of residences and jobs.

Wheeler (1970) tests the hypothesis that higher income households have longer work trips. He develops a linear programming model to determine the lengths of work trips for seven employment categories, six racial–sex combinations, and a male and female dichotomy. The objective function seeks to minimize the commuting distance.

Minimize:
$$Z = \sum_{i=1}^{m} \sum_{j=1}^{n} D_{ij} X_{ij} \qquad (9)$$

where D_{ij} is the travel distance from a residential zone i to a working zone j, and X_{ij} the number of commuters.

One constraint recognizes the labor supply R_i limit at the residential zones.

$$\sum_{j=1}^{n} X_{ij} = R_i \tag{10}$$

A second requires the number of commuters to fill the employment categories E_j.

$$\sum_{i=1}^{m} X_{ij} = E_j \tag{11}$$

Finally, negative commuting is prohibited

$$X_{ij} \geq 0 \tag{12}$$

Data on 4200 work trips for testing the transportation model were obtained from the Tulsa Metropolitan Area Planning Commission. Separate runs were made for each of the 15 different occupation, race–sex and, sex categories noted. Each of 15 objective function values (Z's) is divided by the number of commuters to obtain a mean minimum commuting distance S.

$$S = Z/\sum X_{ij} \tag{13}$$

Wheeler correlated the rankings of the S values with rankings of occupational status and obtained a significant Speaman rank correlation coefficient of 0.64. He concludes that in Tulsa occupational groups reside in accordance with the transport-related hypotheses of land–rent theory.

While the groups tended to minimize their work trips according to their economic status, actual travel behavior could not be accurately modeled by the linear programming transportation approach. The major reason for the failure to accurately model work trips on the basis of distance minimization is the spread of employment to the suburbs. Those in lower status occupations, such as craftsmen and service workers, engage in long-distance reverse commuting, while at the time of the Tulsa survey the higher status professional and manager occupations remained oriented to the central business district. In short, the changing distribution of jobs in the metropolitan region could not be modeled accurately by only a distance-sensitive model.

Brown *et al.* (1970) attempted to define the concept of place utility for intraurban migrants. The authors attempt to model intraurban migration patterns in Cedar Rapids, Iowa, using linear programming.

Minimize:

$$Z = \sum_{i} \sum_{j} C_{ij} X_{ij} \tag{14}$$

where X_{ij} are the migrant households from origin nodes i to destination nodes j, and C_{ij} the costs of migration:

subject to

$$\sum_j X_{ij} \geq m_i \tag{15}$$

Equation (15) requires that all the intraurban migrants be allocated to a new region.

$$\sum_i X_{ij} \leq V_j \tag{16}$$

Equation (16) requires that the supply of vacancies be recognized.

$$\sum_i m_i = \sum_j V_j \tag{17}$$

Equation (17) makes the number of migrant households equal to the number of household vacancies.

$$X_{ij} \geq 0 \tag{18}$$

The dual problem is

Maximize:

$$Z = \sum_j V_j W_j - \sum_i M_i U_i \tag{19}$$

This objective function seeks to maximize the difference between the value of the households in the destination $V_j W_j$ and origin $M_i U_i$ places

subject to

$$W_j - U_i \leq C_{ij} \tag{20}$$

The constraint requires that the value difference of relocating a household is less than the cost of migrating from i to j. The values of W_j and U_i are indicators of the place utility associated with, respectively, a present household at i and a new household at site j.

The C_{ij} values that were calculated from the Cedar Rapids, Iowa, survey data for five neighborhoods represent a combination of a typology of socioeconomic and demographic characteristics weighted by a distance decay function. In essence, the model minimizes the migration distance to neighborhoods that migrants are likely to choose to inhabit. The model test results indicate that too many migrants are allocated to the same zone or the immediately adjacent zones. Brown *et al.* suggest that the model could be improved by recursive linear programming runs. The time intervals would be reduced to 1 month, and migrants would be selected randomly throughout the year. Alternatively, the shadow prices derived from the linear programming model could be used as an input to a probabilistic (Monte Carlo) model. The Monte Carlo approach would define migration rules and allow probabilistic fluctuations to occur in recognition of the complexity of the migration decision.

In recent years, the linear optimization model has given way to nonlinear alternatives, and the early American dominance in the development of large-scale land-use models has passed to other English-speaking, largely European analysts. For example, Wilson (1969, 1971), Senior and Wilson (1974), and Wilson and Senior (1974) have redefined some of the linear programming formulations and have developed more elementary residential models. Senior and Wilson (1974) combined the Lowry (1964) model, which allocates residences around workplaces with a gravity model with Lagrangian multipliers. The Senior and Wilson model was compared with a linear programming formulation for Leeds, England, with similar results. Abadie (1970), Erlenkotter (1972), Noland and Sovereign (1973), Lundqvist (1973), Dickey *et al.* (1972, 1974), Gordon and MacReynolds (1974), Patterson (1972), and Courtney *et al.* (1972) have proposed or developed specialized, heuristic, or nonlinear programming models that may be perceived by some as second-generation land-use models. Lundqvist (1973) develops a quadratic programming model that explicitly considers the interactive impacts of transportation and land-use plans. The interacting model is tested for the aggregated regions within the Stockholm region from 1970 to 1985. Gordon and MacReynolds (1974) devised integer linear programming and quadratic assignment models for land-use assignments. The models are tested for 27 regions and 27 activities in the Los Angeles metropolitan region.

Two of these new approaches—a heuristic algorithm called TOPAZ and a goal programming approach—will be reviewed in some detail. TOPAZ is among the most interesting of the second-generation land-use models. [See Dickey *et al.* (1972, 1974). The acronym TOPAZ stands for *T*echnique for the *O*ptimal *P*lacement of *A*ctivities in *Z*ones. TOPAZ was developed by Brotchie, Sharpe, and Toakley of the Commonwealth Scientific and Industrial Research Organization and applied to Melbourne, Australia. The goal of the heuristic optimization model was to determine how future land uses in the Melbourne region should be designed and arranged to minimize transportation, water supply, and sewer system costs. TOPAZ uses cost–benefit techniques, gravity models, and a nonlinear objective function. A review of TOPAZ will help demonstrate the heuristic approach of many of the new models. The notation used is as follows:

X_{ij}	amount of activity i allocated to zone j, acres
E_{ij}	existing amount of activity i in zone j, acres
A_i	future amount of activity i to be allocated, acres
B_j	area available for development in zone j, acres
C_{sij}	unit establishment benefits or capital costs for service s for activity i in zone j, dollars/acre

C_{ij} total establishment costs–benefits for locating activity i in zone j, dollars/acre
PR_i daily vehicular trip production rate for activity i, vehicles/day/acre
AT_i daily vehicular trip attraction rate for activity i, vehicles/day/acre
S_L speed over link L, mph
L_L length of link L, miles
P_{jk} set of links on the minimum time path from zone j to k,
T_{jk} minimum highway travel time from zone j to k, min
M_{jk} distance over minimum highway travel time path from zones j to k, miles
d number of repetitions of daily trips in a year
y length of planning horizon, years
pm_L vehicular cost to travel over link L, dollars/mile
Z sum total of all travel costs and establishment cost–benefits, dollars
Z' value of the objective function of the linear "transportation problem," dollars
K_{jk} cost over the planning period for a repetitive trip from zones j to k, dollars/daily trip.

First, the establishment costs and benefits for each service s for activity i in zone j in dollars per acre are summed. Second, the minimum time paths between zones are obtained by observation. Using the minimum time paths, the travel distances and times between zones are found by adding the link distances. Next, a gravity model incorporated into TOPAZ is used to calculate travel costs over each minimum path for each year through the planning period.

TOPAZ minimizes the combination of travel costs and net establishment costs. The objective function is

$$\text{Minimize:} \quad Z = \sum_i \sum_j C_{ij} X_{ij} + \sum_j \sum_k K_{jk} \sum_i PR_i (X_{ij} + E_{ij}) \times \frac{\sum AT_i (X_{ik} + E_{ik})(1/T_{jk}^2)}{\sum_n \sum_i AT_i (X_{in} + E_{in})(1/T_{jn}^2)} \tag{21}$$

The first part of the objective function is total net establishment cost. The second part is the gravity model equation (daily trips between zones j and k multiplied by the travel cost K_{jk}). The term $\sum PR_i(X_{ij} + E_{ij})$ adds the allocated amount of activity of type i in zone $j(X_{ij})$ to the existing activity E_{ij}.

Total activity $(X_{ij} + E_{ij})$ is then multiplied by the daily trips for the activity PR_i, which yields the number of trips produced by the activity in the zone. All trips produced by all activities in the zone are then summed to obtain total trip production for each zone j. The trip attraction terms $\sum_i AT_i(X_{ij} + E_{ij})$ are developed in the same manner.

Upon calculation of the productions and attractions, the gravity model is then used to estimate the number of trips between each pair of zones. The calculation is made by dividing the trip productions for zone j according to the trip attractions and squared travel times of a zone k relative to all other zones.

The first constraint requires that all of the future activities are allocated.

$$\sum_{j} X_{ij} = A_i \tag{22}$$

The second constraint requires that all of the available land is used.

$$\sum_{i} X_{ij} = B_j \tag{23}$$

The following constraint is redundant but is needed by the computer algorithm.

$$\sum_{i} A_i = \sum_{j} B_j \tag{24}$$

Finally, no negative activities may be allocated.

$$X_{ij} \geq 0 \tag{25}$$

Equation (21) is nonlinear and cannot be easily solved. TOPAZ requires the user to give an initial intuitive solution. Equation (21) is then assumed to be linear. The linear version is solved as a linear programming model. The X_{ij} values from the first iteration are substituted into Eq. (21) and create another linear programming problem. This process continues until the minimum value for Z is obtained.

Dickey *et al.* (1974) have borrowed and applied TOPAZ to Blacksburg, Virginia (a town of 22,000 in 1970). They report that the cost minimization criteria appear reasonable, and the land-use allocations validate the existing development. On the other hand, the cost models leave a good deal to be desired and implementation mechanisms are weak.

Like TOPAZ, goal programming is a clever variation of linear programming. Goal programming is oriented to solving multiple conflicting objectives [Charnes and Cooper (1961), McGrew (1975), Lee (1972)]. An optimization based solely on a simple objective is avoided by minimizing the deviations between multiple goal levels while subject to a series of constraints. The multiple goals are given priority weightings according to their significance.

A simple example is perhaps the best way of presenting the goals programming approach. A county has m geographic parcels to manage. Each of the m parcels are to receive n different resources. Allocation goals have been established based on past records and county management goals.

Minimum assignment requirements are specified. Maximum availability of resources are defined and an overall budget constraint is present. The objective function is keyed to penalties for underbudgeting or overbudgeting resources in any given zone. The penalties vary by type of resource and by zone.

The notation used is the following:

X_{ij}	the goal level of resource j established for zone i
MR_i	the minimum resource allocation of resource i to all zones
AR_i	the maximum resource allocation of resource i to all zones
C_{ij}	coefficients used to convert resource j in zone i into a standard unit (e.g., trucks to dollars)
O_{ij}	units of resource j overbudgeted for zone i
U_{ij}	units of resource j underbudgeted for zone i
CO_{ij}	unit cost of overbudgeting resource j in zone i
CU_{ij}	unit cost of underbudgeting resource j in zone i
B_{ij}	unit budget equivalent for resource j in zone i
V	budget value limit

Minimize:

$$Z = \sum_i O_i CU_i + \sum_i U_{ij} CU_{ij} \tag{26}$$

The objective function minimizes the penalty costs for overbudgeting and underbudgeting.

$$\sum_i O_i - \sum U_i \geq MR_i - \sum_i X_i \tag{27}$$

Equation (27) defines the minimum resource allocation requirements.

$$\sum_i O_i - \sum U_i \geq AR_i - \sum_i X_i \tag{28}$$

Equation (28) defines the maximum resource allocation requirements.

$$\sum_j \sum_i (X_{ij} + O_{ij} + U_{ij}) B_{ij} \leq V \tag{29}$$

Equation (29) defines budget limitations.

$$X_{ij}, O_{ij}, U_{ij} \geq 0 \tag{30}$$

The basic difference between a standard linear programming formulation and the goal programming model may be seen in the role of the basis and slack variables. In linear programming the weights of the basis variables (activities) drive the solution. In the goal programming approach the slack variables (which minimize the deviations) drive the basis variables.

This role can be summarized by comparing a matrix formulation of a linear programming and a goal programming problem. The general linear programming problem is the following:

Maximize/Minimize: $$Z = \sum_j C_j X_j \tag{31}$$

subject to

$$\sum_j A_{ij} X_j \pm B_{ij} S_j = B_i \tag{32}$$

and

$$X_j \geq 0 \tag{33}$$

where A_{ij}, B_{ij}, and C_j are matrices or vectors of constants, and the X_j's the activities. The term B_{ij} is the constant matrix of the slack variables, and the S_j's the slack variables. In contrast, the general goals programming model is the following:

Maximize/Minimize: $$Z = \sum_j C_j S_j \tag{34}$$

subject to

$$\sum_j A_{ij} X_j \pm B_{ij} S_j = B_i \tag{35}$$

and

$$X_j, S_j \geq 0 \tag{36}$$

The objective function of the linear programming model [Eq. (31)] optimizes the activities, while Eq. (34) pivots around the slack variables. Goal programming may be viewed as one possible solution to the problem of multiple objectives in large-scale land-use optimization models. Frankly, however, this writer believes that a goal programming model would only serve to further complicate large-scale models, which are already too complex. On top of including complex interactions among transportation, place utility, and economic development in the model, a goal programming model incorporates subjective judgments about the penalty of over- or underassigning people, housing units, and transportation facilities. Goal programming appears to be more realistic in the context of small, problem-oriented land-use applications.

Overall, the second generation of large-scale land-use models seem to have added mathematical flexibility, realism with respect to the budgets for modeling studies, and a new European audience. These improvements, however, may not counteract the major criticisms that have been made against the large-scale land-use models.

B. CRITICISMS OF LARGE-SCALE LAND-USE OPTIMIZATION MODELS

In recent years, lashing out at large-scale land-use models has come into vogue. For example, Lee (1973) describes these models as "dinosaurs [which have] collapsed rather than evolved." When viewed with the perfect 20/20 hindsight that many authors possess, the large-scale land-use models have not met the lofty management goals set forth by their proponents. On the other hand, it was not an accident that they evolved, and we have learned from their gradual demise.

Beginning in the late 1950s, two phenomena came together to generate the large-scale land-use model. One was the development of computer hardware that could quickly perform large numbers of iterations. As a result, the application of modeling techniques to a broad range of society's problems, rather than to a few military and business applications, became feasible. Second, the United States manifested sufficient concern about urban development problems to provide money for the answers to important questions such as Why do some cities grow while others decay? and What will future intraregional land use patterns be? In short, while skepticism and cynicism toward the large-scale models are products of the 1970s, the models were products of a period when important advances in hardware and software and concern with urban problems coincided. Moreover, as the third section of this chapter will demonstrate, the death of most of the bigger species of land-use dinosaurs, while dramatic, does not signal the death of all of the species; some can and have thrived.

This section reviews the weaknesses of the large-scale land-use models. The immediate goal is to isolate their important weaknesses. A corollary goal is to prepare a foundation for the third section to demonstrate that many of these weaknesses can be overcome if the models are applied to narrowly defined land-use problems.

The large-scale land-use models were undermined by their attempts to simultaneously be faithful to classical economic theory and to regional planning needs. The results were models that were too general, too dependent on some weak theoretical bases and weak data sets, and too costly. The trap of validating numerous, existing partial theories was probably the most important trap to avoid because it forced the model builders to try to be too comprehensive. Trying to operationalize the intraregional land-use theories of Von Thunen, Hurd, Haig, Alonso and of others have led to the following set of operational problems:

(1) Too many variables (activities) and parameters (constraints) are required.

(2) Many of these variables and parameters are interactive, a fact that is difficult, if not impossible, to remove from the models.

(3) Numerous nonlinear relationships have to be expressed as linear because of the complexity of nonlinear solution algorithms.

(4a) A huge amount of data on the current state of the urban system is required, much of which is not available. This information has to be collected by survey or replaced by dubious substitutes.

(4b) Alternatively, all the myriad of interactions in the intraregional system are superficially characterized by a few variables and parameters.

(5) A large number of forecasts about the projected state of the regional economy are used as exogenous inputs to the intraregional models. The models and data bases used to project these exogenous parameters leave a good deal to be desired.

(6) The large-scale land-use models are sufficiently complex to all but eliminate the possibility of public participation and even agency participation in their formulation.

(7) The output from most of the models is not of the temporal scope and geographical scales required by the numerous private groups and public agencies that can use the projections. Moreover, the output is sufficiently complex so as to render the results and the limitations of the results beyond the clear comprehension of nontechnical audiences. Many users faithfully accepted the model estimates; others completely rejected them. All too few users use the estimates while recognizing their inherent limitations. Summarizing, the large-scale land-use models tried to accomplish too many goals at the same time. The result was that few potential users could be satisfied with the results.

The operational problems are exacerbated by a weak theoretical base. Two of the most frustrating findings for a group interested in modeling a phenomena are that the underlying data bases and theoretical foundations of the model rest on weak supports. A paucity of data is so common as to be accepted as a rule of thumb. The analyst struggles to find substitutes or is forced into a costly survey. Budgets are shifted around, and time schedules are bent or fractured.

When, however, a gap in theory is found, the uneasiness of data shortcomings are replaced by the desperation of a major structural gap that must be quickly filled. The previous chapters have pointed to data and knowledge gaps that have hindered the application of optimization techniques to solid-waste, water resources, and law, health, and law enforcement management. The theory gaps facing land-use modelers are legend. Frankly, this writer believes that the state of our knowledge about urban spatial structure is too weak and uneven to support a fully comprehensive land-use model. We know, in general, how often people move and the proportion of these moves that are due to changes in family status, economic status, and other reasons [Rossi (1955); Speare *et al.* (1975)]. Our knowledge and information fall short of being able to model these at small spatial scales. For example, the Herbert–Stevens model requires that the population be classified into homogeneous preference groups. Each preference group is given a unique bid rent portfolio for each housing unit for each location. The preference groups are classified on the basis of statistically independent factors drawn from economic, geographic, and sociological theory: socioeconomic, family, and ethnic status. Or the entire population is classified with a simpler economic index. The single economic index is simple, but of dubious accuracy because the majority of moves are due to changes in family status. The use of multivariate indicators is an improvement. However, Hughes and Carey (1972) and Greenberg and Hughes (1974) have demonstrated that the statistical independence of the multivariate indicators may be an artifact of the statistical analysis method and is related to the geographical scale of the analysis. In cities like Washington, D.C., and Newark, racial dichotomies so dominate that differentiation on the basis of economic and family status factors is doubtful. Additional theoretical problems exist for those who seek to model residential land-use patterns, and a plethora of similar problems face modelers who must also operationalize rules for the siting of industrial, commercial, transportation, and other activities.

In the absence of sound and validated theory, the numerous behavioral aspects of land-use decisions have been operationalized with rules of thumb (for example, distance minimization with respect to workplace drives residential location, and industrial and commercial activities tend to cluster in groups or along linear thoroughfares). While these generalizations are reasonable, it is a long and dangerous step from generalization to operationalizing generalities as deterministic parameters that will drive land-use patterns a decade or two into the future. A single change, such as the increasing cost of auto fuel, or an unforeseen technological breakthrough that changes the location tendencies of industry, may collapse the underpinnings of the complex interactive edifice upon which the large-scale land-use model is based.

Finally, large-scale land-use models have proven extraordinarily expensive. Most of the expense is the cost of obtaining data and transferring the data to a format that is acceptable to the models.

Overall, the opponents of the large-scale land-use models charge that a good deal of money has been spent to obtain verification of the modeler's preconceived notions of urban spatial structure supported by a few data sets rather than the promised objective and scientifically based blending of computer and theory.

C. PROBLEM-ORIENTED LAND-USE APPLICATIONS

The problems of overgenerality, insufficient theory, and great expense can be mitigated by focusing optimization techniques on specific policy problems and by deciding a tradeoff between theoretical elegance and policy needs in favor of policy needs. Following these guidelines will send analysts to specifically defined problems that can be modeled within existing budgets and may be useful to policymakers. The purpose of this section is to review studies that conform to one or both of these guidelines.

Among the more realistic applications of optimization techniques to land-use decisionmaking are models that center on a specific policy problems or land-use types. Illustrations are provided by the work of Day (1970, 1973) and Smiarowski *et al.* (1975) on floodplain management. The authors confine the problem to the allocation of land, subject to restrictions on flood damage. The Smiarowski study is especially well done and will be used as an illustration.

The objection function maximizes the difference between the selling price of the land and the potential flood damage

$$\text{Maximize:} \qquad Z = \sum_{ijkt} (C_{ijkt} - D_{ijkt}) X_{ijkt} \tag{37}$$

where C is the per acre selling price of land, D the per acre expected flood damage, and X_{ijkt} the number of acres of land devoted to activity i in geographical region j in floodplain frequency zone k during time period t.

Six land-use activities i are defined: single-family, apartments, industry, commerce, agriculture, and open space. Three geographical regions j are defined in Farmington, Connecticut. Three flood frequency zones k are developed: 50-year flood (.02 annual probability), 100-year flood (.01 annual probability), and 500-year flood (.002 annual probability). Finally, five time periods t are defined with 5-year time intervals. At each 5-year interval, only undeveloped land is subject to allocation.

Seven types of constraints are defined:

$$\sum_{i} X_{ijkt} \le b_{jkt} \tag{38}$$

where b_{jkt} is the total land available for development in region j and flood frequency zone k at time period t.

$$\sum_{k} X_{ijkt} \le b_{ijkt} \tag{39}$$

where b_{ijkt} are constraints for each land-use activity i in each region j, flood frequency zone k, and time period t.

Next, a population constraint is defined:

$$a_1 \sum_{jk} X_{1jkt} + a_2 \sum_{jk} X_{2jkt} \leq P_t \tag{40}$$

where a_1 is a coefficient expressing the number of people per single family unit, a_2 a coefficient expressing the number of people per apartment unit, and P_t the population estimated by local planners.

Three nonresidential land-use constraints are developed. The first is for commercial development:

$$\sum_{jk} X_{4jkt} \leq b_{4t} \tag{41}$$

where the commercial land use is estimated in relationship to the estimated population growth at each period.

The second is for industrial development:

$$\sum_{jk} X_{3jkt} \leq b_{3t} \tag{42}$$

where b_{3t} is estimated by community leaders.

The third is for open space:

$$\sum_{i=1}^{4} X_{ijkt} \leq .25 \sum_{i=5}^{6} X_{ijkt} \tag{43}$$

where the community defines an open-space requirement on the basis of developed land. Specifically, for every developed acre allocated during optimization (uses 1–4) four acres are required as open space.

Finally, negative assignments are not permitted.

$$X_{ijkt} \geq 0 \tag{44}$$

An extensive parametric programming analysis was made of three of the most dubious parameters: the flood damage coefficient, the flood depth, and the discount rate. Next, additional constraints were added to take into account political and social factors suggested by local leaders. Overall, four different versions of the model were tested and found useful for planning in the Connecticut River Basin.

While the floodplain applications demonstrate feasible land-use applications, even the floodplain applications have three noteworthy limitations. First, the mathematical precision of the optimization model deserves better flood frequency data than is normally available. Frankly, this writer would not use linear programming unless engineering survey methods had been used to measure the channel size and flood frequency. Second, the data for estimating the damages D are rough, especially for commercial and industrial development. Third, the linear objective function may not be representative of the values of the activities which may vary with scale.

The provision of outdoor recreation has become a problem area that has recently attracted a wide range of optimization technique approaches. Meier (1968) develops a linear programming model that maximizes the recreational value of urban park lands and rural forest preserves. The constraints are the budget, the supply of land, and the demand for recreational land. Meier's article is an excellent place to start because he carefully describes the model and presents a small problem that can be transferred to a computer.

Meier's article, however, is deceptively simple. Papers by Hill and Shechter (1971), Menchik (1973), and Gannon (1974) demonstrate the complexity of applying optimization techniques to recreation planning. Hill and Shechter (1971) concentrate on defining an objective to maximize or minimize. They develop the following six indices:

(1) national economic benefits
(2) regional economic benefits
(3) distributional equity index
(4) index of choice
(5) participation index
(6) index of nature preservation

The programming model objective function weights the six goals in a linear combination. A goal programming approach would have been an alternative. The constraints include the budget, other land and labor resources, mutually exclusive restrictions on activities, and project sequencing contingencies (i.e., some projects cannot be undertaken before others are completed). The authors do not offer test data nor are they sure of an algorithm to be used for solving the nonlinear programming model.

Menchik (1973) sets forth a theoretical quadratic programming model for allocating recreational uses while constrained by carrying capacity limits. The objective function maximizes the benefits (measured monetarily) of users subject to constraints on potential demand, equity constraints, and carrying-capacity restrictions. The equity constraints seek to set minimums for lower income citizens. In this writer's opinion, monetary redistribution policies drawn from welfare economics are too peripherally related to recreational use to be of use in ensuring greater access to outdoor recreation for the poor. This writer suggests that the more direct means of providing access to the poor is to allocate funds for purchasing new and expanding existing recreational sites within easy reach of the poor. In short, equity considerations might be better dealt with under facility siting problems. Menchik defines damage coefficients (scalars) related to site characteristics, user–activity type, party size, and length of stay.

Finally, Gannon (1974) develops a probabilistic dynamic programming model for testing and scheduling alternative open-space policies. As in the Hill and Shechter and the Menchik papers, the model is not tested.

Overall, the purpose of reviewing a sample of the recreational land-use literature is to illustrate the complexity of trying to model land uses that are subject to widely varying utility functions. While most public utility managers would agree about the parameters and variables in solid-waste, air quality and water resources studies, many of which are defined by law, analysts must struggle to determine the general public's perceptions of recreational benefits and to convert these into economic terms and demand constraints. Nevertheless, the confinement of the analysis to a single land-use type problem leads the writer to the conclusion that optimization can be used to make reasonable intraregional allocations. Obvious conflicts with other land uses would be screened out prior to the analysis (to be reflected as a land constraint) or after the analysis by performing parametric programming on the land availability constraints.

The water resources and solid-waste management models presented in Chapters 6 and 7 are land-use models insofar as they directly consume land and, more important, insofar as they influence industrial location and other land uses. An assimilative capacity load allocation model applied to a river basin can severely restrict economic development or direct development to specific sites. The equivalent of the water quality load allocation model can be prepared for air and noise quality. Kohn (1970), Seinfeld and Kyan (1971), Werczberger (1974) develop models and provide case studies of air quality load allocation models. In the first two cases, the objective function determines a minimum cost combination of control measures subject to constraints on air quality. Werczberger's mixed-integer model draws directly from the Herbert and Stevens land-use model and therefore seeks to maximize the net value of urban development subject to air quality constraints. As in the case of their water resources counterparts, the authors' goals are to use repeated runs to determine the tradeoff of added cost or decreased value against improvements in air quality. The analogy to the water resources load allocation models are obvious. For example, each source has a maximum daily control cost that is analogous to the upper biochemical oxygen demand (BOD) removal level. The equation structure for an air quality allocation model can be prepared from the load allocation equations in the water resources chapter. Indeed, this writer finds it difficult to believe that no references to the water quality literature are to be found in the air quality papers. Second, and most important, the model outputs are only as good as the simulation models that are used to diffuse the effluent into the air. Water quality modeling and verification are more advanced than air quality modeling. Accordingly, state-of-the-art air quality optimization models should be used with considerable caution.

In addition to the three studies overviewed above, Teller (1968) used a linear programming model with the goal of determining the minimum cost of fuels that would not violate air quality standards. Overall, in comparison to water resources and solid-waste management, the applications of optimization techniques to air quality management have been minimal. An explanation appears to be that air pollution analysts have devoted nearly all of their attention to establishing a data base and basic diffusion models. While the applications have so far been limited, the potential use of optimization in air quality and land-use management problems should be anticipated.

The applications to flood plain management, outdoor recreation, and air quality are a few of the many new potential applications of optimization techniques to land-use allocation problems with environmental constraints. Realistic applications to urban development plans are possible in which the focus is inner city development or redevelopment as well as environmental management. Laidlaw's (1972) excellent monograph presents four such applications. After developing a model of urban renewal for a hypothetical town, he overviews four actual studies. One is an intensive-housing development evaluation carried out in Jersey City, New Jersey. The goal was to design a varied housing mix while constrained by building and design standards. A linear programming model of 42 activities (housing types, schools, shopping, and others) by 38 constraints was prepared and tested. The result was a set of alternatives that could be evaluated for the 500-housing-unit project.

A second project focused on a large-scale community development plan for an 8-square-mile area of Baltimore. A linear programming problem of 104 activities (housing, retailing, light industry, schools, and open space) and 66 constraints (fiscal, budgetary, land, housing, market, and facility capacity) was developed to derive optimum and suboptimum solutions.

Laidlaw's third case study was a redevelopment project in a 38-acre area of downtown Baltimore. The activities include six types of housing, two types of schools, open space, four kinds of parking facilities, retail space, and commercial office space. Finally, a fourth practical application sought to identify a system of general aviation airports to service all users in metropolitan Baltimore.

Laidlaw's realistic evaluation of the role of linear programming is illustrated by the following statement:

> More likely than not, the initial Baltimore test area experimentation did not identify the best possible development pattern for that area. But it did show that such a search could be started and then move in the direction of attainable improvement. [An overall development plan for the area] would take time, patience and plenty of communication long after the completion of this particular book.

Laidlaw's attitude is one that can only have a healthy effect on nontechnical audiences perceptions of the use of optimization models. Rather than fitting a problem to a method or a theory, optimal and suboptimal solutions are offered to decisionmakers who seek a set of relatively simple and flexible alternatives that can form the base from which to build in nonquantifiable social, political, and other considerations.

Another key to achieving positive feedback from nontechnical audiences in intraregional land-use applications is flexibility in the definition of objective function activities and constraint measures. Given the sorry state of intraregional land use and related data sets and the quite specific needs of users, many criteria including, but certainty not limited to, any of the following might be optimized:

(1) Minimize construction and/or operating cost.
(2) Maximize the number of users, value to users, or user days.
(3) Minimize the amount or impact of wastes discharged.
(4) Minimize distance traveled or cost of travel.
(5) Maximize the value or productivity of a site.
(6) Minimize the public cost.

Flexibility in aggregating the activities is also critical. The activities that are to be maximized or minimized can be subdivided into numerous subactivities. For example, residential activities can be subdivided into numerous categories ranging from single-family detatched to high rise. Because intraregional land-use planning is much more of an art than planning for the other applications in this volume, the analyst must avoid inflexible and preconceived notions about levels of aggregation. Each subdivision of an activity increases the cost of the study and the potential for obfuscation of the results. A few early meetings with informed private and public participants concerning the choice of activities and optimization criteria may result in a study that will gather a readership instead of dust.

Flexibility in the choice of constraints is also a good idea. Common constraints include the following:

(1) financial, for example, budgetary, mortgage, bonding
(2) land availability
(3) minimum space requirements and siting requirements for particular land uses and facilities
(4) legal standards for the construction and the operation of facilities
(5) requirements for minimum user satisfaction
(6) regulations relating to ownsership and control
(7) environmental standards and limitations
(8) public service capacity limitations

(9) social requirements such as a minimum level of low and moderate income units
(10) demand for a land use type
(11) intrazonal land use relationship constraints

In short, the analyst should not necessarily seek to validate and certainly not venerate theory when faced with a question of public policy and with a public that frequently associates veneration with financially wasteful day-dreaming.

D. SAMPLE PROBLEMS

1. A SAMPLE LAND-USE PROBLEM

The following simple model can be used to illustrate some land-use relationships. The objective function minimizes the cost of development:

Minimize: $$Z = C_1X_1 + C_2X_2 + \cdots + C_nX_n \tag{45}$$

Four constraints are defined. One specifies total demands for each type of land use.

$$S_1X_1 + S_2X_2 + \cdots + S_nX_n = R_k \tag{46}$$

where R_k is the requirement for each land use, and S_n are coefficients that provide for supporting service land requirements (e.g., streets, parking). One equation is required for each land-use type. The second constraint defines maximum land-use limits within a region.

$$S_1X_1 + S_2X_2 + \cdots + S_nX_n \leq L_m \tag{47}$$

where L_m is the maximum amount of land use n permitted in zone m. The third constraint requires that at least a small proportion of particular land uses be developed in each region.

$$X_1 + X_2 + \cdots + X_n \geq PX_n \tag{48}$$

where P is a scalar indicating the required fraction. Finally, negative assignments are not permitted.

$$X_1, X_2, \ldots, X_n \geq 0 \tag{49}$$

Table I provides sample data to test the problem.

Table I Data for Sample Problem

Region	Constraints	Residential	Commercial	Open space	Total
1	Cost	$1000X_1$	$750X_2$	$100X_3$	
	Land use L_m				100
2	Cost	$1500X_4$	$1200X_5$	$150X_6$	
	Land use L_m				100
3	Cost	$700X_7$	$500X_8$	$70X_9$	
	Land use L_m				300
	S (coefficient)	1.2 for X_1, X_4, X_7	1.3 for X_2, X_5, X_8	1.1 for X_3, X_6, X_9	
	R_k	35	28	252	

Minimize:

$$Z = 1000X_1 + 750X_2 + 100X_3 + 1500X_4 + 1200X_5 + 150X_6 + 700X_7 + 500X_8 + 70X_9 \quad \text{(P1.1)}$$

$$1.2X_1 + 1.2X_4 + 1.2X_7 = 35 \quad \text{(P1.2)}$$

$$1.3X_2 + 1.3X_5 + 1.3X_8 = 28 \quad \text{(P1.3)}$$

$$1.1X_3 + 1.1X_6 + 1.1X_9 = 252 \quad \text{(P1.4)}$$

$$1.2X_1 + 1.3X_2 + 1.1X_3 \leq 100 \quad \text{(P1.5)}$$

$$1.2X_4 + 1.3X_5 + 1.1X_6 \leq 100 \quad \text{(P1.6)}$$

$$1.2X_7 + 1.3X_8 + 1.1X_9 \leq 300 \quad \text{(P1.7)}$$

$$1.2X_1 + 1.3X_2 - 1/6(1.2X_7 + 1.3X_8) \geq 0 \quad \text{(P1.8)}$$

$$1.2X_4 + 1.3X_5 - 1/6(1.2X_7 + 1.3X_8) \geq 0 \quad \text{(P1.9)}$$

$$X_1, X_2, X_3, X_4, X_5, X_6, X_7, X_8, X_9 \geq 0 \quad \text{(P1.10)}$$

Solution: $Z = 52977$, $X_2 = 6.05787$, $X_5 = 6.05787$, $X_7 = 29.16667$, $X_8 = 9.42271$, $X_9 = 229.09091$.

This simple problem can be expanded by adding ecological, budgetary, and other constraints. Alternatively, you can change the objective function to maximize or minimize a different criteria. The development costs values may be modified. The required fraction P in Eq. (48) may be changed or eliminated to test the impact of intraregional distribution requirements on cost.

2. A SECOND LAND-USE PROBLEM

The second example expands the types of land-use constraints. The general objective of the problem is to maximize the number of units to be constructed on a 1000-acre area subject to housing demand, water supply, and environmental quality constraints.

(1) *The Objective Function* This function seeks to maximize the number of housing units to be built in the tract.

$$\text{Maximize:} \quad Z = \sum_{i=1}^{2} \sum_{j=1}^{3} \sum_{k=1}^{2} X_{ijk} \tag{50}$$

where i are the number of housing unit types ($i = 1$, one-family units; $i = 2$, two-family units) and j are the three lot sizes available ($j = 1$, $\frac{1}{2}$-acre lot; $j = 2$, 1-acre lot; $j = 3$, 5-acre lot). Values of K are developable levels of zones based on environmental criteria ($K = 1$, zones of limited development potential; $K = 2$, zones of normal development potential).

(2) *Housing Type Allocation* Let us assume that housing demand will be a reflection of developments built during the last 5 years. Each unit type can be expressed as a percentage of all housing.

$$\sum_{i=1}^{2} A_i X_i = U_i \tag{51}$$

where A_i is the proportion of units of type i among units built during the last 5 years, and U_i the number of housing units of type i to be built. These equations are developed algebraically.

$$X_1/(X_1 + X_2) = .60$$

This expression states that 60% of the units must be single-family. It is converted into a more simple linear form as follows:

$$\begin{aligned} X_1 &= .60X_1 + .60X_2 \\ X_1 - .60X_1 - .60X_2 &= 0 \\ .40X_1 - .60X_2 &= 0 \end{aligned}$$

Note that this equation may be used to allocate housing or any land-use mix in any manner desired. One equation is required for each housing type.

(3) *Housing Demand and Acreage Demand* We also know that different land uses normally require different-sized plots. Using local records or a specific plan, we can express these relationships as follows:

$$\sum_{i=1}^{2} \sum_{j=1}^{3} L_j X_{ij} = U_{ij} \tag{52}$$

where L_j is the percentage of housing unit of type i on a lot size j. For example, for single-family units

$$X_{11}/(X_{11} + X_{12} + X_{13}) = .50$$

$$X_{11} = .50X_{11} + .50X_{12} + .50X_{13}$$
$$X_{11} - .50X_{11} - .50X_{12} - .50X_{13} = 0$$
$$.50X_{11} - .50X_{12} - .50X_{13} = 0$$

Separate equations are prepared for each housing type on each lot size.

(4) *Environmental Constraints* These constraints define carrying capacity for water quality, runoff and septic tank discharges, air emissions, and other environmental constraints. The constraints define the number of units that will be accepted as proportions of the available acreage.

$$\sum_{i=1}^{2} \sum_{j=1}^{3} \sum_{k=1}^{2} C_i X_{ijk} \leq Q \tag{53}$$

where C_i are multipliers that express the number of acres each unit will require, and Q is the maximum number of available acres.

For example, let us assume that 300 of the 1000 acres are in the severely constrained development zone. This zone has steep slopes, outcrops, and limited ability for septic systems. Theoretically, neglecting roads and other public facilities for this simple example, 600 single-family units could be constructed on $\frac{1}{2}$-acre lots. However, due to environmental problems, for every $\frac{1}{2}$-acre lot that is sold, another $2\frac{1}{2}$ acres of $\frac{1}{2}$-acre lots will have to go undeveloped. Accordingly, 3 acres are required for each single-family unit on a $\frac{1}{2}$-acre lot in this zone. Table II below presents the numbers that were used to prepare the environmental constraints. For example, the second row represents single-family units on 1-acre lots. Given 1 unit per acre, 300 units could be built. However, the area has environmental limitations that restrict the construction to 60 units, or 1 unit for every 5 acres.

Separate carrying-capacity constraints may be prepared for different environmental factors. I would suggest that, if possible, a single constraint be used to save computation time. Savings in computation time might be gained, however, at the expense of important information.

(5) *Minimum Sewer and Water Service* The township operated water and sewerage service will not expand into the area unless \$.65 million in capital service costs are generated:

$$\sum_{i=1}^{2} S_i X_i \geq 650{,}000 \tag{54}$$

where S_i is the cost of providing municipal sewer and water service to housing type i (S_1 = \$7000 for housing type X_1; S_2 = \$5500 for housing type X_2).

(6) *Average and Peak Water Supply* Assuming expansion of the municipal system, the town will be able to deliver .1 million gallons per day above existing levels during peak periods. Assuming 60 (two-family) to 120 (single-family) gallons per capita per day, three persons per dwelling unit for the single-family units, and two person for the garden apartments, we have the following constraints:

$$\sum_{i=1}^{2} P_i X_i \leq 100{,}000 \tag{55}$$

where P_i is the numbers of gallons consumed on a peak day times the number of people in the unit [$P_1 = 360\,(120 \times 3)$ for housing unit type X_1; $P_2 = 240$ (60×4) for housing unit type X_2]. This same constraint can be written for average demands and for other public services.

$$\text{Maximize:} \quad Z = X_{111} + X_{121} + X_{131} + X_{112} + X_{122} + X_{132} + 2X_{211} + 2X_{221} + 2X_{231} + 2X_{212} + 2X_{222} + 2X_{232} \tag{P2.1}$$

subject to

Housing Demand for Single-Family Units:

$$\frac{X_{111} + X_{121} + X_{131} + X_{112} + X_{122} + X_{132}}{X_{111} + X_{121} + X_{131} + X_{112} + X_{122} + X_{132} + X_{211} + X_{221} + X_{231} + X_{212} + X_{222} + X_{232}} = .80$$

or

$$.2(X_{111} + X_{121} + X_{131} + X_{112} + X_{122} + X_{132}) - .8(X_{211} + X_{221} + X_{231} + X_{212} + X_{222} + X_{232}) = 0 \tag{P2.2}$$

Housing Demand for Two-Family Units

$$\frac{X_{211} + X_{221} + X_{231} + X_{212} + X_{222} + X_{232}}{X_{111} + X_{121} + X_{131} + X_{112} + X_{122} + X_{132} + X_{211} + X_{221} + X_{231} + X_{212} + X_{222} + X_{232}} = .20$$

or

$$.8(X_{211} + X_{221} + X_{231} + X_{212} + X_{222} + X_{232}) - .2(X_{111} + X_{121} + X_{131} + X_{112} + X_{122} + X_{132}) = 0 \tag{P2.3}$$

Single-Family Housing Demand and Acreage Demand
For $\frac{1}{2}$-acre plots

$$\frac{X_{111} + X_{112}}{X_{111} + X_{121} + X_{131} + X_{112} + X_{122} + X_{132}} = .50$$

or

$$.5(X_{111} + X_{112}) - .5(X_{121} + X_{131} + X_{122} + X_{132}) = 0 \quad \text{(P2.4)}$$

For 1-acre plots

$$\frac{X_{121} + X_{122}}{X_{111} + X_{121} + X_{131} + X_{112} + X_{122} + X_{132}} = .30$$

or

$$.7(X_{121} + X_{122}) - .3(X_{111} + X_{131} + X_{112} + X_{132}) = 0 \quad \text{(P2.5)}$$

For 5-acre plots

$$\frac{X_{131} + X_{132}}{X_{111} + X_{121} + X_{131} + X_{112} + X_{122} + X_{132}} = .20$$

or

$$.8(X_{131} + X_{132}) - .2(X_{111} + X_{121} + X_{112} + X_{122}) = 0 \quad \text{(P2.6)}$$

Two-Family Housing Demand and Acreage Demand
For $\frac{1}{2}$-acre plots

$$\frac{X_{211} + X_{212}}{X_{211} + X_{221} + X_{231} + X_{212} + X_{222} + X_{232}} = 0$$

or from Table II and above

$$X_{211} + X_{212} = 0 \quad \text{(P2.7)}$$

For 1-acre plots

$$\frac{X_{221} + X_{222}}{X_{211} + X_{221} + X_{231} + X_{212} + X_{222} + X_{232}} = 0$$

or from Table II and above

$$X_{221} + X_{222} = 0 \quad \text{(P2.8)}$$

For 5-acre plots

$$\frac{X_{231} + X_{232}}{X_{211} + X_{221} + X_{231} + X_{212} + X_{222} + X_{232}} = 1.0$$

Table II Carrying Capacity

Housing type zone	300 acres: Maximum (ignoring public use land)	300 acres: Severely constrained area	300 acres: C_i value	700 acres: Maximum (ignoring public use land)	700 acres: Normal development area	700 acres: C_i value
X_{11}	600	100	$\frac{300}{100} = 3$	1400	1000	$\frac{700}{1000} = .7$
X_{12}	300	60	$\frac{300}{60} = 5$	700	538	$\frac{700}{538} = 1.3$
X_{13}	60	60	$\frac{300}{60} = 5$	140	140	$\frac{700}{140} = 5$
X_{21}	0	0	$\frac{300}{0} = 0$	0	0	$\frac{700}{0} = 0$
X_{22}	300	0	$\frac{300}{0} = 0$	700	350	$\frac{700}{350} = 2$
X_{23}	60	15	$\frac{300}{15} = 20$	140	70	$\frac{700}{70} = 10$

or from Table II and above

$$-X_{211} - X_{221} - X_{212} - X_{222} = 0 \qquad \text{(P2.9)}$$

Environmental Constraints (see Table II)

$$3X_{111} + 5X_{121} + 5X_{131} + 0X_{211} + 0X_{221} + 20X_{231} \leq 300 \qquad \text{(P2.10)}$$

$$.7X_{112} + 1.3X_{122} + 5X_{132} + 0X_{212} + 2X_{222} + 10X_{232} \leq 700 \qquad \text{(P2.11)}$$

Minimum Sewer and Water Service

$$7000(X_{111} + X_{121} + X_{131} + X_{112} + X_{122} + X_{132}) + 5500(X_{211} + X_{221} + X_{231} + X_{212} + X_{222} + X_{232}) \geq 650{,}000 \qquad \text{(P2.12)}$$

Average and Peak Water Supply

$$360(X_{111} + X_{121} + X_{131} + X_{112} + X_{122} + X_{132}) + 240(X_{211} + X_{221} + X_{231} + X_{212} + X_{222} + X_{232}) \leq 100{,}000 \qquad \text{(P2.13)}$$

Solution: $Z = 340.9 \simeq 341$; $X_{131} = 45.4 \simeq 45$; $X_{112} = 113.6 \simeq 114$; $X_{122} = 68.2 \simeq 68$; $X_{231} = 3.6 \simeq 4$; $X_{232} = 53.2 \simeq 53$.

E. SUMMARY

This chapter has stressed the need to concentrate on practical intra-regional land-use applications. The large-scale land-use models have tried to test theory, and support planning objectives and to be comprehensive at the same time. Several technical, theoretical, and financial problems have resulted.

The overall result has been widespread discouragement with land-use models. Narrowly defined, problem-oriented applications have and are likely to continue to be more successful. Their success depends on applying optimization only when necessary and when data permits and on the ability of the analyst to successfully interact with the audience of decisionmakers.

REFERENCES

Abadie, J. (1970). "Application of the GRG Algorithm to Optimal Control Problems." *In* "Integer and Nonlinear Programming" (J. Abadie, ed.), pp. 191–212, North-Holland Publ., Amsterdam.

Alonso, W. (1960). A theory of the urban land market. *Papers Regional Sci. Assoc.* **6**, 149–157.

Alonso, W. (1964). "Location and Land Use." Harvard Univ. Press, Cambridge, Massachusetts.

Brown, L., and Longbrake, D., (1969). "On the Implementation of Place Utility and Related Concepts: The Intra-Urban Migration Case." Ohio State University Department of Geography, portions presented at "Behavioral Models in Geography," 1968 Assoc. Amer. Geogr. Meeting.

Brown, L., Horton, F., and Wittick, R. (1970). On place utility and the normative allocation of intra-urban migrants. *Demography* **7**(2), 175–183.

Charnes, A., and Cooper, W. (1961). "Management Models and Industrial Applications of Linear Programming," Vol. 1. Wiley, New York.

Courtney, J., Jr., Klastorin, T., and Ruefli, T. (1972). A goal programming approach to urban–suburban location preferences. *Management Sci*, **18B**(6), 258–268.

Day, J. (1970). A recursive programming model for non-structural flood damage control. *Water Resour. Res.* **6**(5), 1262–1271.

Day, J. (1973). A linear programming approach to flood plain land use planning areas. *Amer. J. Agr. Econ.* **55**(2), 165–173.

Dickey, J., Leone, P., and Schwartz, A. (1972). Use of TOPAZ for generating alternate land use schemes. *Highway Res. Rec.* **422**, 39–52.

Dickey, J., Anthony, J., Burton, D., Gordon, A., Morgan, K., Peters, R., Jr., and Volpe, M. (1974) "Minimizing Water and Sewer System Costs Using TOPAZ." National Technical Information Service, Springfield, Virginia.

Erlenkotter, D. (1972). Economic integration and dynamic location planning. *Swed. J. Econ.* **74**, 8–18.

Gannon, C. (1974). Optimal intertemporal supply of a public facility under uncertainty. *Regional Urban Econ.* **4**, 25–40.

Gordon, P., and MacReynolds, W. K. (1974). Optimal urban farms. *J. Regional Sci.* **14**(2), 217–231.

Greenberg, M., and Hughes, J. (1974). Analysis of human territoriality: alternative factor analytic methodologies. *New York Statistician* **25**(5), 3–4 and 6–7.

Haig, R. (1927). "Regional Survey of New York and Its Environs." New York City Planning Commission, New York.

Hall, P., ed. (1966). "Von Thunen's Isolated State." Pergamon, New York [translated from 1826 original].

Harris, B. (1962). "Linear Programming and the Projection of Land Uses." Penn Jersey Transportation Study, Paper No. 20, Philadelphia, Pennsylvania.

Henderson, J. (1968). "The Utilization of Agricultural Land: A Theoretical and Empirical Inquiry" *In* "Readings in Economic Geography" (R. Smith, E. Taaffe, and L. King, eds.), pp. 281–293. Rand McNally, Chicago, Illinois.

Herbert, J., and Stevens, B. (1960). A model for the distribution of residential activity in urban areas. *J. Regional Sci.* **2**(2), 21–36.

Hill, M., and Shechter, M. (1971). "Optimal Goal Achievement in the Development of Outdoor Recreation Facilities." *In* "Urban and Regional Planning: London Papers in Regional Science" (A. Wilson, ed.), Vol. 2, pp. 110–120. Pion Press, London.

Hughes, J., and Carey, G. (1972). Factorial ecologies: oblique and orthogonal solutions. *Environment and Planning* **4**, 147–162.

Hurd, R. (1903). "Principles of City Land Values." The Record Guide, New York.

Kohn, R. (1970). A linear programming model for air pollution control: a pilot study of the St. Louis airshed. *J. Air Pollution Contr. Assoc.* **22**, 78–82.

Laidlaw, C. (1972). "Linear Programming for Urban Development Plan Evaluation." Praeger, New York.

Lee, D., Jr. (1973). Requiem for large-scale models. *J. Amer. Inst. Planners* **39**(3), 163–177.

Lee, S. (1972). "Goal Programming for Decision Analysis." Auerbach Publ., Philadelphia, Pennsylvania.

Lowry, I. (1964). "Model of Metropolis." Rand Corporation Study RM-4125, Rand Corporation, Santa Monica, California.

Lundqvist, L. (1973). Integrated location–transportation analysis: a decomposition approach *Regional Urban Econ.* **3**(3), 233–262.

McGrew, J., Jr. (1975). "Goal Programming and Complex Problem-Solving in Geography." *In* "Papers in Geography," Department of Geography, Pennsylvania State University. University Park, Pennsylvania.

Meier, R. (1968). Programming of recreational land acquisition. *Socio-Econ. Planning Sci.* **2**, 15–24.

Menchik, M. (1973). "Optimal allocation of outdoor recreational use in the presence of ecological carrying capacity limitations and congestion effects." *Papers Regional Sci. Assoc.* **30**, 77–96.

Morrill, R. (1967). "The Movement of Persons and the Transportation Problem." *In* "Quantitative Geography: Economic and Cultural Topics" (W. Garrison and D. Marble, eds.), pp. 84–93. Northwestern Univ. Press, Evanston, Illinois.

Nolan, R., and Sovereign, M. (1973). A recursive optimization and simulation approach to analysis with an application to transport systems. *Management Sci.* **18B**, 676–690.

Ochs, J. (1969). An application of linear programing to urban spatial organization. *J. Regional Sci.* **9**(3), 451–457.

Patterson, R. (1972). "Applications of Linear Integer Programming to Problems of Land Use Allocation." National Technical Information Service, Springfield, Virginia.

Rossi, P. (1955). "Why Families Move: A Study in the Social Psychology of Urban Residential Mobility." Free Press, New York.

Schlager, K. (1965). A land use plan design model. *J. Amer. Inst. Planners* **31**(2), 103–111.

Seinfeld, J., and Kyan, C. (1971). Determination of optimal air pollution control strategies. *Socio-Econ. Planning Sci.* **5**, 173–190.

Senior, M. L., and Wilson, A. (1974). Explorations and syntheses of linear programming and spatial interaction models of residential location. *Geogr. Anal.* **6**(3), 209–238.

Smiarowski, J., Willis, C., and Foster, J. (1975). "Flood Plain Land-Use Management: An Application of Operations Research Methodology." University of Massachusetts, Department of Food and Resource Economics, Amherst, Massachusetts.

Speare, A., Jr., Goldstein, S., and Frey, W. (1975). "Residential Mobility, Migration, and Metropolitan Change." Ballinger, Cambridge, Massachusetts.

Stevens, B. (1961). Linear programming and location rent. *J. Regional Sci.* **3**, 15–26.

Stevens, B. (1968). Location models and programming models: the Von Thunen case. *Papers Regional Sci. Assoc.* **21**, 19–34.

Teller, A. (1968). The use of linear programming to estimate the cost of some air pollution abatement policies. *Proc. IBM Sci. Computing Symp. Water Air Resour. Management*, IBM, White Plains, New York.

Von Boventer, E. (1961). The relationship between transportation costs and location rent in transportation problems. *J. Regional Sci.* **3**, 27–40.

Werczberger, E. (1974). A mixed-integer programming model for the integration of air-quality policy into land-use planning. *Papers Regional Sci. Assoc.* **33**, 141–154.

Wheeler, J. (1966). Occupational status and work trips: a minimum distance approach. *Social Forces* **45**, 508–515.

Wheeler, J. (1970). Transport inputs and residential rent theory: an empirical analysis. *Geogr. Anal.* **2**(1), 43–54.

Wilson, A. (1969). Developments of some elementary residential location models. *J. Regional Sci.* **9**(3), 377–385.

Wilson, A. (1971). "Generalizing the Lowry Model." *In* "Urban and Regional Planning" (A. Wilson, ed.), Vol. 2, pp. 121–133. London Papers in Regional Science, Pion Press, London.

Wilson, A., and Senior, M. (1974). Some relationships between entropy maximizing models, mathematical programming models, and their duals. *J. Regional Sci.* **14**(2), 207–215.

Chapter 10

Economic Development and Transportation

Chapter 9 was concerned with intraregional land-use and transportation applications. This chapter focuses on interregional economic development and transportation applications. At the intraregional scale, flows are normally measured in periods of hours to weeks. Interregional studies concentrate on flows that persist for months and years. The geographical setting for the intraregional study is usually a group of local governments. The interregional study aggregates local governments into regional centroids. The movements of people, housing, jobs, and services are the concerns of the intraregional study. The interregional study focuses on the impact of capital flows on economic development. Overall, Chapter 10 presents applications concerned with relatively long-duration flows of capital between regions. The chapter is divided into four parts: (1) economic development, the transportation and transshipment problems, (2) economic development and classical applications of the transportation problem, (3) new applications of mathematical programming for studying transportation and economic development, and (4) the gravity model and linear programming.

A. ECONOMIC DEVELOPMENT: THE TRANSPORTATION AND TRANSSHIPMENT PROBLEMS

The process of extracting, processing, and transporting a commodity leads to economic development in the source and receiving regions. The linear programming transportation problem is usually expressed as a problem of distributing a commodity from a group of spatially separated sources to a set of spatially separated markets in such a way as to minimize total transportation costs.

Minimize:
$$Z = \sum_i \sum_j t_{ij} X_{ij} \tag{1}$$

where t_{ij} is the cost of transporting the commodity from source i to destination j, and X_{ij} the flow of commodities from i to j,
subject to

$$\sum_i X_{ij} \geq D_j \tag{2}$$

where D_j are the demands in the j consuming regions.

$$\sum_j X_{ij} \leq S_i \tag{3}$$

where S_i are the supplies available at source i.

$$X_{ij} \geq 0 \tag{4}$$

so that no negative flows are permitted.

$$\sum_i S_i = \sum_j D_j \tag{5}$$

If constraint (5) is added, Eqs. (2) and (3) become equalities.

Versions of the transportation problem have been used in the solid-waste, the health, education, and law enforcement, and the land-use and transportation chapters. In the solid-waste chapter the t_{ij} value was modified to include not only transportation costs, but also capital and net operating costs. Additional constraints were added to allow for the transshipping and processing of the waste through resource recovery plants. In the health, education, and law enforcement chapter, students were allocated to schools on the basis of distance minimization. In the land-use and transportation chapter, models were presented that allocated residents to subregions on the basis of minimizing their daily commute. This section expands the review of the transportation problem to the transshipment or, as it is sometimes called, the warehouse problem.

The transshipment problem first appeared in the literature in Orden (1956) and most recently has received an excellent treatment by Soufflet (1973). In the transportation problem, commodities go directly from the source to the consumer; in the transshipment problem, an intermediate transfer, typically a warehouse, is required. Figure 1 illustrates the difference between the two problems.

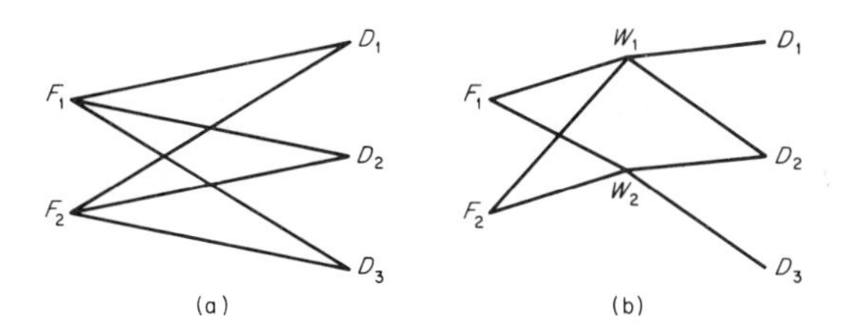

Fig. 1

The most simple version of the transshipment problem seeks to find the least expensive manner of transporting a single, homogeneous product.

Minimize:
$$Z = \sum_j b_j X_{ij} + \sum_{ij} c_{ij} X_{ij} + \sum_{jk} c_{jk} X_{jk} \tag{6}$$

subject to

$$\sum_i X_{ij} \leq P_i \tag{7}$$

$$\sum_j X_{jk} \geq D_k \tag{8}$$

$$\sum_j X_{ij} - \sum_k X_{jk} = 0 \tag{9}$$

where X_{ij} are the goods moved from factory i to warehouse j, X_{jk} the flow of goods from the warehouse j to the buyer k, b_j the costs of the running warehouse j, c_{ij} and c_{jk} the transportation costs, P_i the capacity of factory i, and D_k the demand for products in region k. Equation (7) is the capacity constraint, Equation (8) the demand constraint, and Equation (9) the continuity constraint.

A review of Chapters 6 and 7 will indicate that the warehouse problem is directly analogous to the resource recovery and water network problems. They all have demand, supply, and continuity constraints. The objective functions contain fixed and transportation charges.

The one product warehouse problem may become more complex if any of the following conditions are added: (1) transfer of more than one product, (2) location of facilities for two or more products, and (3) location of facilities for one or more products during more than one time period. Soufflet (1973) reviews these more complex cases. In the multiproduct transfer case he suggests that linear programming is likely to work. However, specialized heuristic algorithms capable of handling this problem are available.

Soufflet argues for the solution of the multiple-product transfer and facility location problem by mixed-integer and linear programming or by heuristic approaches. The reader should note the similarity between these formulations and the approaches suggested for the fixed-facility problem in the solid-waste chapter. Facility siting for future periods is a problem involving dynamic programming and heuristic models that combine integer and linear programming search techniques.

B. ECONOMIC DEVELOPMENT AND CLASSICAL APPLICATIONS OF THE TRANSPORTATION PROBLEM

Applications of linear programming to the solution of interregional transportation problems date back to the early 1950s. A sample of these studies will be reviewed. Fox (1953, 1955) examined the livestock feed industry of the United States. He demonstrated that interregional flows were related to the distribution of demand and of agricultural resources and to price variations. A few years later, Judge and Wallace (1958) published a similar study.

In the mid-1950s, Henderson (1955) analyzed interregional flows of coal. He sought an optimum pattern of coal transfers for the years 1947, 1949, and 1951. The problem included 14 areas of consumption, 11 production areas, and 168 interregional transfer links. Appropriate demand, supply, and cost data were used as right-hand-side values and objective function weights. Land (1957) performed a similar study for coking coal movements over the British railroad system.

Morrill and Garrison (1960) published an analysis of interregional wheat and flour trade patterns in the United States. They used 1954 data to study interregional flows between five regions: the Northeast, Midwest, South, Southwest, and West.

The primal and dual are as follows:

Primal

Minimize:

$$Z_p = \sum_i \sum_j t_{ij} X_{ij} \tag{10}$$

subject to

$$\sum_j X_{ij} = a_i \tag{11}$$

$$\sum_i X_{ij} = b_j \tag{12}$$

$$\sum_j b_j = \sum_i a_i \tag{13}$$

$$X_{ij} \geq 0 \tag{14}$$

where X_{ij} is the set of interregional flows that are to be selected, t_{ij} the transportation costs between the ith and jth region, a_i the surplus of wheat in the ith region, and b_j the deficit of wheat in the jth region.

Dual
Maximize:

$$Z_d = \sum_j b_j u_j - \sum_i a_i v_i \tag{15}$$

subject to

$$u_j - v_i \leq t_{ij} \tag{16}$$

$$u_j \geq 0; \qquad v_i \geq 0 \tag{17}$$

where the u_j and v_i are commodity prices in the exporting and importing regions.

The Morrill and Garrison study produced interesting results, such as the impact of projected population shifts and of a drought on interregional wheat transactions. In general, however, the author would characterize the Morrill and Garrison and other applications of the 1950s and early 1960s as demonstration projects. They demonstrated that variations in supply and demand exist and that they implicitly exert a profound effect on regional economic structure. However, the early studies were limited in size and were restricted in their ability to perform much postoptimal analysis by the lack of automatic computing systems and codes.

By the late 1960s, computers and codes had appeared. The results were some large-scale studies. Howes (1967) developed an interregional transportation model to determine the economic feasibility of water resources investments in the Susquehanna River Basin. The objective function minimizes the total cost of producing and shipping crops and livestock products from sources to market areas.

Four sets of constraints are developed. One restricts the amount of land devoted to agricultural production. A second constructs product mixes that may be used to produce final products. A third set of constraints specifies demands. And a fourth set consists of nonnegativity restrictions.

The study area consists of four agricultural production regions within the Susquehanna River Basin and seven demand regions, including the four regions within the basin. The agricultural land-use pattern estimated by the model is considerably different from the actual basin land-use pattern. Howes concludes that the failure of the linear programming model to be validated is due to the lack of information about how farmers react to opportunities to increase their profits when they face changes in production practices.

O'Sullivan (1972) used a transportation model to predict commodity flows between regions of Great Britain. The problem includes 11 commodities and 78 origin and destination zones. A subset of 24 urban regions was selected from the 78 to serve as a homogeneous sample.

The results support the use of the optimal transportation model as a means of predicting actual interregional transfers. For the 78-region group, the average coefficient of determination between the actual and optimal values was $r^2 = .79$. For the 24-urban-region group, the range of r^2 was .94–.99. O'Sullivan states that the linear programming flows work best for uniform commodity categories such as steel and food and work, worst for heterogeneous products such as miscellaneous manufactures. The linear programming estimates are more accurate when multiple modes of transportation are available and when the regions are homogeneous with respect to population and type of development.

The work of Brokken and Heady (1968) and of Waverman (1972) are the author's two favorite classical interregional studies. Brokken and Heady sought to minimize the national cost of producing and transporting crops and livestock. The United States is divided into 20 consuming regions, 157 crop-producing areas, and 20 livestock-producing regions. Three types of constraints are developed: (1) land supply for pasture and land in different rotation for the 157 crop-producing regions, (2) livestock supply including feed, calfs, yearlings, and hogs for the 20 livestock-producing regions, and (3) consumer demand for fluid milk, manufactured milk products, pork, wheat, food concentrates, soybeans, feeder cattle, two grades of beef, and other products. This enormous study has 4560 activities. The authors carefully used the model to test the impact of different combinations of demand and supply on interregional transactions.

Waverman's (1972) analysis of the economic impact of Canadian restrictions on the use of its western natural gas fields is among the most interesting uses of mathematical programming. Both the Canadian and American governments discourage the flow of many products across their common border by the use of quotas and tariffs. Waverman uses a linear programming model to measure the additional costs borne by the consumers in the two countries because of the added costs of long-distance west-to-east gas shipments in Canada.

First, a free-trade model is written as follows:

Minimize:
$$Z = \sum_i \sum_j c_{ij} X_{ij} \tag{18}$$

subject to

$$\sum X_{ij} \geq D_j \tag{19}$$

$$\sum X_{ij} \leq S_i \tag{20}$$

where c_{ij} is the total cost of supplying natural gas (production at the site and transportation), X_{ij} the flow of gas from field i to market area j, D_j the demand at market j, and S_i the supply at gas field i.

The free-trade problem minimizes the cost of meeting demand subject to demand constraints and gas field capacities. Waverman did not use the pipeline network because it is a manifestation of the Canadian and American governments' restrictive trade policies. If the pipelines had been included, the model would, in essence, be a duplicate of the water supply network model presented in the Chapter 7.

Using a 1966 data base, the free-trade model divides Canada and the United States into north–south trade zones. For example, the western zone contains four Canadian provinces (British Columbia, Alberta, Saskatchewan, and Manitoba) and seven American states (California, Oregon, Washington, Montana, Utah, Wyoming, and Colorado). Transcontinental flows from western to eastern Canada are not identified by the optimal free-trade solution.

The restricted programming problem takes into account the Canadian policy that restricts north–south movements of gas between the United States and Canada and the American response, which is not to encourage the use of Canadian gas in the American northwest, midwestern, and California markets. These trade policies were implemented by splitting Eq. (19) into two parts.

$$X_{kj} \geq D_j \tag{21}$$

where $j = 1, \ldots, h$.

$$\sum X_{ij} \geq D_j \tag{22}$$

where $j = h + 1, \ldots, n$.

Equation (21) requires that gas shipments from kth field in western Canada to the eastern Canadian provinces must at least equal to 1966 interprovincial shipments. Equation (22) requires that demands in the remaining study regions be met.

Nineteen demand and nineteen supply points were chosen. The difference between the free- and the restricted-trade models is sharp. Whereas the free-trade model called for north–south orientations, the restricted trade model recreated the existing west–east Canadian pattern of distributions from Alberta to the eastern Canadian provinces. Of the Canadian gas production, 68% was marketed outside Canada in the free-trade model; only 46% was actually exported in 1966.

Waverman points out that the west–east Canadian constraint affected the pattern of flows throughout the United States (e.g., Louisiana, Texas, and others). The cost of the nationalistic policies amount to about $50,000 per day for the United States and Canada. The Canadians pay 20–30% more for Alberta gas than they would for gas imported through the northeastern United States. Assuming the continuation of restrictive policies for the life of the fields, he calculates that Canada will pay $200 million extra for self-sufficiency in natural gas. The United States, on the other hand, profits to the extent of $18 million because national policies instead of a continental policy allow the U.S. to avoid more expensive shipments to eastern Canada.

C. NEW APPLICATIONS OF MATHEMATICAL PROGRAMMING FOR STUDYING TRANSPORTATION AND ECONOMIC DEVELOPMENT

The transportation and transshipment formats are capable of solving the problem of efficiently moving surplus production to deficit regions. The demands for and supplies of products must be estimated outside the model. During the mid-1960s and 1970s, the emphasis changed from designing the input of the problem to fit the transportation problem to problem-oriented case studies in which the classical transportation model is modified, and to comprehensive planning efforts in which economic development and interregional flows are estimated within a mathematical programming format.

Studies by Casetti (1966), King *et al.* (1971), and Schumaier *et al.* (1974) exemplify the problem-oriented case study. Casetti (1966) developed a linear programming model to determine if and where a steel mill should be located in the St. Lawrence River system of Canada. The objective function seeks to minimize the cost of steel production at four possible sites, the shipment of the steel from one of four plants to markets, the shipment of coal and ore to the plants, and the return of the empty ships. The constraints include demands for steel, raw material requirements at the plants, and ship movement constraints. The model results suggest that as northern Canadian iron ore mines are developed, the number of empty carriers will substantially increase, leading to the justification of a new mill that would utilize raw materials carried by the otherwise empty ore carriers. Summarizing, while transportation costs are critical the linear programming model is not in the classical transportation form.

King *et al.* (1971) examine the efficiency of bituminous coal flows from 15 mining districts to 19 demand areas in the Great Lakes area. The demands represent the requirements of 139 generating stations and 21 steel plants located as far south as southern Kentucky, as far north as the northern Great Lakes, as far east as Harrisburg, Pennsylvania, and as far west as Minneapolis–St. Paul. The authors seek to minimize the shipment cost subject to link capacity and continuity constraints. The authors separate the model runs to fit the seasonal patterns of lake use. Some nodes and links are used to supply one set of regions during the ice-free months and others during the winter months.

The interesting twist to the problem is that it is solved with the out-of-kilter algorithm that was developed by Ford and Fulkerson (1962) and made available by Durbin and Kroenke (1967). This specialized network algorithm has two advantages over a linear programming formulation. First, it is much faster than simplex-based linear programming codes. Second, if costs or constraints are to be changed, the out-of-kilter algorithm has the advantage of being capable of starting with an already existing solution.

In addition to the out-of-kilter algorithm, a number of other specialized algorithms have been developed to solve specific network problems. Several of these problems, the traveling salesman and the Chinese postman problems, have already been mentioned in Chapter 6. Others are the shortest path problem [Dantzig (1963)] and the minimal spanning tree problem [Scott (1971)]. Each of these network algorithms is aimed at saving computation time.

The last of the problem-oriented case studies in which transportation is the critical driving force but in which a specialized model is used is a five-volume study of the economic aspects of petroleum refinery and deepwater port location in the United States. Schumaier *et al.* (1974) developed a heuristic optimization model to determine the relative importance of market, resource, transportation, and technology scale pulls on refinery locations. The authors test different deepwater locations. One of the versions of the model has 10 sources of crude oil, 18 refining centers in the United States, and 484 markets. Another has 13 refining centers, 15 sources of crude oil, and 406 markets. Transportation costs are found to be the driving force. The authors indicate that markets determine where refineries should be. Refineries dictate where deepwater ports should be. Deepwater ports are only partially influenced by overseas sources of crude oil and by economies resulting from the use of supertankers. Suffice it to say that petroleum refineries may be major economic generators of growth.

A second group of applications use mathematical programming as a means of arriving at optimal economic planning goals. Most of these applications are massive multiple-model studies that include input–output models and linear programming as submodels. Input–output analysis was briefly reviewed in Chapter 5. An understanding of the complex programming models that follow requires that the reader understand the mathematics of input–output analysis, of which a brief review follows.

Input–output analysis is a technique for organizing the resource flows that occur in an economy. Interregional flows of products occur because the supply of goods and demand for goods are not equal. Some regions produce vast quantities of many products, other produce little. Some regions are rich in natural resources, others are poor. Each producing unit is characterized by a set of input requirements that is related to technology and to the availability of resources.

Some of the flows are interindustry (steel to automobile plants); others are industry to final demand (automobiles to citizens). Interindustry and industry-to-final-demand flows may be organized as one prepares an orderly accounting sheet. For the moment, assume we are examining a single regional economy.

$$\begin{aligned} x_{11} + x_{12} + \cdots + x_{1n} + F_1 &= X_1 \\ x_{21} + x_{22} + \cdots + x_{2n} + F_2 &= X_2 \\ &\vdots \\ x_{m1} + x_{m2} + \cdots + x_{mn} + F_m &= X_m \end{aligned} \tag{23}$$

The x_{ij}'s are the interindustry flows (iron ore to steel, steel to auto), the F_m column vector is sales to the final consumer, and the X_m vector is the total production of the m industries. The first row indicates that industry 1 consumes some of its own production, sells some to industry 2, to industry n, and to final demand.

The columns tell us each industry's purchases. Industry 1 uses the amount x_{11} from its production, purchases the amounts $x_{21}, \ldots,$ and x_{m1} from other industries.

If we assume that the purchases of each industry are constant, we can make each column element a function of the total output of the industry.

		Industry Use				
		x_1	x_2	x_3	F_i	X_i
	x_1	2	3	2	3	10
Industry	x_2	3	2	1	14	20
Production	x_3	1	1	2	6	10
	x_m	$\vdots$	$\vdots$	$\vdots$	$\vdots$	$\vdots$

In the above matrix, we divide the interindustry shipments by the total outputs to derive a technological coefficient $a_{ij} = x_{ij}/X_i$.

For industry 1

$$\begin{aligned} a_{11} &= x_{11}/X_1 = \tfrac{2}{10} = 0.2 \\ a_{21} &= x_{21}/X_1 = \tfrac{3}{10} = 0.3 \\ a_{31} &= x_{31}/X_1 = \tfrac{1}{10} = 0.1 \end{aligned}$$

For industry 2

$$a_{12} = \tfrac{3}{20} = .15$$
$$a_{22} = \tfrac{2}{20} = .10$$
$$a_{32} = \tfrac{1}{20} = .05$$

The calculations for industry 3 are left to the reader.

The a_{ij} coefficients are invaluable not only as a means of expressing interindustry transactions, but also as a means of converting an accounting sheet into a mathematical model. If we solve for x_{ij} in the expression $a_{ij} = x_{ij}/X_i$, we get $x_{ij} = a_{ij}X_i$. Next, we substitute the $a_{ij}X_i$ values for x_{ij} in Eq. (23). The results are defined by Eq. (24).

$$\begin{aligned} a_{11}X_1 + a_{12}X_2 + \cdots + a_{1n}X_n + F_1 &= X_1 \\ a_{21}X_1 + a_{22}X_2 + \cdots + a_{2n}X_n + F_2 &= X_2 \\ &\vdots \\ a_{m1}X_1 + a_{m2}X_2 + \cdots + a_{mn}X_n + F_m &= X_m \end{aligned} \tag{24}$$

Equation (24) can be mathematically solved as a simultaneous equation set because the number of unknowns equals the number of equations.

Numerous interesting problems have been studied with input–output techniques. For example, a typical application is to determine the impact of a change in final demand F on interindustry flows and total output. In many of the studies that follow, input–output is used to measure the impact of government investments and purchases on regional economies. Please note that this very brief sketch has bypassed the strengths and weaknesses of the huge intput–output literature. The uninitiated reader would be wise to consult some of standard books (see Chapter 5 for references), because input–output analysis in and of itself is a powerful technique for projecting impacts. The six studies that follow are all products of the 1970s. They demonstrate the increasing complexity of the modern wave of interregional programming models.

From a purely mathematical perspective, you may have noticed similarities between the input–output and linear programming equations. Indeed, they are identical with two exceptions. The constraints in the input–output equations are equalities, and the input–output model does not have an objective function.[1]

Golladay and Sandoval (1972) developed a linear programming model from an input–output model of New Mexico. The purpose of the study was to determine New Mexico's best set of responses to changes in export demand, federal government expenditures, and exogenous investments. The authors

[1] See Dorfman *et al.* (1958), Chapters 9 and 10, for the case in which input–output is a simple form of linear programming that has equality constraints and an implicit objective function which seeks maximum production.

use a 20-sector input–output model of New Mexico derived from 1958–1963 data. Their goal is to state economic policies through 1980.

Due to the large number of symbols, Golladay and Sandoval's notation has been retained. Uppercase letters represent activities, and lowercase letters represent parameters. The letter s represents labor skills, and m represents the number of sectors.

1. MODEL PARAMETERS

$X(m \times 1)$	gross state output
$I(m \times 1)$	gross investments
G^{s}(scalar)	level of state and local expenditure
$G^{f}(m \times 1)$	federal government purchases (exogenous)
$E(m \times 1)$	exports
$M^{c}(m \times 1)$	competitive imports
$S^{k}(m \times 1)$	slack capacity
$K_0(m \times 1)$	initial capital stocks (predetermined)
$U(s \times 1)$	unemployment
$L(s \times 1)$	labor supply (exogenous)
F (scalar)	external finance (exogenous)
T(scalar)	transfer payments to households (exogenous)
$E^{u}(m \times 1)$	upper bounds on exports (exogenous)
P(scalar)	state population (exogenous)
$X^{u}(m \times 1)$	upper bounds on production (exogenous)
$\hat{1}$	a vector of ones used to sum vectors into scalar quantities
$a(m \times m)$	input–output coefficients
$b(m \times m)$	input–investment coefficients
$g^{s}(m \times m$, diagonal)	pattern of state and local government expenditures
$k(m \times m$, diagonal)	output–capital ratios
$d(m \times m$, diagonal)	depreciation rates
$l(s \times m)$	labor–output coefficients
$s(1 \times s)$	savings ratios
$m^{x}(1 \times m)$	noncompetitive, intermediate import–export coefficients
$m^{i}(1 \times m)$	noncompetitive, intermediate import–investment coefficients
$v(m \times 1)$	value added per unit of gross output
$e(1 \times m)$	state and local tax rates on gross output
$f(1 \times m)$	federal tax rates on gross output
$c(m \times 1)$	minimum per capita consumption levels
$h(m \times m$, diagonal)	maximum rates of excess capacity

2. MODEL CONSTRAINTS

Resource balance equations

$$X = aX + bI + C + g^s G^s + G^f + E - M^c \qquad (25)$$

Capital constraint

$$X_t + S_t^k = K_0 + k_1 I_1 + \cdots + k_{t-1} I_{t-1} - d_1 X_1 - \cdots - d_{t-1} X_{t-1} \qquad (26)$$

Labor constraint

$$lX + U = \mathrm{L} \qquad (27)$$

Savings–investment constraint

$$\hat{1} I \leq sX + F_t \qquad (28)$$

Interregional resource flows constraint

$$m^x X + m^i I + M^c \leq \hat{1}\{E + G^f\} - fX + F \qquad (29)$$

Bounds on private consumption

$$\hat{1} C \leq (v - s - f - e)X + T; \qquad C \geq cP \qquad (30)$$

Bounds on state purchases

$$\mathrm{G}^s \leq eX \qquad (31)$$

Bounds on exports

$$E \leq E^u \qquad (32)$$

Bounds on production

$$X \leq X^u \qquad (33)$$

Bounds on investment

$$S^k \leq hX \qquad (34)$$

Terminal conditions

$$k_t I_t \geq (1 + d_t)X_t - X_{t-1} \qquad (35)$$

Objective function

$$\text{Maximize:} \quad Z = \hat{1}\{C + bI + g^s G^s + G^f + E - M^c - m^i I - m^u X\} \qquad (36)$$

Equation (25) is a resource balance restriction. Gross state output X must equal the sum of outputs to industries within the state aX, to investments bI, to consumption C, to local and state government purchases $g^s G^s$, to federal government exports G^f, to exports E (not including federal government), and to minus competitive imports M^c.

Equation (26) balances available capital. Current production X_t plus slack capacity S_t^k must equal initial stocks of capital K_0 plus investment capital $(k_1 I_1 + \cdots + k_{t-1} I_{t-1})$ minus depreciation $(d_1 X_1 - \cdots - d_{t-1} X_{t-1})$.

The third equation (27) requires that employed labor lX plus unemployed labor U equals available labor supply L.

The fourth equation [Eq. (28)] requires that investment $\hat{1}I$ minus savings sX be less than available finances F_t.

Equation (29) requires that noncompetitive, intermediate imports–exports $m^x X$, plus noncompetitive intermediate import–investments $m^i X$, plus competitive imports M^c, minus exports E, minus federal government purchases G^f, plus federal taxes on state output fX be less than a set level of external finances F.

The sixth equation [Eq. (30)] has two parts. One requires that consumption $\hat{1}C$ minus net income $(v - s - f - e)(X)$ be less than government transfer payments. The private consumption equation requires that consumption at least equal some minimum level $(C \geq cP)$.

The seventh equation requires that state purchases G^s not exceed state revenues eX. The eighth, ninth, and tenth equations set limits, respectively, on exports E^u, production X^u, and slack capacity hX. The slack capacity requirement means that investments are restricted. The last constraint [Eq. (35)] requires that investments at least meet depreciation.

The objective function [Eq. (36)] maximizes the output of New Mexico's activities for consumption C, investment bI, government expenditures $(G^s + G^f)$, and exports E, minus competitive imports, noncompetitive intermediate import–exports, and import–investments.

The New Mexico programming model contains 342 constraints and 487 activities. The authors developed a trend solution and, through the use of parametric programming, studied the impact on the New Mexico economy of changes in federal government expenditures, externally financed investments, and the demand for exports. The trend solution, with several exceptions, assumes that federal expenditures, investments, and exports will grow at the same rates as the national economy. The major findings of the base level runs is that New Mexico's small internal demand prevents rapid economic expansion.

The effects of a decrease in federal defense expenditures in New Mexico are less than the authors' intuitive expectations. Federal expenditures (largely Vietnam War related) are assumed to be cut 16% between 1970 and 1975 and 31% between 1975 and 1980. The 16% decrease in federal spending leads to a 3.4% decrease in per capita income, and the 31% decrease leads to a 5.6% income reduction. The relatively low impact is due to the recycling of skilled labor into production leading to exports.

The impact of changes in investment by out-of-state residents and in export demand were tested by assuming the worst-case circumstances perceived by New Mexico policymakers. Overall, the Golladay and Sandoval (1972) study illustrates the potential application of linear programming to the numerous input–output matrices that heretofore have been without slack variables an an explicit optimization goal.

The transportation problem has been used for a specific purpose as part of an overall study aimed at forecasting the urban economic structure of the American economy. Hopkins (1972) examined the impact of transportation costs on the American household furniture industry in 1964 and 1966 at the county level. The usual transportation model formulation was used:

Minimize:
$$Z = \sum_i \sum_j t_{ij} X_{ij} \tag{37}$$

subject to

$$X_{ij} \leq S_j \tag{38}$$

$$X_{ij} = D_i \tag{39}$$

$$X_{ij} \geq 0 \tag{40}$$

where t_{ij} is the cost of supplying household furniture from j supply areas to i demand areas, X_{ij} the flow of household furniture, and S_j the supply of household furniture at region j.

The transportation cost coefficients were derived from data supplied by the Interstate Sanitation Commission. Hopkins points out that the household furniture industry is a good industry to analyze. It appears to be highly competitive and potentially mobile, with more than 6000 companies, rather than an oligarchical organization in which industrial location patterns are decided by a few decisionmakers.

Hopkins uses the dual solution to explain employment shifts in the industry. The results support the contention that transportation costs are a meaningful factor affecting industrial location.

Hopkins contention was incorporated into the research design of the multiregional and multiindustry forecasting model [see Harris (1973)]. The forecasting design includes the use of input–output analysis, econometrics, and linear programming. The research design assumes that industrial location is the force that drives regional economic development. A national scale input–output model is used to generate demand, supply, and other control totals for the 219 standard metropolitan statistical areas (SMSAs) that constitute the regions. Econometric models are used to estimate economic characteristics of each SMSA. Linear programming is used to develop transportation costs. Specifically, as in Hopkins (1972), the dual shadow prices are used to represent the transportation costs of shipping and receiving products. Along with the U.S. Environmental Protection Agency's Strategic Environmental Assessment System, the University of Maryland's multiregional and multiindustry study is the most complex interregional economic modeling effort of the decade.

Complex models have also been developed for other nations and regions of the United States. Thoss (1974) developed a complex linear programming model to coordinate regional policies in the Federal Republic of Germany. At best, the tool is intended to help federal planners choose the best overall decisions; at worst, the model would prevent agencies from making decisions that contradict other federal policies. The Thoss paper is worth distinguishing because of the model structure. The model specifies the impact between 1970 and 1985 of projected production on economic characteristics and land use for 38 regions. The linear programming model maximizes the net domestic product of all 38 regions. The equations are ordered so that the output from one run becomes part of the input to the next run. While a linear programming model is used, the author notes that many of the relationships tying together an urban–industrial society's productive sectors are nonlinear. In response to this fact, nonlinear approximations are used and are readjusted at every time period.

Charnes *et al.* (1975) have begun one of the most ambitious American applications of mathematical programming models. They have developed a goal-programming model to evaluate regional economic development on the Texas Gulf Coast. The model allows for explicit consideration of policies at three levels of government (state, county, and local). Accordingly, the impact of local policies on the state economy and state policies on local land-use and environmental characteristics can be evaluated.

At the state scale, population and industrial activity are allocated. At the local government level, housing and industry are allocated, and potable water service, waste treatment, and population distribution occur. A variety of models are incorporated into the formulation, including input–output models for economic development, gravity models for transportation, probability models for aging the population, and mathematical models for water quality. The complexity of the model is increased by its formulation in the goal-programming format. The problem seeks to minimize the weighted sum of deviations from regional growth goals and from unemployment and job vacancies on a local basis.

The final economic development design example [Mastenbroek and Nijkamp (1975)] was prepared for the area around the Rotterdam delta in the Netherlands. It is a linear programming problem in which regional production is maximized. The maximization explicitly includes the cost of environmental damage.

Maximize: $$Z = \sum_r \sum_i V_i^r g_i^r - \sum_r \sum_k C_k^r d_k^r \tag{41}$$

where g_i^r is the production of i manufacturing sectors in region r, V_i^r the value added by production by manufacturing sector i in region r, d_k^r the concentration of pollutant k in region r, and C_k^r the damage caused by pollutant k in region r.

Twelve constraints are defined. Three constraints are drawn from input–output equations. They specify each region's production and the markets for its products. Five constraints are devoted to investments. Two of these five indicate investments in pollution control equipment. Next, one constraint indicates the relationship between pollution emissions and concentrations (see Chapters 7 and 9 for water quality and air pollution examples). A second constraint imposes an upper limit on emissions. Labor demand and supply constraints are written. Finally, nonnegativity constraints are needed because the objective function [Eq. (41)] has some negative weights. The authors also discuss a nonlinear version of the model.

Mastenbroek and Nijkamp's (1975) model is conceptually appealing. However, their case study results and this author's work with economic damage data suggest that damage functions are of dubious accuracy. The best way of using the Mastenbroek and Nijkamp model is to approximate damage functions and run repeated parametric programming tests on the C_k^r values.

One of the great chicken-or-egg problems of regional economics is does the highway cause development or does development necessitate the highway? While we frequently hear hotly contested conversations on the subject, few models were developed to explicitly test the interrelationship until the 1970s. When models such as the models in the previous section were finally created, they included complex interactions between input–output, econometric, and transportation models.

In many countries and regions, these expensive, interactive models are infeasible and indeed may be hitting the proverbial thumbtack with the sledgehammer. In response to the need for relatively simple transportation investment models, some recent literature has explored the question of where should transportation investments go to maximize economic benefits.

In an early paper, Scott (1967) proposed an integer programming model that would seek the minimum distance (transportation network) through sets of nodes (cities). A budget constraint is superimposed on the selection process.

To date, most of the exciting work has been in developing countries. The work of Barber (1975) is exemplary. One of his models is a single-period minimum-operating-cost model. It minimizes the transportation and production costs of meeting demands by a combination of network additions and plant capacity increases while limited by budget restrictions. It uses an input–output model to specify economic interrelationships.

Minimize:

$$Z = \sum_i \sum_j \sum_k \sum_s C^s_{ijk} X^s_{ijk} \tag{42}$$

The objective function minimizes the sum of the transportation and production costs, where C^s_{ijk} is the cost of producing a unit of product k at node i and shipping it over route s to node j, X^s_{ijk} is the amount of product k produced at node i and shipped to node j over route s.

$$\sum_i \sum_j \sum_k \sum_s b^{ls}_{ijk} X^s_{ijk} \leq Z_l K_l \tag{43}$$

The first constraint does not allow any flow over a link unless the link is constructed, $Z = 1$. If the link is constructed, its capacity is set at K. The symbol b_{ijk} is a binary-valued variable that assumes the value of 1 if link l is the sth or shortest route, and 0 if not.

$$\sum_j \sum_s X^s_{jik} - \sum_m a_{km}\left(\sum_j \sum_s X^s_{ijm}\right) - \sum_m \left(Y_{km}\left(\sum_j \sum_s X^s_{ijm}\right) - X_{iom}\right) \geq D_i^k \tag{44}$$

The second constraint is a demand restriction. The production of k at i shipped over s to j ($\sum_j \sum_s X^s_{ijk}$) minus the amounts used for interindustry production and investments must equal the final demand D_i^k requirements at i for product k. In the expression $\sum_m a_{km}(\sum_j \sum_s X^s_{ijm})$, the symbol a_{km} is an interindustry coefficient expressing the per unit input requirements between two products k and m (for example, cents on a dollar of leather per pair of shoes). In the expression $\{-\sum_m [Y_{km}(\sum_j \sum_s X^s_{ijm}) - X_{iom}]\}$, Y_{km} is the amount of product m required to support an increase in the productive capacity of industry k by one unit, and X_{iom} is the initial production of commodity m at node i. The entire expression represents the cost of investing for new productive capacity.

Equation (45) is a budget constraint.

$$\sum_i \left[\sum_k a_i^k\left(\sum_j \sum_s X^s_{ijk}\right) - X_{iok}\right] + \sum_l Z_l F_l \leq B \tag{45}$$

where B is the total amount of available investment funds, a_i^k the increased capital–output ratio for commodity k at node i, X_{iok} the initial production for k at i, and F_l the fixed investment cost of constructing link L. This budget equation sets new investments in transportation links and productive capacity equal to a fixed budget.

The fifth equation requires that production at a node does not decrease.

$$\sum_j \sum_s X_{ijk}^s - X_{iok} \geq 0 \tag{46}$$

Without Eq. (46) the computer can increase investment at one node by decreasing production at another.

The last equations are nonnegativity restrictions and require that $Z = 0$ or 1.

$$X_{ijk} \geq 0; \qquad Z_l = 0 \text{ or } 1. \tag{47}$$

Barber's model is a mixed linear and integer programming model that will simultaneously determine interregional commodity flows, production levels, and approximate network additions. A second model is developed that minimizes the capital investment requirements necessary to meet the same levels of demand used in the first model. Barber notes that the capital-poor situation of many developing countries may make the second model more realistic.

The models are tested using data from the Brookings Institution study of Columbia. The problem consists of 36 nodes, 60 links, over 500 rows, and 1220 variables. The test results lead to two major conclusions. The minimum investment strategy solution leads to higher transportation costs because of the use of more circuitous routes. Some of these costlier routes are not used when more capital is available for new links. The second conclusion is that the existing network in 1956 was important in determining the sites for new investments.

Summarizing, Barber's model is a relatively simple means of explicitly testing the interdependence between new transportation systems and economic development. A modification of it is incorporated into an example later in the chapter.

Ahmed *et al.* (1976) have developed a comprehensive procedure for road investment in developing countries. Using Indonesia as a case study, they use a branch and bound, mixed-integer programming algorithm to determine the minimum network for a developing country.

At this time, the author has not found enough evidence to conclude that the models presented in this section and others like them will be helpful tools or futile mathematical exercises. For such models to succeed, the quality of data must be high, and the decisionmakers will have to be convinced of the efficacy of complex mathematical models. Suffice it to say that neither of these requirements is ensured in most regions.

D. THE GRAVITY MODEL AND LINEAR PROGRAMMING

Another area of considerable interest to students of regional economic planning and transportation is the relative efficiency and accuracy of different mathematical models. Comparisons between the transportation problem of linear programming and the gravity model account for a large proportion of the simulations of people and products between and within regions. In recent years, the literature has reported on the theoretical and empirical similarities of the two techniques.

First, the gravity model will be reviewed. Supposing we wish to estimate the volume of goods that is produced in region i and will be shipped to region j. If we assume that overcoming the distance between regions i and j costs the producer neither money nor time, we can expect that the flow of goods from i to j will depend on the relative size of the market in j. If j has 10% of the multiregional market, then j should be expected to receive 10% of i's production. In general, if the transportation cost is zero, proportional flows F_p between regions should be equal to the ratio $F_p = M_j/M$, where M is the total interregional market and M_j the regional market. Given a representative production unit and a representative consumer, we may estimate the volume of flow between i and j as the multiregional average $\bar{C}$. The volume of flow between a production unit in i and a consumer in j is $\bar{C}(M_j/M)$. If 10% of the market resides in j, production unit i will ship 10% of its products to j. For example, if 50 units are produced in i, then 5 units will be shipped to j.

Thus far we have been discussing the case of one production unit in i. If there are M_i production units located in i, the flow of goods from i to j will be M_i times the average number of trips from a single unit in i to j.

$$F_{ij} = \bar{C}(M_i M_j/M)$$

where F_{ij} is the flow from production units in i to consumers in j. The same formula can be applied to determine the product flow between any two regions.

To be realistic, the cost of distance must be factored into the estimating procedure. The literature contains exponential weights, usually between -3.0 and -0.5. The average is about -2.0. Normally, the analyst collects data about the actual flows between regions. The symbol A_{ij} represents the actual flow. We divide A_{ij} by the hypothetical flow F_{ij}. We also obtain an indicator d_{ij} that is the distance, time, or cost of shipping from i to j. Next, we plot A_{ij}/F_{ij} and d_{ij} on a logarithmically scaled graph. Normally, an inverse straight line results. The straight line can be converted into an equation.

$$\log(A_{ij}/F_{ij}) = a - b \log d_{ij}$$

where A_{ij} is the actual flow of products from i to j, F_{ij} the estimated flow if distance is zero, a a constant that intercepts the y-axis, and b a constant representing the slope of the equation. With some further transformation, the equation can be used to estimate interregional commodity flows.

Hartwick (1972) compared gravity and linear programming estimates of flows for 80 transportation problems. The distance exponents of the gravity model ranged from 0.0455 to -1.703. None reached the -2.00 value. Of the 80, 68 had the expected negative sign. However, Hartwick found the exponents were not stable. He concluded that the flows generated by the gravity model estimates are "relatively poor proxies for those generated in a transportation cost minimizing system."

Evans (1973) presents a theoretical proof of the similarity of the gravity and transportation models. Gordon (1974) criticizes Hartwick's formulation of the gravity model, reformulates the gravity model in an acceptable form, and discusses his empirical results. Using some data from Ireland, he generates a trading pattern with the gravity model that closely approximates a linear programming solution.

Niedercorn and Moorehead (1974) accept Hartwick's conclusion that the simple commodity flow gravity model is not a good predictor of flows generated by linear programming. However, they find that a modified gravity model predicts flows generated from a revenue-maximizing firm. They argue that there is a fundamental difference between predicting flows when you are the manager of a profit-maximizing firm and when you are a planner, geographer, or economist dealing with aggregate flow data between two or more points that represent the decisions of thousands of people. Clearly, they are raising the basic issue of the optimality assumption.

E. SAMPLE PROBLEM

1. BARBER'S SINGLE-PERIOD MINIMUM OPERATION COST MODEL

Barber (1975) ties the construction of a transport network to the operation of the economic system. The problem is reproduced as in the text with a few changes to make it manageable. The objective function minimizes the production and transportation costs of the outputs of different products.

Minimize:
$$Z = \sum_i \sum_j \sum_k \sum_s C_{ijk}^s X_{ijk}^s \tag{48}$$

$$\sum_i \sum_j \sum_k \sum_s b_{ijk}^{ls} X_{ijk}^s \leq Z_l K_l \tag{49}$$

Constraint (49) limits flows to those links that are constructed.

$$\sum_j \sum_s X_{jik}^s - \sum_m a_{km}\left(\sum_j \sum_s X_{ijm}^s\right) - \sum_m \left[Y_{km}\left(\sum_j \sum_s X_{ijm}^s\right) - X_{iom} \right] \geq D_i^k \tag{50}$$

Constraint (50) forces the amount produced minus the products needed for interindustry production and new investment to be greater than the demand for product K at node $i(D_i^k)$.

$$\sum_i \left[\sum_k a_i^k \left(\sum_j \sum_s X_{ijk}^s\right) - X_{iok} \right] + \sum_l Z_l F_l \leq B \tag{51}$$

This budget constraint [Eq. (51)] limits the investment funds B to investments on productive capacity and roadway construction.

$$\sum_j \sum_s X_{ijk}^s \geq X_{iok} \tag{52}$$

Equation (52) requires that final production at a node i not decrease below its original output

$$X_{ijk} \geq 0; \qquad Z_l = 0 \text{ or } 1 \tag{53}$$

Finally, all output must be greater than or equal to zero, while Z_l must be zero or one. This is a mixed linear and integer problem that we will solve with linear programming. To test this complex model, a 4-node, 2-product (agricultural and manufacturing) system was hypothesized. The system and its potential links are shown in Fig. 2.

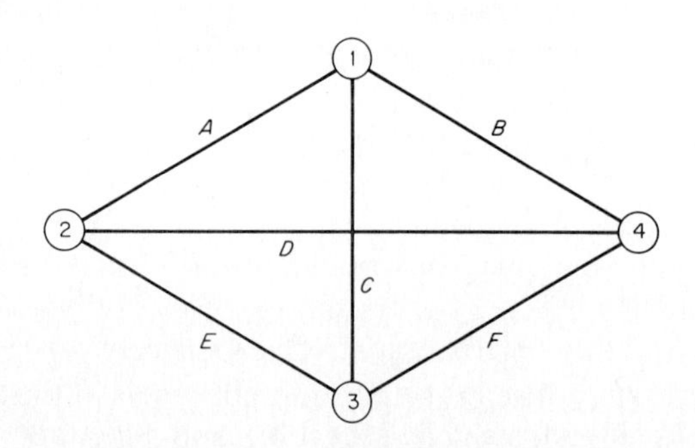

Fig. 2

The demand for product k at node i (in thousand tons) is as follows:

node	D_{agr}	D_{manu}
1	19	0
2	17	24
3	0	16
4	18	22

We add some simplifications:

1. All initial output, X_{iom} [Eq. (50)] is assumed to be 0.
2. All links are assumed to be equal in distance; the value of $\{b_{ijk}^{ls}$ [Eq. (49)] is 1 for all links$\}$.

The budget B is set at $80 million. Three sets of interindustry coefficients are given:

	a_{km} agr	a_{km} manu
agr	.2	.3
manu	.4	.3

	Y_{km} agr	Y_{km} manu
agr	.08	.07
manu	.10	.07

	a_i^k agr	a_i^k manu
1	.100	.150
2	.200	.100
3	.050	.200
4	.100	.125

Production costs ($ million/thousand tons)

	agr	manu
1	100	75
2	50	100
3	100	75
4	50	100

Link costs

Link	Construction costs F_l ($ million)	Transport costs C_l ($ million) 1000 tons)	Link capacity K_l (1000 tons)
A	2	5	23
B	3	6	30
C	7	9	42
D	2	7	18
E	4	5	35
F	8	6	32

The total costs in the objective function, C_{ijk}^s are the sum of the production costs at i plus the transport costs from i to j for product k. For example,

C^A_{12A} will equal the production costs for A agr at 1, \$100, plus the transport costs from 1 to 2 (link A) or \$5. The total cost of option X^A_{12A} is \$105. Using the preceding tables we construct the following objective function.

Minimize:

$$\begin{aligned} Z = {} & 100X_{11A} + 105X^A_{12A} + 109X^C_{13A} + 106X^B_{14A} + 55X^A_{21A} + 50X_{22A} \\ & + 55X^E_{23A} + 57X^D_{24A} + 100X_{33A} + 109X^C_{31A} + 105X^E_{32A} \\ & + 106X^F_{34A} + 56X^B_{41A} + 57X^D_{42A} + 56X^F_{43A} + 50X_{44A} + 75X_{11M} \\ & + 80X^A_{12M} + 84X^C_{13M} + 81X^B_{14M} + 105X^A_{21M} + 100X_{22M} \\ & + 105X^E_{23M} + 107X^D_{24M} + 75X_{33M} + 80X^C_{31M} + 80X^E_{32M} \\ & + 81X^F_{34M} + 106X^B_{41M} + 107X^D_{42M} + 106X^F_{43M} + 100X_{44M} \end{aligned} \tag{P1.1}$$

The objective function has 32 possible activities: 4 agricultural links from each of 4 nodes and 4 manufacturing links from each of 4 nodes.

subject to

$$X^A_{12A} + X^A_{12M} + X^A_{21A} + X^A_{21M} \leq 23Z_A \tag{P1.2}$$

$$X^B_{14A} + X^B_{14M} + X^B_{41A} + X^B_{41M} \leq 30Z_B \tag{P1.3}$$

$$X^C_{13A} + X^C_{13M} + X^C_{31A} + X^C_{31M} \leq 42Z_C \tag{P1.4}$$

$$X^D_{24A} + X^D_{24M} + X^D_{42A} + X^D_{42M} \leq 18Z_D \tag{P1.5}$$

$$X^E_{23A} + X^E_{23M} + X^E_{32A} + X^E_{32M} \leq 35Z_E \tag{P1.6}$$

$$X^F_{34A} + X^F_{34M} + X^F_{43A} + X^F_{43M} \leq 32Z_F \tag{P1.7}$$

The first constraint sets the total flow along a link as less than or equal to the link capacity K_l. Each link has the potential to receive a flow of agricultural and manufacturing products from the node. The demand constraints are rather substantial even with initial production X_{iom} set at 0. For example,

$$\begin{aligned} & X_{11A} + X^A_{21A} + X^C_{31A} + X^B_{41A} - .2(X_{11A} + X^A_{12A} + X^C_{13A} + X^B_{14A}) \\ & -.3(X_{11M} + X^A_{12M} + X^C_{13M} + X^B_{14M}) - .08(X_{11A} + X^A_{12A} + X^C_{13A} \\ & + X^B_{14A}) - .07(X_{11M} + X^A_{12M} + X^C_{13M} + X^B_{14M}) \geq 19 \end{aligned} \tag{P1.8}$$

which fortunately can be reduced to

$$\begin{aligned} & .72X_{11A} + X^A{}_{21A} + X^C_{31A} + X^B_{41A} - .28(X^A_{12A} + X^C_{13A} + X^B_{14A}) \\ & -.37(X_{11M} + X^A_{12M} + X^C_{13M} + X^B_{14M}) \geq 19 \end{aligned} \tag{P1.8$'$}$$

The other demand equations can be similarly reduced.

$$.63X_{11M} + X^A_{21M} + X^C_{31M} + X^B_{41M} - .5(X_{11A} + X^A_{12A} + X^C_{13A} + X^B_{14A})$$
$$-.37(X^B_{12M} + X^C_{13M} + X^B_{14M}) \geq 0 \qquad \text{(P1.9)}$$

$$.72X_{22A} + X^A_{12A} + X^E_{32A} + X^D_{42A} - .28(X^A_{21A} + X^E_{23A} + X^D_{24A})$$
$$-.37(X^A_{21M} + X_{22M} + X^E_{23M} + X^D_{24M}) \geq 17 \qquad \text{(P1.10)}$$

$$.63X_{22M} + X^A_{12M} + X^E_{32M} + X^D_{42M} - .5(X^A_{21A} + X_{22A} + X^E_{23A} + X^D_{24A})$$
$$-.37(X^A_{21M} + X^E_{23M} + X^D_{24M}) \geq 24 \qquad \text{(P1.11)}$$

$$.72X_{33A} + X^C_{13A} + X^E_{23A} + X^F_{43A} - .28(X^C_{31A} + X^E_{32A} + X^F_{34A})$$
$$-.37(X^C_{31M} + X^E_{32M} + X_{33M} + X^F_{34M}) \geq 0 \qquad \text{(P1.12)}$$

$$.63X_{33M} + X^C_{13M} + X^E_{23M} + X^F_{43M} - .5(X^C_{31A} + X^E_{32A} + X_{33A} + X^F_{34A})$$
$$-.37(X^C_{31M} + X^E_{32M} + X^F_{34M}) \geq 16 \qquad \text{(P1.13)}$$

$$.72X_{44A} + X^B_{14A} + X^D_{24A} + X^F_{34A} - .28(X^B_{41A} + X^D_{42A} + X^F_{43A})$$
$$-.37(X^B_{41M} + X^D_{42M} + X^F_{43M} + X_{44M}) \geq 18 \qquad \text{(P1.14)}$$

$$.63X_{44M} + X^B_{14M} + X^D_{24M} + X^F_{34M} - .5(X^B_{41A} + X^D_{42A} + X^F_{43A} + X_{44A})$$
$$-.37(X^B_{41M} + X^D_{42M} + X^F_{43M}) \geq 22 \qquad \text{(P1.15)}$$

The budget constraint is

$$.1(X_{11A} + X^A_{12A} + X^C_{13A} + X^B_{14A}) + .15(X_{11M} + X^A_{12M} + X^C_{13M}$$
$$+ X^B_{14M}) + .2(X^A_{21A} + X_{22A} + X^E_{23A} + X^D_{24A}) + .1(X^A_{21M} + X_{22M}$$
$$X^E_{23M} + X^D_{24M}) + .05(X^C_{31A} + X^E_{32A} + X_{33A} + X^F_{34A}) + .2(X^C_{31M}$$
$$+ X^E_{32M} + X_{33M} + X^F_{34M}) + .1(X^B_{41A} + X^D_{42A} + X^F_{43A} + X_{44A})$$
$$+ .125(X^B_{41M} + X^D_{42M} + X^F_{43M} + X_{44M}) + 2Z_A + 3Z_B + 7Z_C + 2Z_D$$
$$+ 4Z_E + 8Z_F \leq 80 \qquad \text{(P1.16)}$$

Since initial production is 0, the fourth constraint merely requires that all production output be greater than 0. Finally, since Z must be less than 1, and cannot immediately be set equal to 0 or 1, the following constraint is necessary:

$$\begin{array}{ll} Z_A \leq 1; & Z_D \leq 1 \\ Z_B \leq 1; & Z_E \leq 1 \\ Z_C \leq 1; & Z_F \leq 1 \end{array} \qquad \text{(P1.17)}$$

The amount of each product, either agriculture or manufacturing, produced in one node and shipped to another is shown. A few patterns are apparent. The primary determinant seems to be production costs. The product is produced where these costs are less and shipped to more expensive nodes. The transport costs determine the specific shipping node (goods move along cheaper links). For intranodal products, goods are produced where production costs are cheaper (agriculture was produced and remained in zones 2 and 4; manufacturing was produced in zones 1 and 3). The exact magnitude of each output appears to be a function of the link capacity and demand equations. The reasons for the much greater output at node 4 (158.3 manufacturing and 124.2 agricultural) are not readily apparent.

The demands have all been met, and the budget has been used up entirely. A number of links have been constructed to carry the output. Links, A, B, and E have all been fully constructed, $Z = 1.0$, since their entire capacity was necessary. Only .73066 of link F's capacity was necessary. The program reflects this by attempting to construct only .73 of link $F - Z_F = .73066$. Since a link can be either constructed or not with $Z_F = 0$ or 1, some adjustments will be necessary. This is where integer programming is relevant.

Solution: $Z = \$35{,}175{,}516$; $X^A_{21} = 8.96$; $X^A_{22} = 28.70$; $X^A_{23} = 4.14$; $X^A_{41} = 22.62$; $X^A_{43} = 23.38$; $X^A_{44} = 124.26$; $X^M_{11} = 12.58$; $X^M_{12} = 14.04$; $X^M_{14} = 7.38$; $X^M_{32} = 30.86$; $X^M_{33} = 43.52$; $X^M_{44} = 158.33$; $Z_A = 1$; $Z_B = 1$; $Z_E = 1$; $Z_F = .73066$.

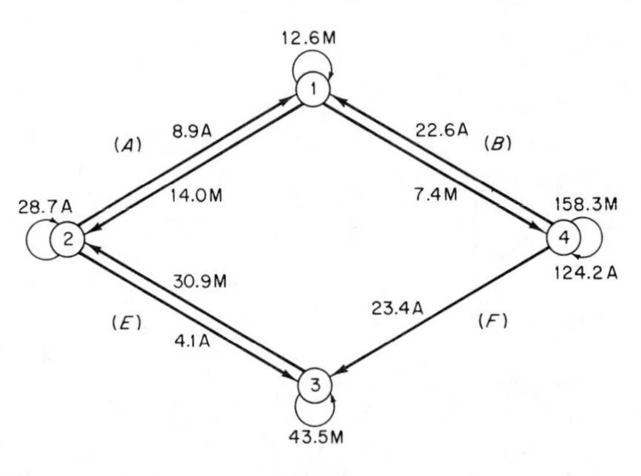

Fig. 3

Because we are using linear programming as a substitute for integer programming, our first option is to set Z_F, Z_A, Z_B, and Z_E each equal to 1.

Solution: $Z = \$35{,}461{,}052$; $X^A_{21} = 5.32$; $X^A_{22} = 25.68$; $X^A_{41} = 26.25$; $X^A_{43} = 22.22$; $X^A_{44} = 132.58$; $X^M_{11} = 12.58$; $X^M_{12} = 17.67$; $X^M_{14} = 3.75$; $X^M_{32} = 21.82$; $X^M_{33} = 38.21$; $X^M_{44} = 172.67$; $Z_A = 1$; $Z_B = 1$; $Z_E = 1$; $Z_F = 1$.

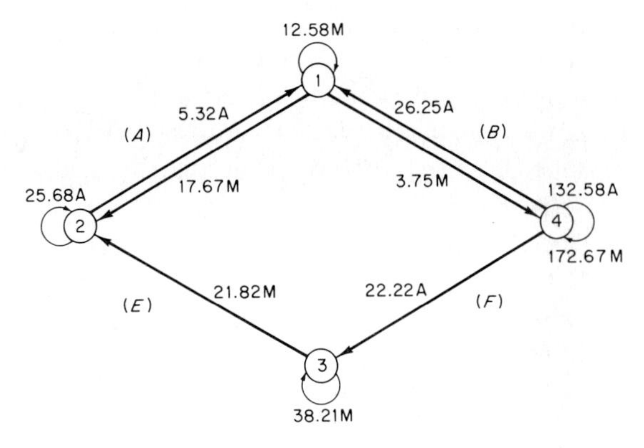

Fig. 4

Next, Z_F is set equal to 0 to test the other possible alternative.

Solution: $Z = \$35{,}691{,}630$; $X^A_{21} = 1.58$; $X^A_{22} = 35.58$; $X^A_{23} = 18.87$; $X^A_{41} = 30.00$; $X^A_{44} = 127.03$; $X^M_{11} = 12.58$; $X^M_{12} = 21.42$; $X^M_{22} = 5.86$; $X^M_{32} = 16.13$; $X^M_{33} = 34.87$; $X^M_{42} = 10.26$; $X^M_{44} = 165.57$; $Z_A = 1$; $Z_B = 1$; $Z_D = .57042$; $Z_E = 1$; $Z_F = 0$.

Clearly, link F must be constructed because the $Z_F = 0$ solution is more expensive and .57042 of D had to be constructed when $Z_F = 0$.

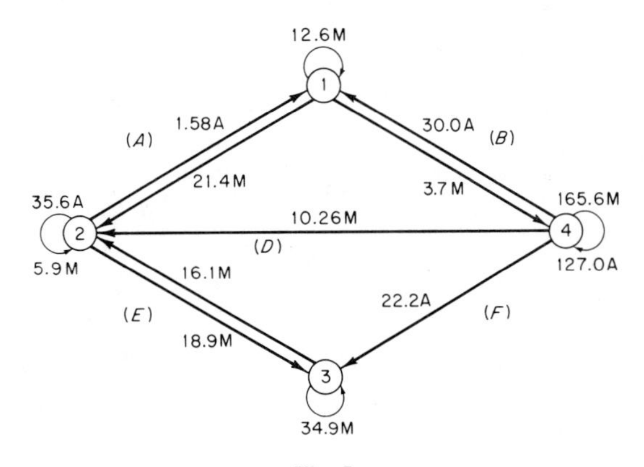

Fig. 5

F. SUMMARY

This chapter has reviewed the use of linear and other programming methods for the projection of interregional development patterns. The early uses focused on commodity flows among surplus and deficit regions. More recent efforts incorporate large-scale input–output, econometric, and transportation models as components of extremely ambitious research designs. The research models reviewed in the later sections of this chapter are among the most complex in the socioeconomic and environmental sciences.

REFERENCES

Ahmed, Y., O'Sullivan, P., Sujono, and Wilson, D. (1976). "Road Investment Programming for Developing Countries: An Indonesian Example." Transportation Center Publ., Northwestern University, Evanston, Illinois.

Barber, G. M., (1975). A mathematical programming approach to a network development problem. *Econ. Geogr.* **51**(2), 128–141.

Brokken, R. F., and Heady, E. O. (1968). "Interregional Adjustments in Crop and Livestock Production." Technical Bulletin No. 1396, U.S. Government Printing Office, Washington, D.C.

Casetti, E. (1966). Optimal location of steel mills serving the Quebec and southern Ontario steel market. *Can. Geogr.* **10**(1), 27–39.

Charnes, A., Haynes, K., Hazleton, J., and Ryan, M. (1975). A hierarchical goal-programming approach to environmental land use management. *Geogr. Anal.* 7, 121–130.

Dantzig, G. (1963). "Linear Programming and Extensions." Princeton Univ. Press, Princeton, New Jersey.

Dorfman, R., Samuelson, P., and Solow, R. (1959). "Linear Programming and Economic Analysis." McGraw-Hill, New York.

Durbin, E., and Kroenke, D. (1967). "The Out-of-Kilter Algorithm: A Primer." Rand Corporation, RM 5472-PR, Santa Monica, California. [The algorithm is available from IBM, IBM Share Distribution No. 3536, "Out of Kilter Network or Transportation Problem Solver," White Plains, New York.]

Evans, S. P. (1973). A relationship between the gravity model for trip distribution and the transportation problem for linear programming. *Transportation Res.* 7(1), 39–61.

Fox, K. A. (1953). A spatial equilibrium model of the livestock feed economy of the U.S. *Econometrica* **21**, 547–566.

Fox, K. A., and Taeuber, R. C. (1955). Spatial equilibrium models of the livestock feed economy. *Amer. Econ. Rev.* **40**, 584–608.

Ford, L., Jr., and Fulkerson, D. (1962). "Flow in Networks." Princeton Univ. Press, Princeton, New Jersey.

Golladay, F. L., and Sandoval, A. D. (1972). Optimal development policy in an open regional economy: a programming analysis. *J. Regional Sci.* **12**(2), 185–198.

Gordon, I. R. (1974). The gravity hypothesis and transportation cost minimization. *Regional Urban Econ.* **4**, 1–9.

Harris, C. C., Jr. (1973). "The Urban Economies, 1985." Heath, Lexington, Massachusetts.

Hartwick, J. (1972). The gravity hypothesis and transportation minimization. *Regional Urban Econ.* **2**(3), 297–308.

Henderson, J. (1955). A short-run model for the coal industry. *Rev. Econ. Stat.*, pp. 336–346.

Hopkins, F. E. (1972). Transportation cost and industrial location: an analysis of the household furniture industry. *J. Regional Sci.* **12**(2), 261–277.

Howes, R. (1967). A test of a linear programming model of agriculture." *Papers Regional Sci. Assoc.* **19**, 123–140.

Judge, G. G., and Wallace, T. D. (1958). Estimation of spatial equilibrium models. *J. Farm Econ.* **40**, 801–820.

King, L., Casetti, E., Odland, J., Semple, K. (1971). Optimal transportation patterns of coal in the Great Lakes region. *Econ. Geogr.* **47**(3), 401–413.

Land, A. H. (1957). An application of linear programming to the transport of coking coal. *J. Roy. Stat. Soc.* **A-120**, 308–319.

Mastenbroek, A. P., and Nijkamp, P. (1975). Environmental protection and spatial allocation of investments. *Int. Reg. Sci. Rev.* **1**(1), 73–86.

Morrill, R. L., and Garrison, W. L. (1960). Projections of interregional patterns of trade in wheat and flour. *Econ. Geogr.* **36**(2), 116–126.

Niedercorn, J., and Moorehead, J. (1974). The commodity flow gravity model. *Regional Urban Econ.* **4**, 69–75.

Orden, A. (1956). The transshipment problem. *Management Sci.* 2, 276–285.

O'Sullivan, P. (1972). Linear programming as a forecasting device for interregional freight flows in Great Britain." *Regional Urban Econ.* **1**(4), 383–396.

Schumaier, C. P., Gezen, A., and Kendrick, M. (1974). "Economic Aspects of Refinery and Deep-Water Port Location in the United States," five vols. Office of the Assistant Secretary for Policy Plans and International Affairs, Department of Transportation, Washington, D.C.

Scott, A. J. (1967). A programming model of an integrated transportation network. *Papers Proc. Reg. Sci. Assoc.* **19**, 215–222.

Scott, A. J. (1969). Combinatorial programming and the planning of urban and regional systems. *Environment and Planning* **1**, 124–142.

Scott, A. J. (1971). "Combinatorial Programming, Spatial Analysis and Planning," Methuen, London.

Soufflet, J. F. (1973). The warehouse problem: a review. *Regional Urban Econ.* **3**(2), 187–216.

Thoss, R. (1974). Resolving goal conflicts in regional policy by recursive linear programming. *Papers Regional Sci. Assoc.* **33**, 59–76.

Waverman, L. (1972). National policy and natural gas: the costs of a border. *Can. J. Econ.* **5**(3), 333–348.

Chapter 11

A Linear Programming Case Study[1]

Having studied the first ten chapters, the reader can solve a linear programming problem and is aware of many applications in the socioeconomic and environmental sciences. Such knowledge is a necessary but not a sufficient educational base from which to use linear and other mathematical programming tools.

Another needed perspective is advice on how to design models for use in the decisionmaking process. Chapter 5 was an introduction to this perspective. Briefly, Chapter 5 presented models, particularly mathematical programming, in the decisionmaking process; it also presented guidelines for using linear programming. The goal of the chapter was to alert the reader to the research design issue prior to a reading of the application chapters.

Now that you have studied many applications, this chapter provides a recapitulation and more detailed treatment of Chapter 5. The case study route is used. The case study is solid waste in northeastern New Jersey. In addition to the fact that the author worked on this project, it was selected because it illustrates numerous research design issues such as relating the model design to goals and outcomes, data adequacy, budget and personnel resources, and many others that affect the utility of the study. In particular, this solid-waste study highlights the data adequacy issue, the issue that frustrates many of the best conceived preliminary research designs. Data in the socioeconomic and

[1] The case study is from Greenberg *et al.* (1976).

environmental sciences are so often inadequate, and the political process is so dominant, that a premium should be placed on those research designs that require a small number of assumptions and relatively little data, and that produce many alternative solutions.

The data adequacy and other research issues are presented in three parts, which have distinctive subsections:

A. Stating the Problem—Northeastern New Jersey's Solid-Waste Management Problems

B. Designing the Facility Siting Research Problem:

1. Consulting Decisionmakers and the Literature
2. Initial Screening
3. Designing the Mathematical Programming Model to Evaluate the Best Alternatives

C. Applying the Linear Programming Model and Interpreting the Results

1. Applying the Model
2. Interpreting the Results from the Analyst's and the Decisionmaker's Perspectives

A. STATING THE PROBLEM—NORTHEASTERN NEW JERSEY'S SOLID-WASTE PROBLEM

From the solid-waste management perspective, northeastern New Jersey (Fig. 1) has been spoiled by the availability of centrally located marshlands. Long before the 1965 federal solid-waste management legislation suggested regional solid-waste solutions, northern New Jersey was landfilling the marshes, fields, and streams of the Hackensack Meadowlands. The Hackensack Meadowlands Development Commission (HMDC) was authorized by the State of New Jersey to manage the Meadowlands; their decision to end inexpensive landfilling has prompted the resource management problem examined in the case study.

1. AN OVERVIEW OF THE REGIONAL SETTING OF THE HACKENSACK MEADOWLANDS

The Hackensack Meadowlands is a unique island of open land amid the densely developed counties of northeastern New Jersey. The approximately 20,000 acres of the Meadowlands are composed of organic silt and marshes that are subject to both freshwater and saltwater flooding. These physical characteristics have retarded development of the area. While the Meadowlands have not been suitable for normal residential and commercial development, they have been seen as eminently suitable for noxious activities legally

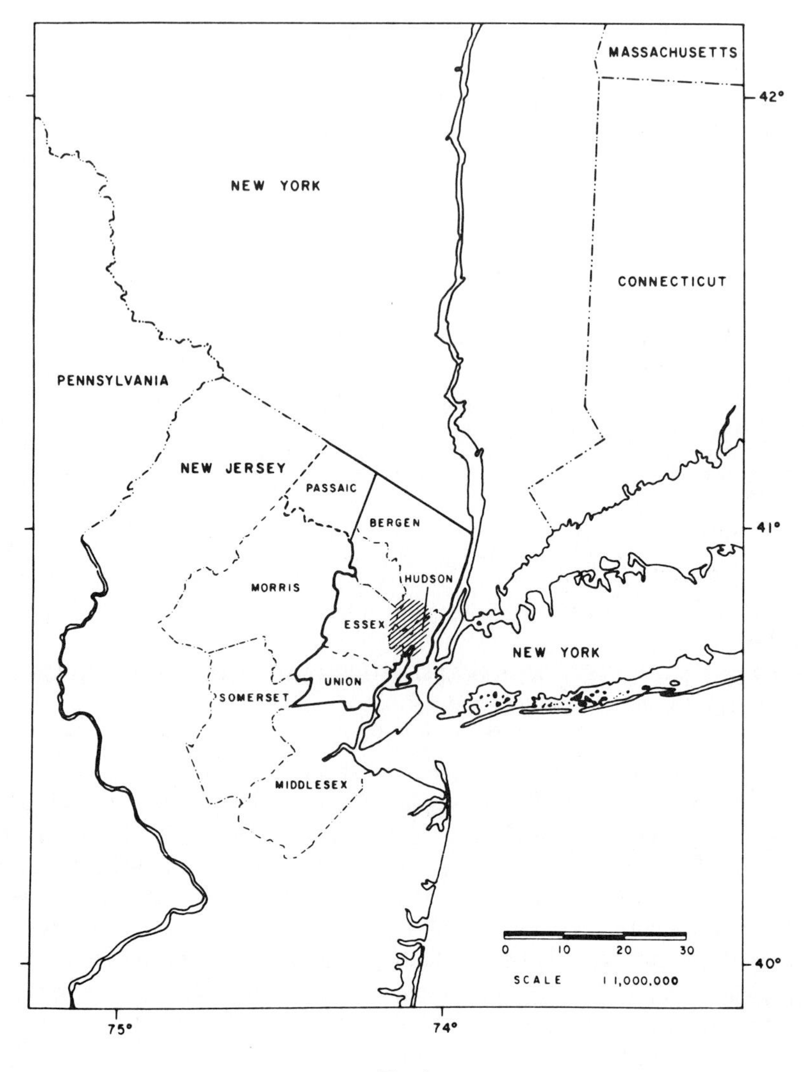

Fig. 1

banned or otherwise unsuitable for other portions of the New York region. Accordingly, most of the approximately 7500 committed acres of the Meadowlands are devoted to such activities as roads, rails and yards, airports, warehouses, transmission lines, storage tanks, and industry. These activities frequently adjoin landfills, which receive about 45,000 tons of solid waste each week, and open space and streams, which have attracted flocks of fowl and fish both rare and widely distributed in the United States.

2. THE ROLE OF THE HACKENSACK MEADOWLANDS DEVELOPMENT COMMISSION

The piecemeal and haphazard development of the Meadowlands led to the creation of the HMDC in 1968. With respect to solid-waste management, the critical powers of the HMDC are to approve and to direct land use in the district. As part of its master plan for the Meadowlands district, the HMDC plans to phase out six remaining inexpensive landfills by the spring of 1979. In addition, to aid in the construction of its 2000-acre recreation area, the HMDC plans to shred, compact, and bale solid waste as building blocks. The charge for baling is estimated at $7.00 per ton compared to the current Meadowlands disposal charge of $2.50 to $4.00 per ton. In the words of William McDowell, executive director of the HMDC:

> People from outside the Meadowlands have been dumping here cheaply for years. And now it's only fair that they help pay the costs of turning it from dumps into parks.[2]

The people from outside the region include more than 100 northern New Jersey communities and communities from New York State. In the 1940s, the combination of the economic legacy of a depression and the material needs of World War II constrained the amount of waste to be landfilled. However, with the conclusion of the war and the beginning of suburbanization and growing affluence, the Meadowlands became a centralized, regional dumping ground. By 1968, it was receiving 29,000 tons per week. Five years later, Meadowlands landfills were receiving 55,000 tons per week of solid waste.

The huge increase in landfilling threatened to wipe out the planned Meadowlands parkland before the master plan could even be adopted. Accordingly, in 1971 the HMDC began to make policy decisions which have inexorably led to northern New Jersey's acute solid-waste management problem. In 1971, the HMDC adopted sanitary landfill regulations and forbade filling virgin wetlands. In 1972, the HMDC closed down a landfill that was receving New York waste and banned liquid waste disposal. In 1973, garbage from outside New Jersey was banned. This policy, which reduced the input by 10,000 tons per week, has been legally challenged. Overall, between 1970 and 1976, the amount of available landfill space in the Meadowlands has decreased from 2500 to about 650 acres.

While it was curtailing landfilling, the HMDC was seeking a substitute for landfilling because it is legally obligated to provide a means of disposal

[2] Remarks made to our group on May 28, 1975, at the office of the HMDC, Lyndhurst, New Jersey.

to 118 northern New Jersey communities that have been using the Meadowlands. Two early attempts (1970–72) were a 6000-ton-per-day incinerator and a 400-acre site in a rural New Jersey county, 50 miles west of the Meadowlands. More recently, the HMDC has begun constructing a baler and is studying a set of resource recovery proposals.

3. THE SOLID-WASTE MANAGEMENT CRISIS IN THE MEADOWLANDS SERVICE AREA

The Meadowlands service area has a legal definition and a functional definition. The legal definition is the obligation of the Meadowlands to provide a means of disposal for the 118 northern New Jersey communities that had been disposing some waste in the Meadowlands. The functional definition we have adopted for use in this study differs from the legal definition in three respects. First, we have included 136 communities in the study area, whereas the Meadowlands is obligated to serve only 118. The 136 communities represent all the minor civil divisions in the approximately 500 square miles of the following northern New Jersey counties: Bergen, Essex, Hudson, Union, and southern Passaic. Legally, 112 of these 136 communities are permitted to use the Meadowlands. The remaining 24 are located in communities that were using other disposal sites, some of which have now been closed, at the time of the survey that established the Meadowlands legal region. It is our view that these 24 communities would seek to use any inexpensive recovery alternative or landfills constructed in the Meadowlands because of their location near the Meadowlands and a lack of other alternatives.

On the other hand, we have eliminated 6 of the 118 legally allowable communities from the functional region. These lie in Middlesex and Monmouth counties (east of Middlesex) and are more likely to use landfills and recovery sites currently under study or operating in their counties. Summarizing, the waste source region for our Meadowlands region consists of 136 communities located in five northern New Jersey counties: 112 are already legally entitled to use the Meadowlands and 24 others are likely to use Meadowlands facilities in the near future.

The third difference between our functional definition of the Meadowlands region and the legal obligations of the HMDC is that we have included some disposal sites in the following three counties that are adjacent to the five waste-source counties in our functional region: Morris, Somerset, and Middlesex. This decision is a reflection of the fact that these disposal sites are sufficiently close to the waste-source counties to be disposal options.

While the HMDC has been eliminating the region's ideal, centralized, and inexpensive landfill solution, a combination of environmental regulations, the increase of waste generation, and the spread of development have eroded most of the other inexpensive landfilling and incineration options available to the Meadowlands service area. Over 100 landfills are available to the region. Of these, 38 are special-use municipal disposal areas that include compost areas and sites limited to bulky items, leaves and trees that were not composted, and demolition wastes. Sixteen are special-use private disposal areas including those limited to demolition wastes, trees, brush, and the disposal of septic tank wastes. Twelve other facilities include sites used by institutions and other public agencies, and those used for onsite industrial disposal.

For management purposes, only 51 conventional landfill sites can be counted on for the disposal of the region's solid wastes. The life expectancy of these 51 conventional landfill sites is not long. Only 13 are expected to be open by 1980. On the surface, the projection of the life expectancy may seem unusually pessimistic. However, it agrees with recent regional trends. Specifically, between 1971 and 1973, 26 landfills in the study area were closed; only 3 new landfills were opened. Five of the six incinerators were closed. Furthermore, several of the Meadowlands landfills were closed while we were making the study. Only 4 of the 13 landfills lie within the waste-generation source region, which implies great political vulnerability on the part of the five waste-source counties to the adjacent counties.

Overall, the Hackensack Meadowlands service region faces an immediate and a long-range solid-waste disposal problem. As early as mid-1977, and certainly by 1979, more than 100 communities with over 3.5 million residents will not have any inexpensive disposal alternatives. Unless a relatively inexpensive resource recovery system is on line within the near future, the residents of the Meadowlands face a series of unpalatable choices: the opening up of new landfills in their communities, the destruction of the remaining open-space sanctuaries in the Meadowlands at a high environmental and economic cost, or long hauls out of the region to nearby regions by truck or to distant regions by rail.

B. DESIGNING THE FACILITY SITING RESEARCH PROBLEM

Preliminary analysis suggested that the essence of the solid-waste management problem was facility siting. To avoid the dual pitfalls of overlooking some obviously important considerations and of reinventing the wheel, interested parties and the relevant literature were consulted.

1. CONSULTING DECISIONMAKERS AND THE LITERATURE

Interested parties include the responsible agency (in this case, the HMDC) state, county, and local officials and special interest groups, especially environmental. Discussions with these parties led to the formulation of three major research questions, each of which had more specific subquestions.

1. How much will it cost the region and its communities to continue the centralized landfilling strategy?
   (a) Are the present landfill charges underpriced?
   (b) If so, how high can the price go before there is an economic incentive to haul the waste out of the region?
2. How much will it cost the region and its communities to convert to a decentralized landfilling strategy?
   (a) If one or two regional landfills are sited in each county, and landfills outside the region are used, how will the centralized and decentralized landfills options compare from an economic perspective?
   (b) How do baling and long-haul strategies compare with other landfilling and resource recovery strategies?
3. How much will it cost the region and its communities to convert to a resource recovery strategy?
   (a) What technologies offer the most economical solution?
   (b) What scales are most appropriate?
   (c) What sites are the most economical?
   (d) What is the relative importance of technology, scale, site, and product revenue?

The preceding questions took many hours to formulate. However, they were worth the effort because the research design was built from them. The composition of hierarchical questions is strongly recommended to the reader.

Facility Siting Literature[3]

The facility siting literature includes everything from locating an electric generating station and airports to siting fire hydrants and traffic lights. Solid-waste facility siting is closer to the first two, and therefore this review will focus on siting controversial facilities. The essence of the problem is finding sites for facilities that are frequently viewed as hazardous, noxious, and certainly undesirable by nearby communities, but which have been judged desirable to the region as a whole. Such circumstances provide an opportunity to completely discard equity as a decisionmaking criteria

[3] The references at the end of this chapter are a good representation of the literature.

and deliberately or inadvertently create or preserve some wealthy communities at the cost of creating or worsening what the author has called "ugly duckling" communities [see Carey and Greenberg (1974)].

The early facility siting literature almost completely ignored the secondary impacts of large-scale facilities. As the facilities have grown in scale and impact and the number of sites has become scarce, the literature has become more concerned with a variety of impacts. While the literature has rapidly expanded, its goal has remained the same—to present technically sound alternatives to decisionmakers who owe their role to the political process, not to technical expertise.

The technical input to the siting process is screening large numbers of proposed sites and carefully evaluating the few most viable sites. Baecher *et al.* (1975) appropriately label this process as "balancing apples and oranges." [See Baecher *et al.* (1975).] Both the initial screening and the more detailed evaluations require the design of methods of weighing and comparing different impacts. Some of these impacts may be quantified, others may not. Some are important to the entire regional population, others to a few city blocks. Some impacts will certainly occur, others are improbable but catastrophic in impact. Some impacts are immediate, others may not occur, if at all, for two or more decades. Some are irreversible and therefore close out future options.

The initial screening is likely to be based on a checklist of thresholds. If these thresholds are violated at a site, the site is eliminated from further consideration. For example, the thresholds for landfills might include location near a surface or underground water supply, inaccessibility, remoteness from a source of clean fill, and presence of vocal citizen's groups.

The feasible sites undergo a more thorough analysis. Three approaches are usually used: (1) estimating people's preferences by evaluating their expenditures (cost–benefit analysis) (2) estimating people's preferences by asking them direct questions about their preferences and (3) judging the impact on the population by direct assessment of economic, social and environmental considerations. Cost–benefit analysis, including our study, has been used extensively in siting evaluations. The advantages of cost–benefit analysis include the following: it expresses all the objectives as one index, it has been used for decades and is therefore accepted, and it is based on general welfare theory. The main disadvantage is that the use of a single monetary index makes it difficult to quantify environmental and social goals and to take improbable but catastrophic impacts into account.

As a result of the many criticisms aimed at cost–benefit methods, two other approaches have begun to be used. One directly measures and creates a statistical representation of people's preferences. The major advantage of this method is the ability of the user to include multiple goals in the objective.

The major problem is the difficulty of measuring and creating the utility function.

The third general approach is the checklist. The checklist approach starts from the assumption that apples and oranges cannot be mixed. Each impact or related impacts are evaluated separately and it is left to the decisionmakers to make the tradeoffs among different impacts. While the matrix techniques are relatively comprehensive, the decisionmaker must subjectively compare the impacts without a comparison of their relative importance. However, major improvements in checklist preparation have been occurring.

Overall, siting is a political decision. The analyst should screen out obviously unacceptable sites and carefully weigh the advantages and disadvantages of the best sites. It is up to the political process to decide if one part of the region is receiving most of the undesirable impacts and not enough of the benefits.

After discussions with interested parties and a literature review, a screening device was developed and a preliminary decision was made to use a mathematical programming model.

2. INITIAL SCREENING

Initial screening of sources, technologies, and sites occurred. First, the 136 source communities in the study area were collapsed into 25 refuse sheds, with the centroid of the refuse shed serving as the point from which distances were measured. This step reduced the number of calculations of travel costs, population, and waste generation by a factor of 5. In addition, we estimate that computer costs were reduced by at least a factor of 3.

Four rules were followed in preparing the aggregations. County boundaries were preserved in recognition of the strong possibility that the county may assume an important management role. All waste sheds were grouped from contiguous communities. The minor civil divisions were grouped along major transportation paths and/or became planning units within the counties for which special data tabulations were available. The projected solid-waste loads were located at the population centroid of each of the waste sheds.

The loss of information about individual communities is more than made up by the ability to test many more technology/location/transportation/price alternatives than would have been possible with a system of 136 waste sheds. Aggregating sites is strongly suggested to the reader. A postoptimizing program can be written to dissaggregate the results to the local community level. The misrepresentation caused by using the centroid of the waste shed can be included in the postoptimization program.

A second accommodation to the realities of the financial and informational constraints was made in selecting disposal sites for the model. A general guideline is to choose a sufficient number of well-spaced sites to determine the impact of different locations on cost. In choosing existing landfills, our objective was to include large sites that might be functioning by 1980. Using these two criteria, the number of existing landfill sites was reduced from 51 to 6. The reduction was achieved in two steps. First, landfills with capacities of less than 450 tons per week that are located in the source region were accommodated by subtracting their waste capacity from the waste generated in the waste shed in which they are located. For example, suppose a waste shed has a landfill that has been receiving 200 tons per week. For the 1975 run, we subtracted 200 tons per week from the waste shed's waste generation. In reality, few of the small landfill sites are expected to be open by 1980. Therefore, their impact on the model runs is negligible.

A more important step was the aggregation of disposal site capacities in the two multiple landfill areas: the Hackensack Meadowlands and Middlesex County. The six remaining Meadowlands sites were merged into one site in Kearny, and the four Middlesex County sites located along the Raritan River in Edison and East Brunswick were aggregated to a single Edison site. This step was taken for three reasons. First, it greatly reduces the number of waste-source-to-ultimate-disposal paths. Second, it recognizes that no one seems to know precisely which landfills are used by which carters. Had we been provided with unlimited research staff and computer time, we would have connected each waste shed to each existing landfill. We did not, however, have these resources. Moreover, travel-time measurements indicated that the sites were located so close to one another that the cost estimates for only the adjacent waste sheds might be biased by the aggregations.

The selection of future landfill sites is literally as well as figuratively a hazardous assignment. None of the state and county officials was willing to share with us possible sites or even site selection criteria. Since this dearth of information had been expected, we devised three criteria for choosing representative sites. One guideline was that whereever possible each county should be assigned at least one major landfill site. The second and third criteria were that regional landfills require at least 200 acres of vacant land and excellent road accessibility. Through a process of elimination, we learned that representative sites could be found in communities that have at least 200 acres zoned for industrial use and access to one or more interstate or county roads. When a county had many communities meeting the industrially zoned land and accessibility criteria, we tested several alternatives to minimize the transportation costs, but found that the model chose the community closest to the mass of the Hackensack service region's population.

Individual sites within communities were not chosen, and the economic, social, and environmental circumstances of individual communities were not investigated. We did not investigate the indirect economic and environmental impacts of the technology/site/scale combinations. We did not conduct field and library investigations of the new sites chosen for landfills or resource recovery systems, nor did we determine if the resource recovery systems would overtax the transportation, water, and other resources and services of their regions. Such investigations are, however, being made in some of the communities. Clearly, evaluation of sites within communities chosen by the model is in itself a major research project. In essence, the 7 dispersed landfill sites and 11 resource recovery sites we chose represent approximations of where regional sites might be placed, on the basis of the preceding general guidelines.

After a preliminary analysis, some management approaches were not investigated because they were considered uneconomical or not politically viable in the near future; these include large-scale incineration and oil pyrolysis and legislation that would greatly reduce the solid-waste stream. Hindsight suggests that incineration might be feasible in specific locations.

3. DESIGNING THE MATHEMATICAL PROGRAMMING MODEL TO EVALUATE THE BEST ALTERNATIVES

With the screening phase in mind, we divided the next phase of the research into four parts: (1) estimating solid-waste generation, (2) transporting the waste from sources to sites, (3) market and technology choices, and (4) mathematical programming choices.

Estimating Solid-Waste Generation

Most of the studies we have read adopt someone's population projections, and some adopt both population and per capita solid-waste generation estimates. We do not agree with these policies, particularly if the projections are for more than a decade. Our disagreement stems from the observation that population projections and per capita solid-waste estimates are developed for different purposes by agencies with different levels of expertise and information. The resultant solid-waste generation estimates may be grossly overestimated in some communities and underestimated in others.

The study area is illustrative. Five sets of population projections were available. One was developed from a modified exponential model of population density at the community scale. A second was prepared by a regional planning agency, but was badly out-of-date. A third was developed by a public utility; while methodologically sound, it could not be used because of company restrictions. A fourth was prepared by the state but was not available below the county scale. The fifth was composed of projections made for the communities by the counties. These projections varied considerably in credibility when judged from the perspective of when and how they were made.

Since the completion of our study, cooperative efforts among the federal and state governments lead the author to believe that population projections will eventually be produced that take into account the federal, state, county, and local perspectives. Until such time, and during the time of our study, the available projections were not considered adequate even for our 10-year period. The difference between the highest and lowest projections was 7% for the year 1985. However, differences at the community scale were frequently in excess of 50%. Having compiled the population projections of the five-county study area, it became apparent that the county projections varied so much in assumptions and methods that they could not be used as absolute numbers. However, it would be an error to disregard completely the counties' own projections, since they reflect local knowledge and are sensitive to economic and social trends in their communities.

In an effort to preserve the advantages inherent in using projections prepared at the county scale, and at the same time in order to maintain a uniformity that would lend credibility to our population projections, a step-down method was utilized. The method uses a chain of alternative techniques, which allows variables or models most appropriate for each geographic scale to play their role in the projection. For example, cohort-survival techniques are inappropriate for a township, but appropriate for the nation. Landholding-capacity techniques are unfeasible for the nation but appropriate for the township. Each technique has its level of greatest competence, and the chain model permits each to play its role with a minimum of distortion to the others.

The approach we used is illustrated by Fig. 2. It begins by adopting the most recent population projections at that time made by the State of New Jersey for the counties. It then adopts projections made by the counties for the communities and uses these as proportions to distribute the projections made by the state for the counties. The projections for the 136 communities are then reallocated to the 25 waste sheds.

Next, in contrast to studies that use a national estimate of per capita solid-waste generation for their regions, current and future estimates of daily per capita waste generation rates for each of the 25 subregions in the study area were developed. These rates were then multiplied by the estimated population of each region to arrive at an estimate of total waste production.

For the purposes of this analysis, solid waste was defined as comprised of the following: (1) household waste, which includes putrescible waste, rubbish, and nonputrescible waste normally generated by individuals or families, (2) household bulky waste referring to large items such as stoves

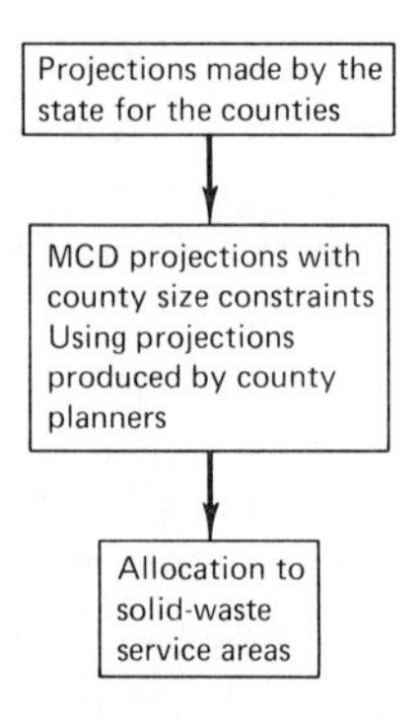

Fig. 2 A flow diagram of the procedure used to estimate the population.

and refrigerators disposed of by households, (3) yard debris including lawn clippings, tree trimmings, and the like coming from ordinary yard maintenance, (4) street sweepings, which refers to the waste collected from maintenance of public roads such as dirt and catch-basin cleanings, and (5) small commercial waste coming from small retail stores, restaurants, and the like that may be located on regular collection routes of regular household waste haulers. For convenience and clarity, the foregoing categories will be designated as municipal solid waste.

Solid wastes produced by large commercial establishments and industrial facilities were excluded due to the specialized nature of these wastes, which are normally collected and disposed of by private, contracted haulers.

The first task was to estimate current municipal waste generation rates. Data were available in computer printout from the Bureau of Solid Waste Management of the New Jersey State Department of Environmental Protection. This information included the total amount of solid waste collected by each hauler from every municipality in the state.

The information was reported as total solid waste, and there was no breakdown as to what proportion constituted municipal waste. An attempt was made to factor out all haulers that appeared to service major industries or businesses. All haulers collecting less than 2500 cubic yards per year were excluded because they would be delivering waste to the disposal site no more than three times per week. Since municipal waste requires a more intensive effort, an assumption was made that these haulers were either specific industrial/commercial waste haulers or small operators hauling negligible amounts of waste.

Next, the amounts of waste collected by haulers that were identified as serving only industrial or commercial customers were subtracted from the total municipal waste figure. It was assumed that municipal waste would be the largest figure left for each town. Dividing this by the population of each region produced the per capita generation rate.

The results of this approach are extremely erratic. Generation rates were produced ranging from less than 2 pounds per capita to greater than 50 pounds per capita per day! A few municipalities were reported as producing no solid waste at all, while other municipalities were not even listed.

Conversations with officials in the bureau also gave the impression that much of the information was not accurate because most collectors keep rather poor records of the actual weight of the waste that they haul. Thus, amounts reported to the bureau are mainly the product of guesswork done to satisfy legal requirements rather than a product of accurate accounting. In addition, it was believed that underreporting might be practiced to avoid providing regular information that could be used in tax investigations. Further analysis of this data set was considered fruitless due to its dubious quality.

Next, models that assume relationships between per capita generation and residential density, income, type of housing, family size, and age of the population were tried. None of these expected relationships could be verified. This negative research finding may have three causes. There may not be any systematic relationship between the variables we chose and waste generation. Other variables such as price of waste removal and frequency of removal may be the causative agents. Third, the dependent variable measurements of per capita residential solid waste may be inaccurate.

Having failed to obtain from records or estimate per capita generation rates, we obtained the best estimates from county and local solid-waste personnel. The lack of other credible available data left little choice but to use these rates.

The final task was to develop future municipal waste generation rates from the base year rates. In the absence of a predictive model, a survey of the literature was made to determine the growth rates used by other studies. Seventeen studies were examined. Annual growth rates in per capita waste generation in these studies varied from .164 to .000 pounds per year. The median rate was .070 pounds per year; the median rate for studies in New Jersey was .082. The .082 was used to normalize all the base year rates and then to project the rates for 1980 and 1985. Municipal waste totals were then projected by multiplying per capita municipal waste rates by the projected populations.

The results (Table I) indicate the region's municipal solid waste will rise 29% or 13,000 tons per week between 1975 and 1985. In most public facility development analyses, the size of the projected demand is a critical decision-making parameter. The size of the projected increase may be the determining factor as to whether a sewage treatment plant, a public water supply, a

Table I Estimated Municipal Waste Generation: 25 Waste Sheds, 1975–85

County	Waste shed	Tons/week 1975	1980	1985
Bergen	1	592	741	915
	2	737	923	1143
	3	307	402	507
	4	894	1083	1308
	5	3291	3892	4564
	6	1165	1355	1583
	7	1353	1612	1896
	8	1234	1474	1742
Hudson	9	1552	1914	2301
	10	2002	2328	2642
Passaic	11	573	717	882
	12	429	512	597
	13	1566	1828	2102
	14	398	486	586
	15	1508	1784	2083
Essex	16	752	1007	1346
	17	4261	4746	5222
	18	1086	1308	1538
	19	5598	6027	6410
	20	7997	8351	8587
Union	21	792	923	1062
	22	1942	2224	2535
	23	1911	2198	2492
	24	2979	3420	3917
	25	1507	1740	2004
Total		46,426	52,995	59,964

school, or some other facility should be expanded. In this case, the 13,000-ton-per-week increase would require approximately one 2000-ton-per-day plant or a number of smaller facilities. Whether the region generates 46,000 or 60,000 tons of municipal waste per week should not, however, be the decisive factor in the decision to use landfill or resource recovery methods. The 46,000 tons generated in the region in 1975 is sufficient to require a number of new facilities or sites, or both, in the Meadowlands or elsewhere. The additional 13,000 tons projected for the period 1975–85 would require a modular expansion of the basic decision about what to do with the wastes.

In addition to the magnitude of the solid waste generated in the region, two critical considerations are the composition of the waste and the spatial distribution of the waste sources. In essence, we have assumed that the composition of the waste in 1980 and 1985 will not be significantly different from the composition of the waste in 1975. We have assumed that in this study area there will not be massive paper recovery or legislation that will drastically reduce the amount of ferrous metal products arriving at the processing facilities.

Finally, with respect to the development of resource recovery facilities, the dense concentration of population in the immediate vicinity of the Meadowlands and the electric generating stations in northern New Jersey strengthen the economic case for resource recovery. We estimated that in 1975 the nine waste sheds within a 10- to 15-minute drive of the Meadowlands produced almost 30,000 tons of waste per week. This huge quantity of concentrated waste provides a powerful economic argument for a regional solution involving the economies of scale of large facilities.

Summarizing, solid-waste generation estimates were produced for 25 waste sheds. The potential important error in the per capita estimates was one factor in our conclusion that a dynamic model would produce results of dubious validity. Overall, I urge the reader not to accept estimates of solid-waste, water systems, land use, and the like without careful scrutiny and consideration of the impact of these estimates on the potential credibility of your study. Having participated in facility siting hearings, the author can state with certainty that opponents who might lose as a result of your study's recommendations will go directly for the study's jugular vein through population, employment, solid waste, water, and other estimates of service need.

Transporting the Waste from Sources to Sites

The hauling of waste to disposal sites is usually more expensive than paying for disposal. Yet our review of solid-waste management studies almost invariably reveals that transportation of the waste has not received careful attention. This section reviews methods used to obtain the travel costs between the sources of waste, the intermediate facilities, and the final facilities. The review is divided into three sections: (a) estimation of travel times, (b) selection of the most relevant paths in the system for incorporation into the model, and (c) estimation of transportation costs via direct haul and by transfer stations.

(a) Travel-Time Measurement

Although it is easily determined, crow-fly distance is generally a poor choice because it ignores the savings of time possible on high-speed, limited-access roads. Some facilities, while closer, are not chosen by carters because of poor access and because traffic congestion causes long delays. Travel-time data is a better choice, and consultation with drivers is an even better choice.

Without data such as those produced from the COLMIS system, information from drivers was not available. County, regional, and local data sets were not comprehensive.

The New Jersey Department of Transportation (NJDOT) offered their recent travel-time study of the entire New Jersey arterial system. They had used the floating car method in which the test car driver attempts to "float" with the stream of traffic while one observer times the journey; thus there is no correction factor for the driver's accuracy in gauging the average speed of the flow of traffic.

In the New Jersey case, each link had been driven at least once and frequently twice. Sometimes, however, the runs were in opposite directions. While off-peak-hour trips were the rule, peak-hour trips were often included, and there were too few runs to justify averaging off-peak and peak-hour runs together.

The form of the data, a computer map displaying four-digit numbers and an accompanying link time printout, implies that a computer simulation capacity restraint study was the objective of this agency. It quickly became apparent when we began working with the data that, while the form of a computer simulation study had been followed, the study itself was grounded in techniques of the 1950s.

The four-digit numbers, which for the most part signified road intersections (sometimes they represented state, county, or municipal boundaries, other landmarks, or abrupt turns in the road), could not be identified as specific highway intersections. The numbers themselves were chosen in sequence from 2000 up, but they have no significance as to the type of road, volume of traffic, signalization, or other considerations. Most significantly, the arterial system was not mapped by the computer.

The absence of this essential feature meant that we could not identify the intersections of the New Jersey network and fix them in space. Consequently it was impossible for us to locate the sources and sinks of our network on the NJDOT computer map. Furthermore, the map had no identifiable boundaries or directional indicators, and the scale of 1 inch to every 8642 feet was one which no other map of New Jersey seemed to have. This scale was an accommodation to the computer plotting device. No map could be found to overlay the NJDOT map, partly because of the odd scale and partly because the nodes themselves had been systematically displaced so that the numbers could be read clearly in areas of dense highway development. Our problem is not unique. Be prepared to encounter similar problems.

The NJDOT data set was used by accepting travel times on links that we knew from personal experience had been accurately timed and then correlating these times with crow-fly distance modified by assuming different truck speeds along different points. The standard error of the estimate of the truck travel time equation was 3.4 minutes; the correlation with the actual travel times was .961.

It was concluded that even though the modified crow-fly method for calculating the speed of the garbage trucks was imprecise, it probably mirrored fact pretty well in a study area, which has few outstanding geological features that would distort traffic flows.

(b) Selection of Important Paths for Incorporation into the Model

The choice of links for inclusion in the model runs is a critical step. The analyst must decide if the added computer and labor costs are worth the possible information to be gained from including all or most of the possible links. On the one hand, we have seen optimization studies in which the results were, in essence, predetermined because so few alternatives were available. For example, a solid-waste study that is programmed for 25 sources and 45 links probably does not warrant a modeling effort of the kind developed in this volume. On the other hand, a system of 25 sources and 800 links will be informative, but will also be extremely expensive to run. If the analyst has unlimited computer time and a large staff, an 800-link system can be modeled. Few studies are so well endowed. In the case at hand, our system had 719 potential links. The 719 potential links were reduced to 319 links (Table II). In order to recognize the legal responsibility of the Meadowlands region to accept the waste from most of the communities in the five-county study area, every source was linked to at least one Meadowlands facility of each type: existing landfill, dry fuel, and gas pyrolysis.

Table II Potential Links in the Solid-Waste Management Model

From	To	Number of possible links	Number of final links
25 waste sheds	6 existing landfills	150	71
25 waste sheds	7 proposed landfills	175	72
25 waste sheds	7 dry-fuel facilities	175	83
25 waste sheds	4 gas pyrolysis facilities	100	54
7 dry-fuel facilities	7 electric generating stations	49	27
7 dry-fuel facilities	6 existing landfills	42	8
4 gas pyrolysis facilities	7 electric generating stations	28	4
Total potential links		719	319

Second, every source was given links to at least two facilities of each type. A few waste sheds were given four links.

The precise selection of links to choose other than Meadowlands links was made on the basis of geography, tempered by political reality. The geographical consideration was cost minimization for the hauler. For example, preliminary calculations of transportation costs indicated that it was not necessary to link central and northern Bergen County communities in the northern part of the study area, to landfills or recovery facilities in the southern portion. Likewise, we did not link waste sheds in the southern portion of the study area to sites in central and northern Bergen County. It makes little sense to link a community to a facility 30 miles away when the same priced facility or facilities are available 10 miles away. The possibility that county solutions to the management problem may be feasible led us to choose at least one link in the county, wherever possible. The county assignments were nearly always the minimum cost link. The summary of 319 links indicates that each waste shed was assigned between 9 and 14 links. The more centrally located waste sheds were assigned more links than the peripheral sheds.

The intermediate-to-final facility links were chosen to minimize cost. In the case of the dry-fuel system, the amount of recovered energy assigned to a particular electric generating station was constrained. Therefore, multiple links to electric generating stations were assigned. With one exception, all the residue from the dry fuel plants was sent to the Meadowlands. The one exception was the Elizabeth facility, which was given the option of the Meadowlands or the Edison landfills. Each of the gas pyrolysis facilities was linked to a single electric generating station in recognition of the high costs of moving this energy product.

(c) Estimation of Transportation Costs via Direct Haul and by Transfer Stations

Given a set of 319 transportation links, we estimated the cost of moving the waste along 315 of these links by direct truck haul and sometimes by a transfer station. The four cost estimates for moving the gas recovered by the pyrolysis process were calculated separately because they should depend on pipelines.

(i) Direct-Haul Costs Our review of local conditions and the literature isolated six possible direct-haul modes involving different truck sizes and crew sizes. Municipal collection systems rely to a large extent on 20-cubic-yard collection vehicles and to a lesser degree on smaller 18-cubic-yard vehicles. For example, in Middlesex and Union counties, New Jersey, more than three-fourths of the municipal collection vehicles are of the 20-cubic-yard size. Only 8% use the large 25-cubic-yard trucks. Conversely, more than half the private collection firms use the larger 25-cubic-yard-capacity vehicles. Forty percent use 20-cubic-yard trucks. Only 10% use small trucks.

By 1980 we expect the present pattern to change in the direction of larger trucks. Generally, private carters amortize equipment more rapidly (3–5 years) than their municipal counterparts (10 years) and replace equipment more frequently. Private firms appear to select the more efficient 25-cubic-yard trucks to replace older equipment. Therefore, in modeling the municipal solid-waste system for 1980 and 1985, we have assumed that private haulers will be using 25-cubic-yard trucks.

Municipal waste collection agencies replace collection vehicles less frequently because of the difficulties they face in obtaining appropriations for routine vehicle replacement. As a result they tend to operate older and smaller trucks, many more than 10 years old. While recent purchases of new equipment by municipal agencies have tended to be of the 25-cubic-yard type, we have assumed that by 1980 the modal vehicle unit for the municipal agencies will be 20 cubic yards.

The 25-cubic-yard trucks cost more than the smaller vehicles. However, they more than compensate for the added 10–25% cost by being more efficient. The 25-cubic-yard trucks have much higher average payloads than their smaller counterparts. Average payloads for 18-cubic-yard vehicles in the study area tend to be low because the majority are older and less efficient than the new equipment. Average payloads for the larger, 25-cubic-yard vehicles are closer to the maximum reported payloads. The average has been chosen to compensate for less intensively used municipal equipment. Overall, the 40–65% payload advantage of the large truck can reduce the number of trips or trucks, or both, required by a community.

The other major choice we observed is crew size. Our review suggests that private contractors tend to use a combination of large truck and two-man crew, while the municipal agencies use smaller trucks and three-man crews. The additional man further increases the unit cost of the municipal operations.

The differences between the six cost combinations of two crew sizes and three truck sizes are considerable, especially when the annual costs are divided by the amount of operational time of the truck crew (about 120,000 minutes per year) and the average payloads of the trucks. It is this cost-per-ton-per-minute figure that is the critical transportation cost variable.

Direct-haul transportation costs were developed for all 315 links in two stages. First, we had to establish the haul cost per ton per minute for each of the 25 waste sheds. Towns were phoned to establish whether they were served by private carters or by municipal agencies. Pure cases (an entire waste shed served by either a municipal or a private carter) were assigned either a

$.098 per-ton-per-minute (municipal) or a $.054 per-ton-per-minute (private) cost. Most of the waste sheds had some communities served by municipal agencies and others by private contractors. In those mixed cases, a per-ton-per-minute cost was calculated by weighing the modal cost ($.098 or $.054) by each community's population (a measure of waste generation). Summarizing, each of the mixed-case waste sheds was assigned transportation costs between $.098 and $.054 per ton per minute. The average direct-haul transportation cost used in the study is $.067 per ton per minute, and the range is $.098 to $.054.

Next, the travel time for each link was calculated by adding twice the one-way travel time in minutes and a 20-minute turnaround time.

The final direct-haul cost estimates were obtained by multiplying the per-ton-per-minute cost for each waste shed times the travel times.

(ii) Transfer Stations Transfer stations can produce significant savings when a large amount of waste is to be hauled long distances. On the other hand, transfer stations are an added expense to the already high cost of waste disposal. Briefly, a transfer station is a fixed facility at which waste is transferred from small vehicles (18–25 cubic yards) to larger (75-cubic-yard) trucks or rail cars. The facility, which usually covers about 5 acres, contains storage areas, scales for weighing the waste, and compactors for reducing its volume.

The final decision to develop new transfer station capacity was made on a link-by-link basis. The direct-haul cost was compared with a haul cost assuming a transfer station. If the cost of direct haul was 50% greater than the cost of haul with a transfer station, then transfer station cost links were established. The decision to use a 50% added cost threshold reflects our judgment that, unless major savings are apparent, few communities will be willing to raise capital funds for a transfer station to be jointly operated with neighboring communities. Frankly, conversations with decisionmakers in some communities lead us to conclude that perhaps the 50% ratio is not high enough. Overall, the 26 existing transfer station links were augmented to 68 for the 1980 and 71 for the 1985 computer runs.

Summarizing, this section has demonstrated the complexity of developing an acceptable transportation cost matrix. Long before the computer model chooses the optimum links, the analyst has screened out choices that are not likely to be selected. Unless and even if the best available data are gathered and judgements are exercised, the model must greatly simplify the real-world options. If the analyst leaves inputs such as waste load generation, travel-time estimation, and link selection to people who do not know the big picture and who have little at stake in the study, then the results may be undermined.

Market and Technology Choices

The need for accurate marketing and technological data can be demonstrated in great detail. For the sake of brevity, however, we will treat only briefly these inputs to the model.

(a) Markets

The major markets for products derived from solid waste are materials industries and utilities and industries that can use the organic fraction as a fuel. Our analysis concluded that about 45% of the refuse could be converted to a fuel with about another 45% available to materials industries (paper, ferrous metal, glass, and aluminum). However, interviews with local industries revealed a definite conservatism toward the prospect of building or modifying facilities which would handle solid waste. While this conservatism has softened, we gave materials only minimum revenue credit in pricing the alternative resource recovery technologies. Fuel for utility boilers was seen as the major source of revenue and means of disposing of the bulk of solid waste in the study area.

(b) Technologies

While the federal and state governments may provide the direction toward resource recovery and regionalization and some of the financial and technical assistance, the burden for implementation has fallen largely on local governments, which have three basic choices:

(1) to continue landfilling by increasing the landfill area, through alteration of the material by compaction, or by raising the heights of the fills,
(2) to reduce the amount of potential waste through legislation designed to discourage throwaways, and
(3) to move toward resource recovery or long-haul.

The second of these, unfortunately, had to be dismissed in the screening process. The costs of the other two options were researched in the literature and through discussions with vendors. A summary of net costs at the 1000-ton-per-day scale for the three major contenders is found in Table III.

Table III clearly indicates that, assuming the resource recovery technologies work, their success will depend on the revenue. The high revenue represents the market value of the fuel that can be derived from solid waste minus costs for utility boiler adaptation, air pollution control, and other utility costs. The low credit is our best guess of what the utilities are willing to pay. In the absence of government intervention, the market for products derived from solid waste is a buyer's market. Accordingly, 1000-ton-per-day

Table III Estimated Net Cost of Alternative Solid-Waste Disposal Technologies[a]

Technology Cost	New landfills	Refuse-derived fuel	Gas pyrolysis
Base operating cost	5.01	9.01	14.58
Low revenue credit	—	4.05	4.83
Net operating cost	5.01	4.96	9.75
High revenue credit	—	13.77	10.60
Net operating cost	5.01	−4.76	3.98

[a] At 1000-ton-per-day operating level and different price levels, late 1974 dollars.

resource recovery plants are barely feasible from the economic perspective in a low-revenue buyer's market. When the technical perspective is added, resource recovery is much less viable than landfilling because neither the refuse-derived fuel nor the gas pyrolysis technologies has an operating prototype even a small fraction of 1000 tons per day. Indeed, several of the most recent experiences with financing and operating refuse-derived fuel and pyrolysis facilities have been very discouraging. At the time of this writing, only incineration with electric and steam production has been proven to work at the 1000-ton-per-day scale.

If, however, the technical problems can be overcome without greatly adding to the cost, then proponents of large-scale recovery facilities argue that ecconomies of scale will bring down the net unit price of waste disposal through resource recovery. Whereas giant landfills are rarely proposed, the literature contains citations and cost data for resource recovery facilities in excess of 5000 tons per day. Such facilities are assumed to provide important economies of scale. Discussions with vendors, however, convinced us that the economics of these massive facilities were very uncertain. In contrast to other studies, therefore, we chose to limit facility size to a maximum of 3000 tons per day and to focus on 1000- and 2000-ton-per-day scale facilities.

Choice of a Mathematical Programming Model

The preceding lengthy description of the data and the needs of the users led us to the conclusion that a linear programming model was better than nonlinear alternatives and that the study should focus on a 5-year horizon. The decision, particularly the first, has been questioned by those who believe that facility siting requires a nonlinear model because of economies of scale and fixed capital costs. We respond that the users needed answers to specific questions that, from a computer cost perspective, were far more economically answered by linear programming. Specifically, while you can design runs on the basis of research questions, the entire application cannot be designed in advance because questions will arise from solutions to early runs. Time, personnel, and budget considerations will dictate how many questions may be answered. The linear model was judged to be a more economical tool for answering many questions. And given the data deficiencies encountered in all the data sets, the concave objective function of the nonlinear model would produce results that give the false impression of precision which does not exist in the input data. The decision to use a simple model is not to be taken lightly. If possible, try to explain the options to the users before a decision is made. In this case the first equation set in Chapter 6 was used. It is repeated for your review.

Minimize: $$Z = \sum_i \sum_j C_{ij} X_{ij} + \sum_i \sum_k C_{ik} X_{ik} + \sum_j \sum_k C_{jk} X_{jk} \tag{1}$$

subject to

$$\sum_j X_{ij} + \sum_k X_{ik} = S_i \qquad \text{for} \quad i = 1, \ldots, m \tag{2}$$

$$\sum_i X_{ij} \leq K_j \qquad \text{for} \quad j = 1, \ldots, n \tag{3}$$

$$\sum_i X_{ik} + \sum_j X_{jk} \leq K_k \qquad \text{for} \quad k = 1, \ldots, u \tag{4}$$

$$\sum_i aX_{ij} - \sum_k X_{jk} = 0 \qquad \text{for} \quad j = 1, \ldots, n \tag{5}$$

$$X_{ij}, X_{ik}, X_{jk} \geq 0 \tag{6}$$

where C is the total net cost in dollars per ton associated with all possible travel links in the system (travel, capital, revenue, operating) and X the amount of waste or recovered product transported from a source to an intermediate facility, from a source to a disposal facility, and from an intermediate facility to a disposal facility. The subscript i denotes a source of waste (a waste shed), j an intermediate facility for waste processing (gas pyrolysis, incinerator, dry fuel plant), and k an ultimate disposal facility. S_i is the amount of waste generated in tons per week at each of m sources of solid waste, K_j the capacity in tons per week for each of n intermediate facilities, K_k the capacity in tons per week for each of u ultimate facilities, and a the percentage of waste remaining at the n site j's after intermediate processing has been completed.

C. APPLYING THE LINEAR PROGRAMMING MODEL AND INTERPRETING THE RESULTS

This section reviews the runs that lead us to the conclusion that, assuming the technology works and the economic data are reasonable, resource recovery is a preferred technology to landfilling. The previous section has considered the sources of residential solid waste, alternative transportation strategies for moving this waste, markets for recovered products, and technologies for making the best economic use of the solid waste. A brief summary of the most important information should help set the stage for the major conclusions and the plethora of detail presented in this section.

The five-county study area produces about 46,000 tons of residential solid waste per week. Projections suggest that the weekly load will increase to about 53,000 tons in 1980 and 60,000 tons in 1985. Nearly all of this waste is hauled directly by trucks. Few communities have transfer stations. Our analysis of about 300 waste-source-to-facility paths indicates that transfer stations should become an important factor in regional solid-waste management planning only if communities feel a need to make long hauls outside the region to the west and south in order to avoid increased disposal charges in the Hackensack Meadowlands.

Four major solid-waste disposal technologies were found possible, if not acceptable, alternatives to meeting the study area's disposal problems from 1975 through 1985. The most obvious alternative is to continue to haul about 46,000 tons of waste each week to the Hackensack Meadowlands landfills. This practice has been recently curtailed due to the closing of some landfills in the Meadowlands. The HMDC plans to all but halt landfilling by 1980.

An alternative to hauling nearly all the waste to a centralized landfilling site is to landfill the solid waste at sites outside the Meadowlands. We have chosen representative sites in communities with excellent road access and a large amount of undeveloped industrial space. On the one hand, this landfill solution would greatly reduce the burden on the Meadowlands area, but would undoubtedly face opposition of the harshest form from the chosen communities.

The third alternative is a dry-fuel technology that recovers an energy product and ferrous metals from the solid waste. Sixty percent of the waste is converted into a fuel that can be fired in utility boilers. About half of the remaining solid waste is marketable recovered products, and the remaining half should be landfilled. Both centralized and dispersed locations for these dry-fuel facilities were chosen. The dry fuel option requires utility boiler alterations and may therefore not be an acceptable alternative to the utilities.

Gas pyrolysis is a fourth option. It does not require major utility alterations but is more expensive to construct and to operate. Pyrolysis recovers a gas product that can be piped to a utility generating station and a frit that can be sold. Best, medium, and worst revenue pictures are developed for both of the resource recovery options.

1. APPLYING THE MODEL

Our analysis of the current centralized landfilling strategy indicates that the current landfill charge of between $2 and $4 per ton is grossly underpriced. The five most important centralized landfilling simulations are summarized in Table IV. We started by modeling the 1975 system with the hope that the results would validate that the Meadowlands have been receiving between 40,000 and 46,000 tons of waste per week. Run 1 confirms this assumption by sending more than 40,000 of the region's 46,000 tons to the Meadowlands at a weekly, total system cost of almost $300,000. Run 1 also identifies the importance of transportation in the management of the solid-waste system. Of the 25 sheds, 21 send all their waste to the Meadowlands landfills (Kearny, Fig. 3). Four do not. These four waste sheds lie at the periphery of the region or have existing transfer stations, or both. While the optimization procedure assigns most of their waste to non-Meadowlands landfills, in reality the waste of three waste sheds is just as likely to go to the Meadowlands. The Mahwah waste shed lies in northwestern Bergen County. Our calculations find that it is $.27 per ton cheaper (3 %) for the communities in this waste shed to send as much of their waste as possible to the Roxbury landfill and the remaining waste to the Meadowlands. The small difference between the Roxbury and Meadowland hauls implies that these communities are in reality just as likely to send their waste to the Meadowlands as to Morris County.

The cases of the Elizabeth and Westfield waste sheds (Figure 3) are even better examples of the indifference of the Meadowlands versus non-Meadowlands choices. The Elizabeth waste shed sends all its waste to the Mount Olive landfill via a transfer station on the basis of a $.02-per-ton difference in cost. The Westfield shed's waste is allocated to Edison because of a $.03-per-ton cost difference. Clearly, these fractions of 1 % difference in cost are unlikely to be recognized by carters and, indeed, may be due our inexact estimates of unit costs and transportation times. In addition, the Mahwah waste shed should, according to the computer, split its waste. In reality it would send the waste to a single site.

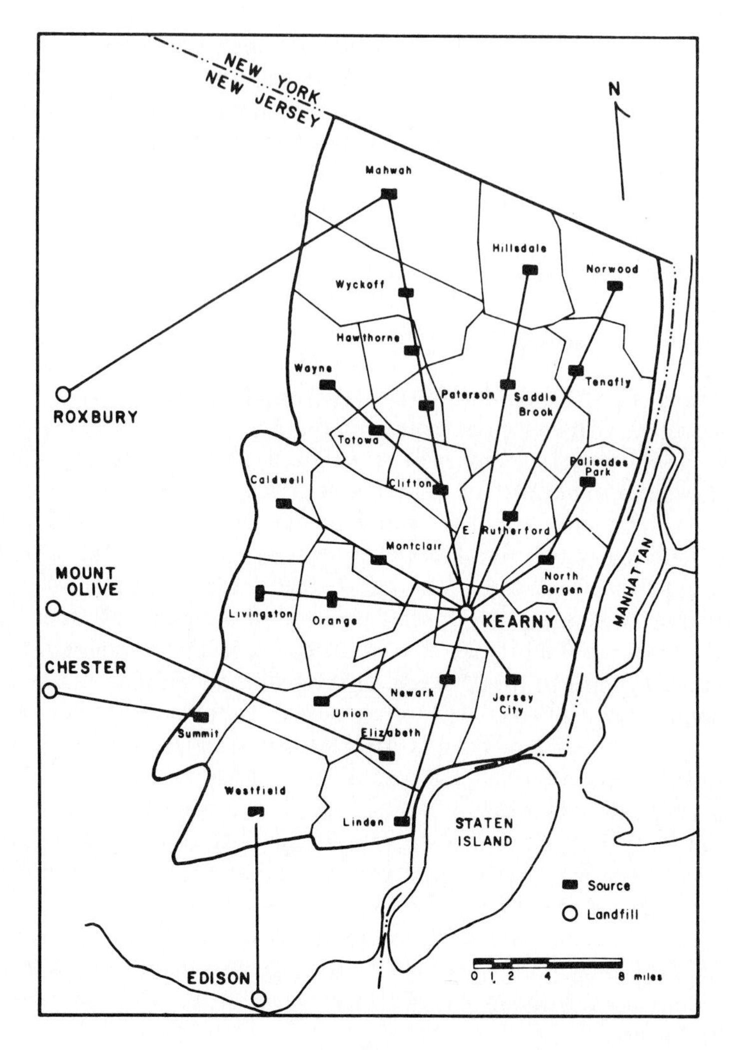

Fig. 3

It is a common result for the computer to assign waste on the basis of little marginal cost difference and split the waste. The analyst has to carefully analyze every link or important management alternatives will be missed.

Overall, in this case only one of the 25 waste sheds (Summit) appears to have an obvious economic motivation to use its existing transfer station to avoid hauling waste to the Meadowlands.

The conclusion reached in the first run is mirrored by the 1980 run. The same geographical assignments are made in run 2 (Fig. 3) as in run 1 (Table

IV). In short, the first two runs point to the unmistakable conclusion that the low price ($2.70 per ton) charge in the Meadowlands results in the hauling of between 40,000 and 45,000 tons per week of solid waste to Meadowlands landfills.

Runs 3–5 pose the following question: How much can the disposal charge be raised before many waste sheds have a clear incentive to haul their waste to other landfill sites? This question is not academic because new landfill sites in the Meadowlands will be more expensive to operate, and the proposed baling operation or rail haul would raise the disposal charge. Runs 3–5 were made with our first planning year data: 1980.

First, we raised the Meadowlands charge to $4.58 per ton, while maintaining the present charge at the other sites (run 3). The optimization model made a negligible change in the hauling pattern. About 2000 tons that had been allocated to the Meadowlands from the peripheral Linden and Hillsdale sheds were sent to other landfill sites. While only two additional waste sheds were given non-Meadowlands allocations, raising the charge from $2.70 to $4.58 per ton provides a clear economic incentive for six instead of only one waste shed not to haul to the Meadowlands. All of these waste sheds lie at the southern, southwestern, and northwestern extremes of the study area, are the farthest removed from the Meadowlands, and have relatively nearby disposal sites available in Middlesex or Morris counties.

Summarizing, at $4.58 per ton, more than four-fifths of the region's waste should come to the centralized Meadowlands landfills. It would appear that, in the absence of new landfills, the disposal charge in the Meadowlands could

Table IV Estimated Costs of Continuing the Centralized Landfilling Strategy

Model results	1	2	3	4	5
Year	1975	1980	1980	1980	1980
Meadowlands,[a] cost, $ per ton	2.70	2.70	4.58	7.00	10.00
Total cost ($1000)	298	341	424	523	619
Haul cost ($1000)	173	198	198	209	261
% of total	58	58	47	40	42
Disposal cost ($1000)	125	143	226	314	358
Tons per week total	46,426	52,995	52,995	52,995	52,995
Tons per week to Meadowlands[a]	40,294	46,004	43,962	39,743	28,576
% of total	87	87	83	75	54
Tons per week to non-Meadowlands landfills	6132	6991	9003	13,252	24,419

[a] The Hackensack Meadowlands.

be raised to \$4.58 per ton without much change in the region's waste-hauling pattern.

At \$7 per ton, the estimated charge for a baling system, the waste of 8 of the 25 waste sheds is allocated to non-Meadowlands landfills. In addition, 15 of the remaining 17 waste sheds have an economic incentive to haul to at least 1 of the other landfills. About half require transfer stations to justify these long hauls. Unfortunately the 13,000 tons per week from the eight other regional waste sheds uses up the disposal capacity of the Morris County sites. Clearly, however, at \$7 per ton in the Meadowlands, one should expect extremely strong pressures for the development of new landfills or other means of disposal from Bergen and Passaic County communities.

Finally, a \$10-per-ton charge (equivalent to rail haul) provides economic incentive for every waste shed to find an alternative to Meadowlands disposal. The Middlesex County landfills and Morris County landfills are allocated about 26,000 tons of waste per week (almost half the region's projected total).

In conclusion, the five-county region is likely to pay dearly for the continuation of the centralized landfilling strategy. At a modest \$4.58-per-ton disposal charge, the disposal cost to the 25 waste sheds rises 20% (\$80,000 per week) above the base price 1980 estimate. Yet, nearly all the communities will have to continue to haul to the Meadowlands because other less expensive alternatives do not currently exist. At \$7 per ton, the price the HMDC has quoted for its proposed baling operation, the cost to the region rises 55% (\$180,000 per week)! And yet, with the exception of spatially peripheral agencies, three-fourths of the waste is still allocated to the Meadowlands because of a lack of inexpensive alternatives. In short, at \$7-per-ton disposal charges, enormous pressures are likely to be generated to develop alternatives to hauling to the Meadowlands for landfilling.

A Dispersed, Regional Landfill Strategy

In the face of substantially increased disposal charges in the Meadowlands, an obvious alternative is the development of regional landfill sites dispersed throughout the region. We made more than 20 model runs to determine where new regional landfill sites could be economically justified. To conserve space, these results will only be summarized. At a \$7 disposal charge in the Meadowlands, between two and four large landfills are warranted in central Bergen (Saddle Brook area), western Essex (Caldwell and Livingston areas), and eastern Morris counties (Roxbury and Mount Olive areas). These landfills would hold the fort against a possible increase in total landfill cost of between 25 and 45%. At the same time, they would precipitate monumental political struggles in the affected communities, and they do not take advantage of solid waste as a possible source of revenue.

A Resource Recovery Alternative: Dry Fuel

From an economic perspective, landfilling seems to be a stable solid-waste management strategy at best and a much more expensive approach at worst. We therefore tested two resource recovery technologies: dry fuel and gas pyrolysis. In each case three factors were systematically varied: (1) location of facilities—centralized in the Meadowlands or dispersed, (2) size of recovery facility—1000, 2000, or 3000 tons per day and (3) revenue for recovered products—high, medium, or low. All runs were made with 1980 projected waste generation rates.

In the case of dry fuel, 16 optimization runs were made. The most important are summarized in Table V.

(a) Low Revenue for Recovered Waste Products

We began by trying to model some of the worst possible circumstances from the perspective of dry-fuel technology. Low price revenue for energy products was assumed, and the Meadowlands was allowed to receive waste for landfilling at $4.58 per ton. A single 2000-ton-per-day dry-fuel facility survived and lowered the regional waste treatment cost from $424,000 (run 3) to $408,000 per week (run 6).

Second, we again stipulated the low price revenue circumstances combined with the presence of dispersed regional landfill sites, but no landfilling in the Meadowlands. As in run 6, a single, centralized facility of about 2000 tons per day survived (run 7). Other runs were made with different variations of the same low-revenue theme. For example, a run was made with a combination of dispersed 2000-ton-per-day facilities, low revenues, and Meadowlands landfilling at $4.58 per ton. Run 8 indicated that even these highly conservative assumptions justified a dry-fuel plant in Newark.

When the price for landfilling in the Meadowlands was assumed to be raised from $4.58 to $7 per ton, three 2000-ton-per-day plants survived (in Newark, Lyndhurst, and Saddle Brook) and one 1500-ton-per-day plant (in Elizabeth) (run 9). Each facility was allocated waste from adjacent communities. The Saddle Brook Facility was allocated waste from the northern and central Bergen and Passaic waste sheds. The Lyndhurst facility received waste from southern Bergen, Hudson, northern Essex, and Passaic counties. The Newark plant was allocated the Newark solid waste and most of the western Essex county waste. Finally, the Elizabeth waste facility received waste from adjacent communities in Union and western Essex counties.

Overall, these low-revenue runs strongly argue that authorities in the Meadowlands or in nearby communities such as Newark should, from an economic perspective, move toward negotiating the development of a

Table V Estimated Costs of a Dry Fuel Strategy[a]

Model results	6	7	8	9	10	11	12	13
Meadowlands, cost $ per ton	4.58	—[b]	4.58	7.00	4.58	4.58	4.58	4.58
Total cost ($1000)	408	385	384	386	178	−79	145	−112
Haul cost ($1000)	214	192	176	220	263	263	230	230
Disposal cost ($1000)	194	193	208	166	−85	−342	−85	−342
Tons per week, total	52,995	52,995	52,995	52,995	52,995	52,995	52,995	52,995
Tons per week to Meadowlands landfills from waste sheds	21,744	—	10,437	0	0	0	0	0
Total per week to Meadowlands existing landfills	19,712	19,712	11,361	11,418	0	0	0	0
Tons per week to proposed dry-fuel plants	11,539	12,997	10,575	41,577	52,995	52,995	52,995	52,995

Tons per week to proposed landfills	—	20,286	20,622	—	0	0	0	0
Tons per week to Meadowlands landfills from dry-fuel plants	1730	1950	1586	6237	7949	7949	7949	7949
Proposed landfills in tons per day	—	SDB-2000 HAN-1500 MAH-150	SDB-2000 HAN-1500 MAH-150	—	0	0	0	0
Proposed dry-fuel plants and size in tons per day	LYE-2000	KRN-2000	NEW-2000	NEW-2000 LYE-2000 SDB-2000 ELZ-1500	KRN-10,000	Same as 10	NEW-2000 LYE-2000 SDB-2000 ELZ-2000 JC-1500	Same as 12
Weekly dry fuel product received by electric generating stations	BER-3873 KRN-3050	MAR-1123 KRN-3050 ESS-3625	HUD-2720 ESS-3625	HUD-7171 MAR-332 KRN-3959 ESS-3625 LIN-4866 BER-5900	HUD-9575 MAR-1123 KRN-3050 ESS-3625 LIN-7700 BER-5900 SEW-597	Same as 10	Same as 10	Same as 10

[a] All runs assume 1980 generation rates.

[b] Dashes mean not relevant to the run.

SDB, Saddle Brook; HAN, Hanover; MAH, Mahwah; LYE, Lyndhurst

NEW, Newark; BER, Bergen; MAR, Marion; HUD, Hudson

SEW, Sewaren; FRF, Fairfield; SBR, South Brunswick; BRW, Bridgewater

KRN, Kearny (Meadowlands); ELZ, Elizabeth; ESS, Essex

LIN, Linden; TOT, Totowa; JC, Jersey City

large dry-fuel resource recovery facility. Even with an extremely modest revenue for the recovered products, at least one major plant is justified. If low-priced, dispersed landfill sites are not provided elsewhere in or near the study area, then between two and four dry fuel facilities are warranted.

(b) Medium and High Revenue for Recovered Products

While it is appropriate to analyze the survivability of a facility under the low-revenue conditions, the low fuel cost assumption used in the previous section seems from the mid-1970s perspective to be long gone, like the 5-cent cigar and the 20-cent hot dog. Therefore, we modeled the system assuming a revenue for No. 6 oil equivalent to the late 1974 and early 1975 price per barrel (our high price) and a price midway between the low and high prices (our medium price).

The medium—and high—price simulations provide strong arguments for a dry-fuel alternative. First, we assumed that the Meadowlands would, through sets of 2000-ton-per-day plants, provide enough capacity to take the entire 1980 waste load of the region (about five 2000-ton-per-day plants). We did not assume that centralized dry-fuel facility capacity would be provided by huge 5000–7000-ton-per-day plants. These large facilities should theoretically lower the disposal charge because of better economies of scale than 2000-ton-per-day plants. However, as yet they have not been demonstrated to be feasible to our satisfaction from the technical, economic, or traffic perspectives.

Run 10 included a 55,000-ton-per-week, dry-fuel capacity in the Meadowlands, the availability of existing landfill capacity, and the development of new, dispersed landfill sites by 1980. The model assigned all of the waste to the Meadowlands dry-fuel facilities at a total weekly regional cost of $178,000 (run 10). This regional cost is less than 60% of the estimated 1975 cost (run 1, $298,000) and 42% of the estimated base price, 1980 cost (run 3, $424,000).

When the revenue for the energy product was raised to current levels and the same set of assumptions were retained about Meadowlands capacity, the unthinkable happened—a revenue was produced (run 11). If, indeed, revenue of almost $80,000 per week were produced, the region would be in the comforting position of arguing whether the funds should be allocated to building parks in the Meadowlands, used to finance education, given back to the waste shed communities, or used in numerous other activities. In short, we have found that, depending on the revenue for fuel, at least one and perhaps five 2000-ton-per-day dry-fuel plants are warranted in the Meadowlands or vicinity by as early as 1980.

The Meadowlands area organizations may not, however, be able or willing to raise the revenue for the facilities. In addition, opposition from local residents to the continued influx of all the region's waste may be high. Accordingly, we modeled the low, medium, and high revenue circumstances for dispersed 2000-ton-per-day dry-fuel plants.

Earlier we concluded that a single 2000-ton-per-day facility could survive in Newark, even if faced with competition by Meadowlands landfills, with a $4.58-per-ton disposal charge and new, dispersed landfills (run 3). When the Meadowlands price was increased to $7 per ton, three 2000-ton-per-day and one 1500-ton-per-day dispersed facility were justified.

The medium and best price, dispersed, dry-fuel simulations were made with the following facility assumptions: (1) six 2000-ton-per-day dry-fuel facilities located in Newark, Lyndhurst, Saddle Brook, Totowa, Jersey City, and Elizabeth, (2) available capacity in existing landfills, and (3) available dispersed landfill capacity. The optimization run procedure allocated all of the waste to the dry-fuel facilities at a weekly cost of $145,000 (run 12). The $145,000 cost is less than half the estimated 1975 cost (run 1, $298,000) and 34% of the estimated base price, 1980 cost (run 3, $424,000). In addition, the dispersed system cost was $33,000 per week, about 20% cheaper than the medium revenue, centralized dry-fuel system. This $33,000-per-week advantage of the dispersed system should be viewed as one indicator of the opportunity cost of locating all the region's waste disposal capacity in the Meadowlands. The geographical distribution of solid-waste movements for the dispersed dry-fuel system is illustrated by Fig. 4. Most of the hauls are relatively short and fall within the county in which the waste shed is located.

The same opportunity cost was obtained when the dispersed dry-fuel system was modeled assuming current (our high) energy price revenues. The model estimated a weekly revenue of $112,000, which is $33,000 per week more than the $79,000 revenue generated by the centralized dry-fuel alternative.

Overall, if the communities in the study area wish to force the Meadowlands to absorb all of their solid waste, then as a whole they will have to incur an additional $33,000 per week in transportation costs. Individual waste shed and community costs will, of course, vary depending on the length of the travel and the locations of facilities.

In conclusion, if reasonable contracts for dry-fuel products can be negotiated with the utilities and if communities can be persuaded to continue consistently to bring their waste to the Meadowlands, then the dry-fuel option is, from an economic perspective, clearly preferable to centralized or dispersed landfilling. If groups of communities, probably county or large cities, choose to take the responsibility of waste disposal away from the Meadowlands, then the cost to most of such communities will be further reduced. The political costs of county or large city management of the final

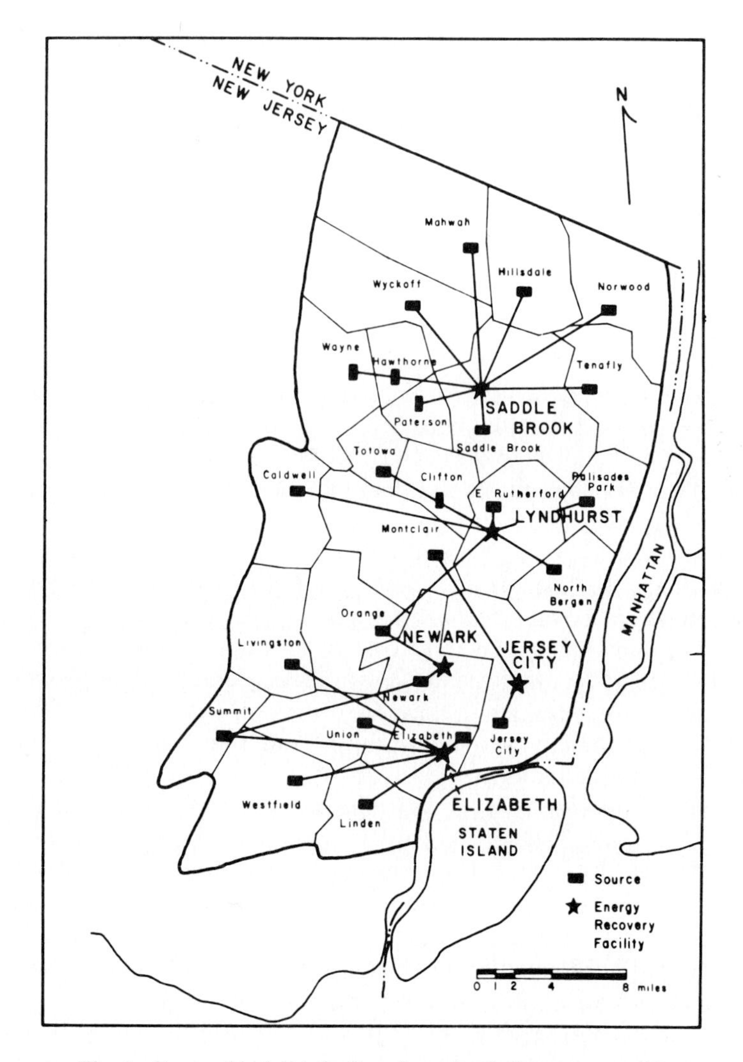

Fig. 4 Geographical distribution of one dry-fuel strategy, run 12.

stage of solid-waste disposal are uncalculable but probably unacceptably high in many areas.

Gas Pyrolysis: A Second Resource Recovery Alternative

Gas pyrolysis is a new and expensive technology. It has been suggested that the gas product is more adaptable to the region's utility boilers and to industrial uses than the dry-fuel technology. Therefore, we tested gas pyrolysis as a possible alternative. Facility locations, plant capacity, and revenue for the recovered products were varied in over a dozen optimization runs. A centralized and a dispersed regional pattern were examined. The recovery facilities were set at 2000 and 3000 tons per day, and the revenue for the gas product was tested at high, medium, and low levels. For brevity's sake, the gas pyrolysis runs will not be reviewed.

2. INTERPRETING RESULTS FROM THE ANALYST'S AND THE DECISIONMAKER'S PERSPECTIVES

Summarizing, from an analyst's perspective, neither the dispersed nor the centralized landfilling strategies are economically competitive with the dry-fuel or gas pyrolysis technologies when the resource credits are high. From an economic perspective, the dry-fuel technology is the preferred alternative. It has clear economic advantages over any other system both as a centralized system in the Meadowlands and at dispersed sites. The 2000-ton-per-day facilities can either be clustered in the Meadowlands or dispersed to sites in each of the counties. If it can be operated at 1000 or 2000 tons per day without technical problems, it can be a breakthrough, large-scale, resource recovery system in our study area.

Gas pyrolysis is preferable to centralized landfilling. However, under the low-revenue conditions, it suffers in competition with a dispersed landfill strategy. Under medium- and high-price revenue circumstances it is clearly superior to centralized or dispersed landfilling. At the present time it is restricted to sites near energy users, a fact which in our study area limits its adaptability. In addition, it has not been successfully demonstrated at the scales modeled in the study.

From the decisionmaker's perspective, while the results are favorable, they are by no means conclusive. The decisionmakers have to be willing to take some important risks and enter into the political process to bring the results to fruition. First, the electricity producing utilities have to be willing to burn refuse-derived fuel and pay for the fuel. Since the utilities are in the energy supply, not the solid-waste business, they are reluctant to expose their expensive boilers, air quality and ash handling systems to relatively untested fuels. Those fuel contracts between utilities and solid-waste suppliers that the author has read or read about do not pay very much for the fuel product, place the legal burden on the solid-waste managers for technical problems resulting from fuel delivery and use, and give the utility the right to reject the solid-waste-derived fuel.

A second source of risk for the solid-waste manager is that communities which are supposed (according to the model) to bring their waste to specific sites will refuse to cooperate. Every community's cost can be displayed by the model. However, convincing over 100 communities to cooperate in a regional project is obviously a Herculean diplomatic effort. At the time of this writing, the HMDC has not contracted for a major refuse-derived fuel facility principally because it has not convinced the legal authorities that it needs a franchise authority to ensure itself of a continuing supply of solid waste.

Next, investors and governments have to be convinced that an unproven technology is a worthy capital risk. This has proven to be a serious problem during a period of investor conservatism.

In the study area, the HMDC is constructing a baler as an interim step and is continuing to press for a refuse-derived fuel facility. The City of Newark which was identified as a resource recovery facility site in the computer analysis, has signed a contract with a private vendor. The vendor will construct and operate the resource recovery facility. The City of Newark is in a position to take the risk because its landfilling options are costly and because its population of about 400,000 can deliver about 700 tons of waste per day, which is enough to guarantee the vendor the initial input it needs.

Summarizing, while the analyst sees these results as a green light, many decisionmakers fear going past the light because of the political and financial risks involved.

D. SUMMARY

The case study presented in this chapter overviews northern New Jersey's solid-waste management problems. Briefly, the study area's solid-waste problems seem to be far more serious than those faced by other metropolitan regions. They emanate, however, from the same sources that afflict other regions: increasing waste generation, decreasing availability of disposal sites, and stricter environmental regulations.

Northeastern New Jersey has been brought to an early, critical, and unavoidable decision point because of the waste generated by its massive population concentrated in a relatively small area amid the largest megalopolitan region in the world. On the one hand, the huge amount of solid waste generated in the region will soon overwhelm conventional solid-waste disposal practices. On the other hand, the severity of the problem offers a fertile and perhaps unparalleled ground for testing the efficacy of new solid-waste technologies at their most efficient scales.

Because of these potential changes in the region's present solid-waste management systems and the choice between land disposal and resource recovery, the study team undertook an evaluation of alternative landfill strategies for northeastern New Jersey. This evaluation was designed to assess the impacts of selected changes in the present regional landfill system, particularly those resulting from the following: the application of higher pollution control standards, the addition of volume reduction techniques to landfilling, and the choices between the continuation of the present pattern of regional landfilling in the Hackensack Meadowlands or the development of alternative regional landfills in other areas of the study region. The study was especially designed to provide economic benchmarks for the continuation of this maximum landfill strategy against which resource recovery strategies could be compared.

More generally, this chapter has attempted to recapitulate and illustrate the perspectives offered in Chapter 5. All five of the analysis steps reviewed in Chapter 5 are needed for this solid-waste problem. The production of solid waste and the demand for refuse-derived products had to be estimated. Solid waste had to be transported from the sources to the disposal sites and from the disposal sites to the markets. Preliminary screening of sites, technologies, and scales was done in order that the problem be kept to a manageable size. The environmental, economic, and social impacts of the alternative solid-waste management strategies were not investigated, but will have to be analyzed if the strategies are to be accepted. Finally, linear programming was used to evaluate the economic efficiency of the alternative strategies that passed through the preliminary screening. Overall, linear programming was one of the important methods in the analysis; it was not the only method, and, moreover, its selection rather than the selection of a more complex mathematical programming model was predicated on results of the other analysis steps.

The guidelines presented in Chapter 5 were also illustrated by the solid-waste case study. Discussions with decisionmakers and the other interested parties were held and led to the formulation of two levels of research questions. Mathematical programming had to be used because of the large number of choices; however, linear programming rather than a more complex programming model was used. The study team was left with the responsibility of defining the study's activities, variables, and parameters. The quality and quantity of data were quickly investigated and found to be deficient, which led to the selection of more simple models, including linear programming. Finally, summary tables and maps were desired as inputs to the decisionmakers. Accordingly, these were designed and are illustrated in this chapter.

REFERENCES

Baecher, G., Gros, J., and McCusker, K. (1975). "Balancing Apples and Oranges: Methodologies for Facility Siting Decisions." Research Report 75-33, International Institute for Applied Systems Analysis, Laxenburg, Austria.

Burnham, J. B. (1974). "A Technique for Environmental Decision-Making Using Quantified Social and Aesthetic Values." National Technical Information Service, Springfield, Virginia.

Carey, G., and Greenberg, M. (1974). Toward a geographical theory of hypocritical decision-making. *Human Ecology* **2**(4), 243–257.

Dee, N., Baker, J., Drobny, N., Duke, K., Whitman, I., and Fahringer, D. (1973). An environmental evaluation system for water resources planning. *Water Resour. Res.* **9**, 523–539.

de Neufville, R., and Keeney, R. (1972). "Use of Decision Analysis in Airport Development for Mexico City." *In* "Analysis of Public Systems" (A. Drake, R. L. Keeney, and P. M. Morse, eds.). MIT Press, Cambridge, Massachusetts.

Fisher, A. C., and Krutilla, J. V. (1974). Valuing long run ecological consequences and irreversibilities. *J. Environ. Econ. Management* **1**, 96–108.

Greenberg, M., Bottge, M., Caruana, J., Horowitz, D. Krugman, B., Masucci, N., Milewski, A., Nebenzahl, L., O'Neill, T., Skypeck, J., Valente, N., (1976). "Solid Waste Planning in Metropolitan Regions." Center for Urban Policy Research, Rutgers University, New Brunswick, New Jersey.

Heath, J. B. (1971). Cost–Benefit Analysis and Airport Location." *In* "Cost–Benefit Analysis" (M. G. Kendall, ed.), pp. 295–304. English Univ. Press, London.

Hill, M. (1973). "Planning for Multiple Objectives." Regional Science Research Institute Monograph 5, Philadelphia, Pennsylvania.

Leopold, L. B., Clarke, F. E., Hanshaw, B., and Balsley, J. R. (1971). "A Procedure for Evaluating Environmental Impact." Geological Survey Circular No. 645. U.S. Geological Survey, Washington, D.C.

Lichfield, N. (1970). Evaluation methodology of urban and regional plans: a review. *Regional Studies* **4**, 151–165.

Maass, A. (1962). "Design of Water Resource Systems." Harvard Univ. Press, Cambridge, Massachusetts.

Ross, M. (1974). London's third airport: quantitative decision-making. *Ekistics* **218**, 27–44.

Sewell, W. (1973). Broadening the approach to evaluation in resources management decision-making. *J. Environ. Management* **1**, 33–60.

Index

8
9
0
1
2
3
4
5